Combinatorial Algorithms: Theory and Practice

Combinatorial Algorithms: Theory and Practice

EDWARD M. REINGOLD

Department of Computer Science
University of Illinois at Urbana-Champaign

JURG NIEVERGELT

Department of Computer Science
University of Illinois at Urbana-Champaign
and
Swiss Federal Institute of Technology, Zurich

NARSINGH DEO

Department of Electrical Engineering and Computer Science Programme
Indian Institute of Technology, Kanpur

PRENTICE-HALL, INC., ENGLEWOOD CLIFFS, NEW JERSEY 07632

Library of Congress Cataloging in Publication Data

Reingold, Edward M (date)
    Combinatorial algorithms.

    Includes bibliographical references and index.
    1. Combinatorial analysis—Data processing.
I. Nievergelt, Jürg, joint author. II. Deo, Narsingh,
joint author. III. Title.
QA164.R43   511'.6   76-46474
ISBN 0-13-152447-X

10 9 8 7 6 5 4 3 2 1

Printed in the United States of America

PRENTICE-HALL INTERNATIONAL, INC., *London*
PRENTICE-HALL OF AUSTRALIA PTY. LIMITED, *Sydney*
PRENTICE-HALL OF CANADA, LTD., *Toronto*
PRENTICE-HALL OF INDIA PRIVATE LIMITED, *New Delhi*
PRENTICE-HALL OF JAPAN, INC., *Tokyo*
PRENTICE-HALL OF SOUTHEAST ASIA PTE. LTD., *Singapore*
WHITEHALL BOOKS LIMITED, *Wellington, New Zealand*

Contents

Preface

The field of combinatorial algorithms concerns the problems of performing computations on discrete, finite mathematical structures. It is a new field, and only in the past few years has it started to emerge as a systematic body of knowledge instead of a collection of unrelated tricks. Its emergence as a new discipline is due to three factors:

An increase in the practical importance of computation of a combinatorial nature, as compared to other computation.

Rapid progress, primarily of a mathematical nature, in the design and analysis of algorithms.

A shift in emphasis from the consideration of particular combinatorial algorithms to the examination of properties shared by a class of algorithms.

The combination of these factors has promoted combinatorial algorithms as an important new discipline on the border between computer science and mathematics. Courses in combinatorial algorithms and related courses in the analysis of algorithms are now being taught in colleges and universities in computer science, mathematics, electrical engineering, and operations research departments.

Combinatorial algorithms can be presented in different ways, and a course or textbook can be directed toward different audiences. This book is aimed at a reader who can best be characterized as having more of a computing background than a mathematics background, a reader who is interested in combinatorial algorithms because of their practical importance. Thus our main goals in writing this book were:

To choose the topics according to their relevance to practical computation (however, we have included some not very practical but mathematically interesting topics).

To emphasize those aspects of algorithms that are important from the point of view of their implementation without belaboring details that any competent programmer can supply.

To present mathematical arguments, where they are necessary, with emphasis on insight.

Prerequisites.   The prerequisites required by this book vary somewhat from chapter to chapter. The minimal knowledge of programming required is that which can be obtained in a first course on computers in which students write several rather extensive programs. This background should suffice to understand the algorithms discussed, which are presented in a notation similar to current high-level programming languages. The additional background obtained in a second computer course on data structures and list processing is highly desirable. The mathematical maturity required is that typical of a student who has taken several mathematics courses beyond calculus.

Starred Sections.   Two sections of this book have been marked by a ★. Section 3.4 is so marked because it requires some familiarity with advanced mathematical concepts, and Section 8.6 is so marked because it is intrinsically complicated. These sections can be considered optional. Exercises related to the starred sections are also starred, as are other exercises that require advanced mathematics or whose solution is unusually complex.

Overall Organization.   The dependencies among the nine chapters are indicated by the following diagram in which "strong" dependencies are shown by solid arrows and "weak" dependencies by dashed arrows:

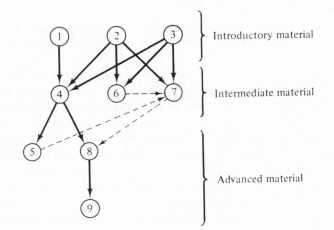

Chapter 1 is designed to display the scope of the material in the book. It also introduces some topics and techniques that reappear in later chapters.

Chapters 2 and 3 discuss data structures and counting techniques, respectively. With the exception of Section 3.4, all the material contained in these two chapters is basic to the rest of the book. It may be more appropriate, however, to cover the topics in these chapters as they are needed in presenting the material in later chapters. The book is not written that way in an effort to make Chapters 4 through 9 as independent of each other as possible.

The material in Chapter 4 on exhaustive search is important in understanding Chapter 5, which is actually a specialization of techniques from Chapter 4 to simple combinatorial objects. For the same reason, the material in Chapter 4 is crucial to understanding many parts of Chapter 8 on graph algorithms. Chapter 9 continues the sequence of material in Chapters 4 and 8 by examining some theoretical questions raised by the failure of computer scientists and mathematicians to find "efficient" algorithms for some of the problems discussed.

Chapters 6 and 7 describe the most common of all combinatorial algorithms: searching and sorting. These chapters rely most heavily on material in Chapters 2 and 3.

ACKNOWLEDGMENTS

The entry "illogicalities" in H. W. Fowler's *A Dictionary of Modern English Usage* (second edition, revised and edited by Sir Ernest Gowers, Oxford University Press, New York and Oxford, 1965) begins with

> The spread of education adds to the writer's burdens by multiplying that pestilent fellow the critical reader. No longer can we depend on an audience that will be satisfied with catching the general drift and obvious intention of a sentence and not trouble itself to pick holes in our wording.

We have been fortunate in having many such critical readers examine all or part of the manuscript for this book, and they have helped to find various instances of "illogicalities." We gratefully acknowledge James R. Bitner, Allan B. Borodin, James A. Fill, W. D. Frazer, Brian A. Hansche, Wilfred J. Hansen, Ellis Horowitz, Alon Itai, Donald B. Johnson, John A. Koch, Der-tsai Lee, Karl Lieberherr, C. L. Liu, Prabhaker Mateti, Yehoshua Perl, David A. Plaisted, Andrew H. Sherman, Robert E. Tarjan, Daniel S. Watanabe, Lee J. White, Thomas R. Wilcox, and Herbert S. Wilf. Without their invaluable assistance this book would be poorer.

We owe an extra debt of gratitude to James A. Fill for his careful preparation of the solutions manual.

We are indebted to the University of Illinois at Urbana-Champaign, Eidgenössische Technische Hochschule, Los Alamos Scientific Laboratory, Weizmann Institute of Science, Washington State University, and Indian Institute of Technology, Kanpur, for providing facilities for the preparation of this book. Their support made our task much easier.

Finally, we would like to acknowledge and thank Connie Nosbisch and June Wingler, our secretaries at the University of Illinois, for their patient, conscientious, and expert typing, retyping and sometimes even re-retyping of the manuscript.

EDWARD M. REINGOLD

JURG NIEVERGELT

NARSINGH DEO

chapter 1

What is Combinatorial Computing?

The subject of combinatorial algorithms, frequently called *combinatorial computing*, deals with the problem of how to carry out computations on discrete mathematical structures. It is a new field: only in the past few years has a systematic body of knowledge about the design, implementation, and analysis of algorithms emerged from a collection of tricks and unrelated algorithms.

An analogy with a more established field may be useful. Combinatorial computing bears to combinatorial (discrete, finite) mathematics the same relationship that numerical analysis bears to analysis. We are witnessing today in combinatorial computing the same development that numerical analysis went through in the 1950s—namely,

new algorithms are being invented at a rapid rate.

rapid progress, primarily of a mathematical nature, is being made in our understanding of algorithms, their design, and their analysis.

emphasis is shifting from consideration of a particular algorithm to properties shared by a class of algorithms, and thus unifying patterns are emerging.

The increasing practical importance of computation of a combinatorial nature has undoubtedly contributed to the recent surge of activity in combinatorial computing. There is every reason to believe that the amount of computation of a combinatorial nature that occurs in applications programs will increase faster than the amount of numerical computation. This is because, outside the traditional areas of applications of mathematics to the physical sciences, discrete mathematical structures occur more frequently than continuous ones, and the fraction of all computing time spent on problems that arise in the physical sciences is decreasing. Hence computer users and applications programmers will probably be called on to solve problems of a combinatorial rather than numerical nature.

Unlike some other fields, combinatorial computing does not have a few "fundamental theorems" that form the core of the subject matter and from which most results can be derived. The entire subject may seem, at first, merely a collection of unrelated, specialized techniques and tricks. Clever tricks do indeed play a role, and in this chapter we will see some examples of "bit pushing" techniques that convey this flavor. However, after examining many combinatorial algorithms, some general principles become apparent. It is these principles that unify the field and make combinatorial computing a coherent subject that can be presented in a systematic way. The purpose of this chapter is to illustrate some of these important principles by means of examples, as well as to serve as an introduction into some of the topics and techniques that will be treated in greater depth in later chapters.

The chapter is organized so as to go from the concrete aspects of computation to more abstract principles, some of which require advanced mathematics for their application to the analysis of algorithms. Therefore the abstract sections of this chapter are more difficult than the others, and in this respect Chapter 1 is a faithful mirror of the field it introduces. Sections 1.1 and 1.2 show examples of the detailed thinking required for efficient implementation of algorithms. Sections 1.3 and 1.4 present general principles of algorithm design. Finally, Sections 1.5 and 1.6 show how algorithms are analyzed. The design, analysis, and implementation of algorithms form the core of combinatorial computing.

1.1 AN EXAMPLE: COUNTING THE NUMBER OF ONES IN A BIT STRING

Bit strings—that is, sequences of zeros and ones—are the basic carriers of information in virtually all modern computers. Most programmers, however, rarely handle information at the detailed level of bit strings. This is certainly true in numerical computation, in which a programmer usually expresses himself in terms of arithmetic operations on numbers and is seldom concerned with the internal representation of these numbers. On the other hand, in areas that are not as well established as numerical computation, certain important operations on data may not be built into computers or high-level programming languages. So, in order to program efficiently, an applications programmer must be familiar with algorithms that operate at the bit level; this is the case with a number of operations that occur frequently in combinatorial computing. Eventually, as these operations become better known, they are likely to be incorporated into computers and programming languages; but until then they are necessary tools of the trade for anyone who programs combinatorial algorithms.

As an example, we consider the problem of counting the number of ones in an n-bit string $B = b_n b_{n-1} \ldots b_2 b_1$. It is a natural operation if we imagine B to represent a subset S of a set U of n elements, the ones indicating which elements of U are in S. This operation, called the *bit sum*, then determines the number of elements in S.

The first algorithm that comes to mind for computing the bit sum of B is to inspect each bit in turn and, if it is a one, to increment a counter.

$$c \leftarrow 0$$
for $i = 1$ **to** n **do if** $b_i = 1$ **then** $c \leftarrow c + 1$

Algorithm 1.1   Computing the bit sum by looking at each bit.

On some computers, Algorithm 1.1 may be the most reasonable one to use. Most computers, however, have features that permit much faster bit sum algorithms. Assume that the memory consists of cells that can hold an n-bit word and that the computer has logical or boolean operations that operate in parallel on each bit of a word, plus arithmetic operations that interpret these words as unsigned nonnegative integers written in base 2.

$$c \leftarrow 0$$
while $B \neq 0$ **do** $\begin{cases} c \leftarrow c + 1 \\ B \leftarrow B \wedge (B - 1) \end{cases}$

Algorithm 1.2   A tricky way to compute the bit sum.

Consider Algorithm 1.2. The statement "$B \leftarrow B \wedge (B - 1)$" contains an interesting operation that makes use of the assumptions stated above. $\wedge$ is the logical "and" operation that operates in parallel on each pair of bits in corresponding positions of its two arguments, B and $B - 1$. $-$ is the arithmetic operation of subtraction on integers represented in base 2. An example best shows that when this operation is executed, the rightmost 1 of B is replaced by 0.

B	11001000
$B - 1$	11000111
$B \wedge (B - 1)$	11000000

The loop in Algorithm 1.2 is executed until $B = 0$, that is, until B contains only zeros. Thus whereas the loop in Algorithm 1.1 is always executed n times, the loop in Algorithm 1.2 is only executed as many times as there are ones in B. The effect is as if Algorithm 1.2 were able to look only at the ones in B, without knowing their positions *a priori*. This algorithm is obviously efficient when applied to "sparse" words—that is, to bit strings that contain few ones and many zeros. It is "tricky" in the sense that it depends intricately on the way that numbers are represented in the computer (in particular, negative numbers!), something entirely foreign to the original problem of counting the number of ones in a word.

The next algorithm is even more interesting. It shares with Algorithm 1.1 the important property that the loop is executed a fixed number of times, dependent on n but independent of the particular value of B. But unlike Algorithm 1.1, which repeats its loop n times, Algorithm 1.3, which we do not give explicitly, runs through its loop only $\lceil \lg n \rceil$ times.[†] For a typical word size of $n = 32$ (or 64), the loop is executed five (respectively six) times, which may result in a significant speedup compared to Algorithm 1.1. The assumptions required for Algorithm 1.3 to work on a given computer are about the same as those required for Algorithm 1.2, and there must be a fast way of shifting words by $1, 2, 4, 8, \ldots,$ places.

Algorithm 1.3. is best explained by means of an example,

	b_8	b_7	b_6	b_5	b_4	b_3	b_2	b_1
B	1	1	0	1	0	0	0	1

1. First, extract the odd-indexed bits b_7 b_5 b_3 b_1 and place a zero to the left of each bit to obtain B_{odd}.

		b_7		b_5		b_3		b_1
B_{odd}	0	1	0	1	0	0	0	1

Next, extract the even-indexed bits b_8 b_6 b_4 b_2, shift them right by one place into bit positions b_7 b_5 b_3 b_1 and place a zero to the left of each bit to obtain B_{even}.

		b_8		b_6		b_4		b_2
B_{even}	0	1	0	0	0	0	0	0

(The newly inserted zeros are shown in small type to distinguish them from the zeros that are extracted from B.)

Then, numerically add B_{odd} and B_{even}, considered as integers written in base 2, to obtain B'.

	b_8'	b_7'	b_6'	b_5'	b_4'	b_3'	b_2'	b_1'
B_{odd}	0	1	0	1	0	0	0	1
B_{even}	0	1	0	0	0	0	0	0
B'	1	0	0	1	0	0	0	1

[†]The symbol $\lg x$ represents $\log_2 x$. $\lceil x \rceil$ is the *ceiling of* x: the least integer $k, k \geqq x$. Similarly $\lfloor x \rfloor$ is the *floor of* x: the greatest integer $k, k \leqq x$.

2. Extract the alternate pairs of bits b'_6 b'_5 and b'_2 b'_1 and place a pair of zeros to the left of each pair to obtain B'_{odd}.

		b'_6	b'_5			b'_2	b'_1
B'_{odd}	0 0	0	1	0	0	0	1

Next, extract the other pairs b'_8 b'_7 and b'_4 b'_3, shift them right by two places into bit positions b'_6 b'_5 and b'_2 b'_1, respectively, and insert a pair of zeros to the left of each pair to obtain B'_{even}

		b'_8	b'_7			b'_4	b'_3
B'_{even}	0 0	1	0	0	0	0	0

Numerically add B'_{odd} and B'_{even} to obtain B''.

	b''_8	b''_7	b''_6	b''_5	b''_4	b''_3	b''_2	b''_1
B''	0	0	1	1	0	0	0	1

3. Extract bits b''_4 b''_3 b''_2 b''_1 and place four zeros to the left to obtain B''_{odd}.

					b''_4	b''_3	b''_2	b''_1
B''_{odd}	0	0	0	0	0	0	0	1

Also, extract bits b''_8 b''_7 b''_6 b''_5, shift them right four places into bit positions b''_4 b''_3 b''_2 b''_1, and place four zeros to the left to obtain B''_{even}.

					b''_8	b''_7	b''_6	b''_5
B''_{even}	0	0	0	0	0	0	1	1

Finally, numerically add B''_{odd} and B''_{even} to obtain $B''' = (00000100)$.

B''' is the representation in base 2 of the bit sum of B (4 in this example).

The generalization of this algorithm to arbitrary n is easy if we imagine that B is padded with zeros to the left until its length is equal to the first power of 2 greater than or equal to n.

The reader is encouraged to prove that the algorithm (in its general form) is correct. In Section 1.4 we describe a general principle of algorithm design from which the algorithm and a correctness proof follow directly. This principle is a good illustration of our claim that combinatorial computing does have general concepts and techniques from which many special cases follow.

Are there still faster algorithms for computing the bit sum of a word? Is there an "optimal" algorithm? The question of optimality of algorithms is an important one, but it can be treated only in special cases. To show that an algorithm is optimal, one must specify precisely the class of algorithms allowed and the criterion of optimality. In the case of bit sum algorithms, such specifications would be complicated and largely arbitrary, involving specific details of how computers work. We will discuss optimality of algorithms in Section 1.5, in more simplified settings.

We can however, make a plausible argument that the following bit sum algorithm (Algorithm 1.4) is the fastest possible, since it uses a table lookup to obtain the result in essentially one operation. The penalty for this speed is an extravagant use of memory space (2^n locations), thereby making the algorithm impractical except for small values of n. The choice of an algorithm almost always involves tradeoffs among various desirable properties, and generally the better an algorithm is from one aspect, the worse it is from another.

Algorithm 1.4 is based on the idea that we can precompute the solutions to all possible questions, store the results, and then simply look them up when needed. As an example, for $n = 3$, we would store the information

Word	Bit Sum
000	0
001	1
010	1
011	2
100	1
101	2
110	2
111	3

What is the fastest way of looking up a word B in this table? Under assumptions similar to those used in the preceding algorithms, we may assume that B can be interpreted as an address of a memory cell that contains the bit sum of B, thus giving us an algorithm that requires only one memory reference.

In concluding this example, we notice the great variety of algorithms that exist for computing the bit sum, each one based on entirely different principles. Algorithms 1.1 and 1.4 solve the problem by "brute force": Algorithm 1.1 looks at each bit and so requires much time; Algorithm 1.4 stores the solution for each separate case and thus requires much space. Algorithm 1.3 is an elegant compromise.

1.2 A REPRESENTATION PROBLEM: DIFFERENCE-PRESERVING CODES

A recurring problem of great importance in combinatorial computing is to find efficient representations of the objects to be manipulated. These objects can be as simple as the bit strings of the preceding example or as complicated as networks of roads or organic molecules. The representation problem arises because there are usually many potential ways of representing complex objects in terms of the simpler structures provided directly in a computer or programming language; but not all of these representations are equally efficient in terms of time and memory requirements. Moreover, the ideal representation depends on the kind of operations to be performed.

This situation is illustrated by the example of this section in which we want to perform a rather special operation on integers. Integers are provided as primitive data types in almost all computers and programming languages, and so a representation problem normally does not occur; the available representation (the one chosen by the computer designer) is almost always best. There are some noteworthy exceptions, however, where it is advantageous (or even necessary) to use a representation of integers other than the one built into the computer. These exceptions arise when

1. integers larger than those directly available in the hardware are needed.

2. only small integers are needed, and it is necessary to preserve memory space by packing several of them into one memory cell.

3. the integers will be operated on in unusual ways, not by means of the conventional arithmetic operations.

4. the integers are used to represent other types of objects, in which case it may be important to be able to convert the integer into its corresponding object easily, and vice versa (this is an important special case of the previous exception that will arise in Chapter 5).

The problem of difference-preserving codes discussed in this section has aspects (2) and (3).

The following technique is standard in pattern recognition and classification problems in order to decide whether two objects X and Y are equivalent. X and Y are represented by feature vectors $(x_1, x_2, \ldots, x_f)$, $(y_1, y_2, \ldots, y_f)$, respectively, in which each component represents a feature of the object that is measured by an integer value. X and Y are accepted as being equivalent if and only if

$$\sum_{i=1}^{f} |x_i - y_i| \leq t,$$

where t is an integer called the *threshold*.

In typical applications (such as measuring the gray levels in digitized pictures, counting the number of characters in English words, or counting the number of vertices in a graph that is abstracted from a handwritten character), the components as well as the threshold can assume only small integer values. In such cases, several components of a feature vector can be packed into a memory location, assigning to each component a field of sufficient length to hold any of the values that component can assume. Figure 1.1 shows how a field can even be broken across memory location boundaries.

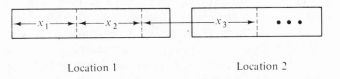

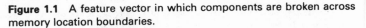

Location 1   Location 2

Figure 1.1 A feature vector in which components are broken across memory location boundaries.

At first it appears that, in order to compute $\sum |x_i - y_i|$, it is necessary to extract one pair of corresponding components of the feature vectors X and Y at a time and then shift them into proper position for the subtraction to be performed. This section shows that if the integers are represented in a special way (a difference-preserving code), the operation of comparing the sum of differences to the threshold can be greatly speeded up by carrying out all operations on an entire memory location, ignoring field boundaries. The speedup is particularly effective on computers with long words and a built-in bit sum operation to count the number of ones in a word.

A *difference-preserving code* $\mathcal{D}$ is a mapping $\mathcal{D} : i \to D_i$ from the set $\{1, 2, \ldots, N\}$ into the set $\{0, 1\}^n$, the binary sequences of length n, with the following property. For all integers i, j such that $1 \leq i, j \leq N$,

   1. $|i - j| \leq t$ implies $H(D_i, D_j) = |i - j|$ and

   2. $|i - j| > t$ implies $H(D_i, D_j) > t$,

where $H(D_i, D_j)$ is the *Hamming distance* between the two codewords D_i and D_j, that is, the number of bits in which they differ, and t is the *threshold*. Intuitively, the code $\mathcal{D}$ preserves small differences but lumps all large differences together. $\mathcal{D}$ is called an *n-bit DP-t code* of *range N* or sometimes an (n, t)-code of range N.

It is easy to verify that when the components of two feature vectors X, Y are represented in a DP-t code, then the inequality $\sum_{i=1}^{f} |x_i - y_i| \leq t$ holds if and only if the Hamming distance between the representations of X and Y, the number of bit positions in which the vectors X and Y differ, is at most t. This property allows the efficient comparison of feature vectors as mentioned earlier.

As an example, the following code has range $N = 8$, length $n = 4$, and threshold $t = 1$. It is *optimal* in the sense of having maximum range among all codes of length 4 and threshold 1.

Integer	Codeword			
	b_4	b_3	b_2	b_1
1	0	0	0	0
2	0	0	0	1
3	0	0	1	1
4	0	1	1	1
5	0	1	1	0
6	1	1	1	0
7	1	1	0	0
8	1	1	0	1

A difference-preserving code can be visualized as a path along the edges of an n-dimensional cube, with vertices corresponding to codewords. Figure 1.2 shows this 4-bit code by labeling certain vertices of a four-dimensional cube with the corresponding integer. Notice that the only codewords that are adjacent to each other correspond to integers that differ by 1.

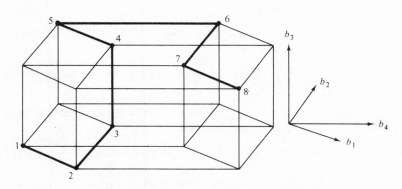

Figure 1.2   An optimal (4, 1)-code depicted on a four-dimensional cube.

Table 1.1 shows the maximal ranges that can be obtained for $n \leqq 6$. Corresponding optimal codes can be constructed fairly easily by hand for $n \leqq 5$ and by computer for $n = 6$. We do not know the ranges of optimal codes for $n = 7$, since the program used to construct the optimal codes for $n = 6$ could not possibly settle the question for $n = 7$ in a reasonable time. The details of the computation for $n = 6$ are given in Section 4.1.5.

This discussion illustrates a rule of thumb that holds surprisingly often in combinatorial computing. Many interesting combinatorial problems contain a parameter that, in a natural way, indicates the size of this problem. (In the case of

Table 1.1   Maximum ranges for (n, t)-codes.

$n \backslash t$	1	2	3	4	5
1	2	2	2	2	2
2	3	3	3	3	3
3	5	4	4	4	4
4	8	6	5	5	5
5	14	8	7	6	6
6	27	14	9	8	7

difference-preserving codes, the codeword length n is such a parameter.) The rule of thumb states:

1. For a sufficiently small value of n, the problem is trivial ($n = 4$ in this case).

2. For the next higher value of n, the problem can be solved by hand ($n = 5$).

3. For the next higher value of n, the problem can be solved with the help of a computer ($n = 6$).

4. For the next higher value of n, the solution is impossible to obtain.

We are not quite ready to claim the optimal DP-codes for $n = 7$ will never be obtained; but the reader who experiments with this problem will soon realize that the amount of work involved in the construction of optimal codes shows a fantastic growth rate (as a function of n). This frequently occurring phenomenon is called the *combinatorial explosion*.

1.3 COMPOSITION TECHNIQUES

The two preceding examples had some of the flavor of clever "bit-pushing" tricks—the kind of highly specialized techniques that work well on one problem but do not generalize or transfer to another. Lest the reader infer that combinatorial computing is merely a collection of isolated tricks, we will discuss two general principles here and in the next section and show how they can be used in such specialized problems as the bit sum algorithm and the construction of difference-preserving codes. The two principles, composition of solutions and decomposition of a problem, are so general that their applicability is by no means restricted to combinatorial computing; they are useful in attacking any kind of mathematical problem.

The composition principle, the subject of this section, leads us to construct complex objects by putting simpler ones together. We illustrate this principle by composing large difference-preserving codes from smaller ones.

In trying to construct difference-preserving codes by hand, we notice that the problem is easy if both the range N and the threshold t are small (preferably $t = 1$). To construct codes of large range and/or large threshold, it is possible to combine small codes by means of two composition techniques, one of which extends the range while keeping the threshold fixed and the other of which increases the threshold while extending the range only slightly. Both composition techniques are based on the idea of juxtaposing, or concatenating, a codeword of one code and a codeword of another so as to form a codeword of the composed code.

Threshold Addition Technique. Let $D_1, D_2, \ldots, D_N$ be an (n, t)-code and let $E_1, E_2, \ldots, E_M$ be an (m, u)-code. Then the sequence

$$D_1E_1, \qquad D_2E_1, \qquad \ldots, D_tE_1, \qquad D_{t+1}E_1, \qquad D_{t+1}E_2, \qquad \ldots, D_{t+1}E_u,$$

$$D_{t+1}E_{u+1}, \quad D_{t+2}E_{u+1}, \quad \ldots, D_{2t}E_{u+1}, \quad D_{2t+1}E_{u+1}, \quad D_{2t+1}E_{u+2}, \quad \ldots, D_{2t+1}E_{2u},$$

$$D_{2t+1}E_{2u+1}, D_{2t+2}E_{2u+1}, \ldots, D_{3t}E_{2u+1}, D_{3t+1}E_{2u+1}, D_{3t+1}E_{2u+2}, \ldots, D_{3t+1}E_{3u},$$

$$D_{3t+1}E_{3u+1}, D_{3t+2}E_{3u+1}, \ldots$$

is an $(n + m, t + u)$-code obtained by *alternatingly* keeping the subscript of one code sequence constant while increasing the subscript of the other code sequence; the range of the compound code is

$$R = \begin{cases} N + \left\lfloor \dfrac{N-1}{t} \right\rfloor u & \text{if } \left\lfloor \dfrac{M-1}{u} \right\rfloor \geq \left\lfloor \dfrac{N-1}{t} \right\rfloor, \\[2ex] M + t + \left\lfloor \dfrac{M-1}{u} \right\rfloor t & \text{otherwise.} \end{cases}$$

If $N/t \approx M/u$, the range of the composed code is approximately $N + M$.

For example, if we compose the optimal $(3, 1)$-code and the optimal $(4, 2)$-code, we obtain a $(7, 3)$-code of range 9.

$(3, 1)$-code: range $N = 5$	$(4, 2)$-code: range $M = 6$	$(7, 3)$-code: range $R = 9$
$D_1 = 000$	$E_1 = 0000$	$D_1E_1 = 0000000$
$D_2 = 001$	$E_2 = 0001$	$D_2E_1 = 0010000$
$D_3 = 011$	$E_3 = 0011$	$D_2E_2 = 0010001$
$D_4 = 111$	$E_4 = 0111$	$D_2E_3 = 0010011$
$D_5 = 110$	$E_5 = 1111$	$D_3E_3 = 0110011$
	$E_6 = 1110$	$D_3E_4 = 0110111$
		$D_3E_5 = 0111111$
		$D_4E_5 = 1111111$
		$D_4E_6 = 1111110$

It turns out that since we have not used the codeword D_5 and since D_1, D_2, D_3, D_4 form a (3, 3)-code, the composed code is actually a (7, 5)-code of range 9, the optimal for (7, 5)-codes.

Range-Extension Technique. Let $D_1, D_2, \ldots, D_N$ be an (n, t)-code and $E_1, E_2, \ldots, E_M$ be an (m, t)-code. Then the sequence

$$D_1 E_1, \quad D_1 E_2, \quad \ldots, D_1 E_M, \quad D_2 E_M, \quad \ldots, D_{t+1} E_M$$

$$D_{t+2} E_M, \ D_{t+2} E_{M-1}, \ \ldots, D_{t+2} E_1, \ D_{t+3} E_1, \ \ldots, D_{2t+2} E_1$$

$$D_{2t+3} E_1, \ D_{2t+3} E_2, \quad \ldots, D_{2t+3} E_M, D_{2t+4} E_M, \ldots, D_{3t+3} E_M$$

$$D_{3t+4} E_M, \ D_{3t+4} E_{M-1}, \ldots$$

is an $(n + m, t)$-code of range

$$R = M\left\lceil \frac{N}{t+1} \right\rceil + t\left\lfloor \frac{N}{t+1} \right\rfloor \approx \frac{NM}{t+1}. \qquad (1.1)$$

For example, if we compose the optimal (3, 2)-code, which has range 4, with itself, we obtain a (6, 2)-code of range 10.

(3, 2)-code: range $N = 4$	(6, 2)-code: range $R = 10$
$D_1 = E_1 = 000$	$D_1 E_1 = 000000$
$D_2 = E_2 = 001$	$D_1 E_2 = 000001$
$D_3 = E_3 = 011$	$D_1 E_3 = 000011$
$D_4 = E_4 = 111$	$D_1 E_4 = 000111$
	$D_2 E_4 = 001111$
	$D_3 E_4 = 011111$
	$D_4 E_4 = 111111$
	$D_4 E_3 = 111011$
	$D_4 E_2 = 111001$
	$D_4 E_1 = 111000$

1.4   DECOMPOSITION TECHNIQUES

The decomposition principle illustrated here is the dual of the composition principle of the previous section; each emphasizes different aspects of the same central idea. The decomposition principle leads us to solve a problem about a complex object by decomposing the object into several smaller ones, recursively solving the same problem for each of the smaller objects, and then composing the solutions found. To ensure that the procedure terminates, sufficiently simple objects are not further decomposed, but instead the problem is solved directly for them. It is this principle that leads to the efficient and elegant bit sum algorithm (Algorithm 1.3) of Section 1.1.

Let $B = b_n \ldots b_2 b_1$ be a bit string of length n and let $S(B)$ be its bit sum. Split B into two parts at some index h and denote its left part by $B_l = b_n \ldots b_{h+1}$ and its right part by $B_r = b_h \ldots b_2 b_1$. The bit sum obviously satisfies the recursive equation

$$S(B) = S(B_l) + S(B_r).$$

This means that the computation of $S(\mathrm{B})$ can be achieved by a single addition if we know the bit sum of the strings shorter than B. Repeating the same argument on B_l and B_r and, in turn, on the substrings they create, and so on, we arrive at a process to compute $S(B)$, provided that we can compute the bit sum on sufficiently short strings. If we take "sufficiently short" to mean "of length 1" and define $S(B) = B$ if B is of length 1, it is clear how the bit sum can be computed by a sequence of additions that depends on the method used to split B repeatedly.

Consider the following two extremes among splitting strategies. First, choose $h = n - 1$, that is, $B_l = b_n$ and $B_r = b_{n-1} \ldots b_2 b_1$. This strategy gives us the left-to-right scan (Algorithm 1.1). It clearly takes $n - 1$ additions. Second, choose $h = \lceil n/2 \rceil$, that is, $B_l = b_n \ldots b_{\lceil n/2 \rceil + 1}$ and $B_r = b_{\lceil n/2 \rceil} \ldots b_2 b_1$. This approach of treating both parts of B symmetrically and of always splitting B in half (approximately) leads to Algorithm 1.3, but a clever idea is still needed before this becomes obvious. Notice that this algorithm still takes $n - 1$ additions and, indeed, that any splitting strategy will take $n - 1$ additions (Exercise 5). Therefore, in order to obtain the logarithmic performance of Algorithm 1.3, we must arrange things so that a number of the additions on short strings are carried out by a *single* addition on long strings.

Figure 1.3 shows how to do this. $S(B)$ now denotes not only the bit sum but also, in particular, the string that is the binary representation of the bit sum, padded with zeros on the left so as to have the appropriate length.

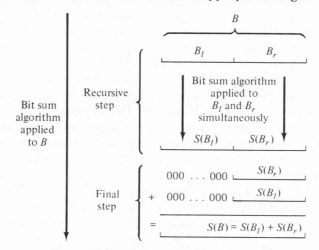

Figure 1.3 The idea behind Algorithm 1.3.

Since the same algorithm is being applied to B_l and B_r, and since B_l and B_r are of the same length, exactly the same operations are performed at each stage on B_l and its parts as on B_r and its corresponding parts. Thus if we have operations (like addition, shifting, etc.) that operate on strings of length n (i.e., the length of B), then one such operation can be interpreted as performing many of the same operations on the shorter parts into which B has been split.

Much can be learned from studying this example thoroughly. It shows how an apparently unmotivated, tricky algorithm can actually be derived from a general principle.

1.5 CLASSES OF ALGORITHMS

We have stated that one of the trends responsible for the rapid progress in combinatorial computing is a stronger emphasis on the study of classes of algorithms as opposed to individual algorithms. In order to make statements like "All algorithms for doing such and such must have property so and so" or "There is no algorithm satisfying thus and such", it is necessary to consider a precisely defined class of algorithms. With such a definition it is possible to say that a particular algorithm is optimal with respect to a certain property if it performs at least as well (with regard to this property) as any other algorithm in the class being considered. Knowing that an algorithm is optimal can prevent searching for a better one that does not exist!

How can a (possibly infinite) class of algorithms be rigorously defined? Let us examine this question by means of an example, the false-coin problem. The class of algorithms considered in this example generalizes to a larger, more important class of algorithms—namely, decision trees.

One form of the false-coin problem is as follows. We have n coins, of which we know that $n - 1$ are identical, "true" coins. At most, one coin can be "false"; that is, it is either lighter or heavier than the true coins. In addition to the n doubtful coins, we are given one coin that is known to be true. We are also given a balance scale with which we can compare the total weight of any m coins against the total weight of any other m coins. The balance will tell us whether the two groups of m coins are each of equal weight, whether the first group is heavier, or whether the first group is lighter. The problem is to identify the false coin (if any) in as few weighings, or comparisons, as possible. The number of comparisons can be counted in several ways, the two most reasonable ways being the average and the worst cases. We will discuss this subject further after defining the class of algorithms considered.

Let the doubtful coins be named $1, 2, \ldots, n$, let the coin known to be true be named 0, and let $S = \{0, 1, 2, \ldots, n\}$ be the set of coins. If S_1, S_2 are disjoint, nonempty subsets of S, we denote by $S_1 : S_2$ a comparison of the total weight of sets S_1 and S_2. The three possible outcomes of the comparison are denoted by $S_1 < S_2, S_1 = S_2$, or $S_1 > S_2$, depending on whether the total weight of S_1 is less than, equal to, or larger than the total weight of S_2.

The algorithms considered can be expressed in the form of a (ternary) *decision tree*. The tree in Figure 1.4, for four coins, serves to explain the concepts and the notation involved. We omit brackets in denoting sets when no confusion arises.

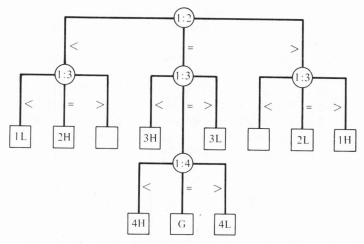

Figure 1.4   A decision tree for the false coin problem with four coins.

The root of this tree, the circle labeled 1 : 2, indicates that the algorithm starts by comparing the weights of coins 1 and 2. The three branches emanating from the root lead to subtrees that indicate how the algorithm continues after each of the three possible outcomes of the first comparison. The squares, called the leaves of the tree, indicate that the algorithm terminates and what the outcome is: "1L" means that coin 1 is light, "1H" means that it is heavy, "G" means that all coins are good, and an empty square means that this case cannot arise under our assumptions.

The algorithm shown in Figure 1.4 requires two comparisons in some cases and three in others. We say that it requires "three comparisons in the worst case." It is usually important to know how much work an algorithm requires on the average, but this knowledge requires an assumption about the probabilities with which different outcomes occur. If we assume that all nine outcomes 1L, 1H, 2L, 2H, 3L, 3H, 4L, 4H, G are equally likely, then this algorithm requires 7/3 comparisons on the average.

We could have put more than one coin on each side of the balance. For example, we could have started by comparing coins 1 and 2 on one side against 3 and 4 on the other (see Figure 1.5). With luck, the problem might be solved with one comparison—that is, if all the coins are good. However, it is clear that no matter how this decision tree is completed, it will still take three comparisons in the worst case, since a single ternary decision cannot possibly identify one among the four outcomes that flow along the < branch, nor among the four outcomes that flow along the > branch. Furthermore, no matter how this decision tree is

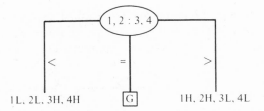

Figure 1.5   The root of another decision tree for the four coins problem.

completed, it will require at least 7/3 comparisons on the average (why?), thereby being no better in this respect than the tree of Figure 1.4.

Using coin 0, known to be true, we can obtain the decision tree of Figure 1.6 (a complete two-level ternary tree), which requires two comparisons both in the worst case and on the average.

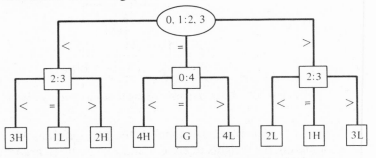

Figure 1.6   The optimal decision tree for the four coins problem.

The class of algorithms we consider for solving the false-coin problem is the set of *ternary decision trees*, as in Figures 1.4 and 1.6, having the properties below.

1.   Each node is labeled with a comparison $S_1 : S_2$, where S_1, S_2 are disjoint, nonempty subsets of the set $S = \{0, 1, 2, \ldots, n\}$ of all coins.

2.   Each leaf is either unlabeled (corresponding to an outcome that is impossible under the assumption that at most one coin is false) or is labeled with an outcome of the form iL, iH, or G, indicating that coin number i is light or heavy or that all coins are true, respectively.

Having precisely defined the class of algorithms under consideration, we can now investigate properties that must hold for any tree in this class, and we can ask how to find algorithms that are optimal in some respect. We will do so first for the special case of four coins and then for the n-coins problem in general.

Since the four-coins problem requires us to distinguish between nine possible outcomes, any decision tree for solving it must have at least nine leaves and hence at least two levels. Therefore the tree of Figure 1.6 is optimal with respect to both the worst case and the average case. Are there other optimal trees? To answer this question, we must consider the space of all decision trees for the four-coins problem; any argument that allows us to eliminate portions of this space from further consideration will make our task easier.

First, observe that other optimal trees can be obtained from the one in Figure 1.6 by any permutation of the set $\{1, 2, \ldots, n\}$ of doubtful coins. Such trees are isomorphic to the one in Figure 1.6, and we make our question more precise by asking only for trees that are not isomorphic to each other.

Next, we ask whether there exists an optimal tree among those that do not use coin 0 at the root. With this restriction, essentially only two different comparisons can be made at the root—namely, $1:2$ and $1, 2:3, 4$. Observe the flow of outcomes along the three branches leading out of the root, as in Figure 1.5. In order to obtain a complete two-level ternary tree, such as the one in Figure 1.6, the nine possible outcomes would have to be split according to the partition $(3, 3, 3)$; instead they are split according to $(2, 5, 2)$ and $(4, 1, 4)$, respectively. Thus we conclude that it is impossible to solve the four-coins problem in two comparisons without using an additional coin known to be true.

Finally, we consider those decision trees that use coin 0 at the root, and we observe that there are essentially only two possible comparisons at the root, $0:1$ and $0, 1:2, 3$. The flow of outcomes for the first one is $(1, 7, 1)$, disqualifying all algorithms that start that way, and $(3, 3, 3)$, leading to the optimal tree of Figure 1.6. A similar check establishes that, for an optimal tree, the comparisons at the level below the root of the tree in Figure 1.6 are uniquely determined. So we conclude that essentially only one optimal tree exists for the four-coins problem.

It is interesting to see what happens when the ideas used to analyze the four-coins problem are applied to the false-coin problem for arbitrary n. All the ideas generalize to some extent, but some are no longer practical when n is significantly larger than 4. In principle, optimal decision trees can always be found by a systematic search of the space of decision trees because, for any given value of n, only a finite number of decision trees need be considered as candidates. In Chapter 4 we will discuss techniques for exhaustively searching through such finite sets. However, even if such a search is cleverly organized and considers only trees that are essentially different (not isomorphic), it is not a practical way to find optimal decision trees. The number of trees grows exponentially with n, and so exhaustive search techniques are practical only for small values of n.

The obvious idea that the number of leaves of a decision tree must be at least as large as the number of possible outcomes of the problem ($2n + 1$ for the n-coins problem) immediately yields a lower bound on the number of comparisons needed (or, equivalently, an upper bound on the number of coins for a given number of comparisons).

Theorem 1.1

It is impossible to solve the n-coin problem with l comparisons in the worst case if $n > \frac{1}{2}(3^l - 1)$.

In Section 2.3.3 we will use the same idea to prove a stronger result:

Theorem 1.2

It is impossible to solve the n-coin problem with l comparisons on the average if $n > \frac{1}{2}(3^l - 1)$.

Observe that the flow of outcomes along the branches leading out of a node allows global conclusions (that hold for the entire decision tree) from local data (that pertains to a single node). We will use this observation to obtain three general results:

1. A negative result concerning algorithms that do not use the true coin 0.

2. An elegant converse to Theorem 1.1, which states that l comparisons suffice to solve the n-coins problem when $n \leq \frac{1}{2}(3^l - 1)$, with an explicit construction of optimal decision trees when $n = \frac{1}{2}(3^l - 1)$.

3. A heuristic rule that allows us to efficiently obtain good (but not necessarily optimal) decision trees when the elegant solution for finding optimal trees does not apply.

Theorem 1.3

It is impossible to solve the n-coins problem with l comparisons, *without using coin* 0, if $n \geq \frac{1}{2}(3^l - 1)$.

Proof: Because of Theorem 1.1, we need only show that when $2n + 1 = 3^l$, an l-level decision tree for this restricted form of the n-coins problem cannot exist. At the root of any such tree, we would compare j coins (say coins $1, 2, \ldots, j$) against another set of j coins (say coins $j + 1, \ldots, 2j$), for some j, $1 \leq j \leq n/2$, as shown in Figure 1.7. Notice that $2j$ outcomes flow along each of the $<$ branch and the $>$ branch, and the remaining

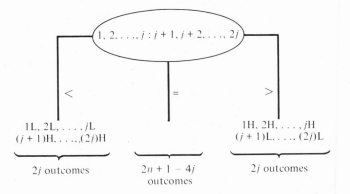

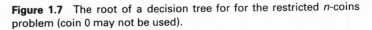

Figure 1.7 The root of a decision tree for for the restricted n-coins problem (coin 0 may not be used).

$2n + 1 - 4j$ cases flow along the $=$ branch. In order to complete this decision tree in such a way that it has $l = \log_3 (2n + 1)$ levels, one-third of the cases would have to flow along each of the three branches (why?). Thus we would have to be able to solve the equation $2j = 2n + 1 - 4j$ or, equivalently, $6j = 2n + 1$ for integers j and n, but this is impossible because $6j$ is even whereas $2n + 1$ is odd.

The converse argument to the one used in the proof of Theoreom 1.3 is worth remembering: If a tree can be found such that at every node the set of outcomes that flow through the node is partitioned into three groups of exactly equal size, then the tree must be optimal. This idea is the key for the proof of the following theorem.

Theorem 1.4

It is possible to solve the n-coins problem with l comparisons if $n \leq \frac{1}{2}(3^l - 1)$.

Proof.   Let $K_l = \frac{1}{2}(3^l - 1)$ and consider the case $n = K_l$ (the case $n < K_l$ is left as Exercise 7). Since the numbers K_l play an important role in the following calculations, it is useful to observe the two identities

$$2K_l + 1 = 3^l$$

and

$$K_l = 3K_{l-1} + 1.$$

Notice that $K_1 = 1$ and that the one-coin problem $n = K_1$ can indeed be solved by the single comparison $0:1$. We must show, in general, that the K_l-coins problem can be solved in l comparisons, and we will do so by exhibiting an algorithm that can be used to construct a decision tree for any given value of l.

During execution of this algorithm we carefully keep track of all the knowledge we have about the set of coins, and this knowledge will always be of one of three types. As the algorithm is executed and i ranges from l down to 1, the algorithm is always in one of the following three states.

State 1:   We have K_i doubtful coins (i.e., if there is a false coin in the set, then it is among these), and we are allowed i comparisons to complete the solution. This is the initial state of the problem when $i = l$.

State 2:   We have a set of $K_i + 1$ coins and another set of K_i coins, we have i comparisons to go, and we know that there is either a heavy coin in the set of $K_i + 1$ coins or a light coin in the set of K_i coins.

State 3:   Same as State 2 but with "heavy" and "light" interchanged.

For each of these three states of the algorithm, we prescribe what the next comparison is to be according to the transition function f shown in Figure 1.8, and we verify that this action will always lead us to one of these same states, with i decreased by 1. When i reaches

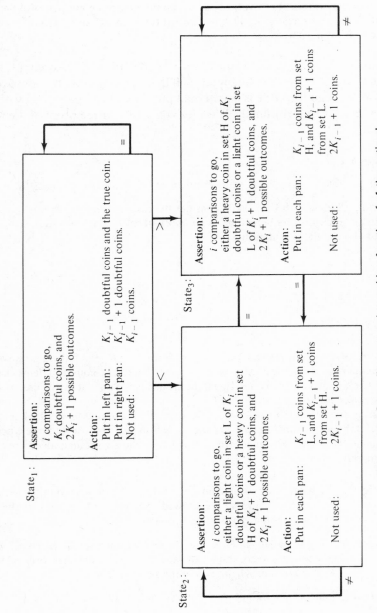

Figure 1.8 State diagram and transition function f of the optimal decision tree algorithm, Algorithm 1.5.

1, we solve the problem by a single comparison, as mentioned before. The algorithm has the following structure.

$$i \leftarrow l$$

$$s \leftarrow \text{State}_1$$

while $i > 1$ **do** $\left\{ \begin{array}{l} s \leftarrow f(s) \\ i \leftarrow i - 1 \end{array} \right.$

Solve the one-coin problem
with a single comparison.

Algorithm 1.5   Optimal algorithm for solving the false-coin problem with $K_l = \frac{1}{2}(3^l - 1)$ coins. The transition function f is given in Figure 1.8.

Completion of the proof is a matter of detail. As you follow each transition in the state diagram of Figure 1.8, you must simply verify that the sets of coins mentioned in the source and destination states match. By using the recurrence $K_i = 3K_{i-1} + 1$, you should also verify that the outcomes possible at each state flow along the three branches leading out of this state in equal-sized groups.

There is no elegant and efficient way to find an optimal solution for most combinatorial problems, as Algorithm 1.5 does for the false-coin problem. In this respect, the literature on combinatorial algorithms is misleading, for it naturally tends to concentrate on the elegant solutions even if they apply only to specialized problems. Elegant solutions often suffer a major flaw: if the problem is changed slightly, the solution becomes inapplicable. This situation would occur in the solution of Theorem 1.3, for example, if we changed the assumptions in the false-coin problem to allow at most two false coins instead of one. The ideas that survive minor changes in the problem statement have an air of generality, like the idea of observing the flow of outcomes along the branches of a decision tree. Let us illustrate another use of this idea.

An important principle of practical computing states that if a problem is hard to solve exactly, one should try to discover heuristics that allow approximate solutions to be found efficiently. In the case of decision trees for a problem with a given number of possible outcomes, an obvious heuristic is to try to keep the tree shallow—that is, to keep the number of levels as small as possible. This can be achieved by making the tree as wide as possible. A ternary decision tree with a given number of leaves can be made to be wide if, at every node, the outcomes that flow through the node are partitioned into three groups of approximately equal size.

In the case of the four-coins problem, with its nine possible outcomes, we would like to choose the comparison at the root in such a way that three outcomes flow along each of the branches leading out of the root. However, as we have seen, it is impossible if we are not given an additional true coin. In this restricted form of the four-coins problem, only two essentially different comparisons can be made at the root—namely, $1:2$ and $1, 2:3, 4$. They split the nine outcomes into the partitions $(2, 5, 2)$ and $(4, 1, 4)$, respectively. Which one is more uniform—that is, closer to our ideal partition $(3, 3, 3)$? Let us introduce an important measure of uniformity of such partitions. The motivation for choosing this particular measure cannot be fully presented here; additional justification for its importance will be given in Chapters 6 and 7.

Consider a probability vector $(p_1, p_2, \ldots, p_n)$, where each p_i is a nonnegative real number and $\sum_{i=1}^{n} p_i = 1$. In our example of Figure 1.4, the partition $(2, 5, 2)$ gives rise to the probability vector $p_1 = 2/9, p_2 = 5/9, p_3 = 2/9$; these are the probabilities of following each of the three branches leaving the root, assuming that the nine outcomes are equally likely. We want a function $\mathcal{H}(p_1, p_2, \ldots, p_n)$ with the following properties:

1. It is symmetric in $p_1, p_2, \ldots, p_n$; that is, it is independent of the relative order of the p_i.

2. It achieves its maximum on the most uniform vector $p_1 = p_2 = \cdots = p_n = 1/n$.

3. It achieves it minimum, 0, on a least uniform vector, $p_1 = 1, p_2 = p_3 = \cdots = p_n = 0$.

4. It increases monotonically from the points at which it achieves its minima to the point where it achieves its maximum: as α goes from 0 to $1/n$, $\mathcal{H}[1 - (n - 1)\alpha, \alpha, \ldots, \alpha]$ increases monotonically.

The well-known *entropy function* of information theory,

$$\mathcal{H}(p_1, \ldots, p_n) = \sum_{i=1}^{n} p_i \lg \frac{1}{p_i},$$

satisfies all these properties (and others that are desirable from different points of view).

Using this function to measure the degree of uniformity of the two vectors corresponding to the roots of the two decision trees of the four coin problem, we find

$$\mathcal{H}\left(\frac{2}{9}, \frac{5}{9}, \frac{2}{9}\right) = \frac{2}{9} \lg \frac{9}{2} + \frac{5}{9} \lg \frac{9}{5} + \frac{2}{9} \lg \frac{9}{2} \approx 1.44,$$

$$\mathcal{H}\left(\frac{4}{9}, \frac{1}{9}, \frac{4}{9}\right) = \frac{4}{9} \lg \frac{9}{4} + \frac{1}{9} \lg \frac{9}{1} + \frac{4}{9} \lg \frac{9}{4} \approx 1.39.$$

The same measure applied to the ideal partition $(3, 3, 3)$ yields

$$\mathcal{H}\left(\frac{1}{3}, \frac{1}{3}, \frac{1}{3}\right) = \lg 3 \approx 1.58.$$

Thus, according to this measure, the partition (2, 5, 2) is closer to uniform than (4, 1, 4). The heuristic principle of choosing the comparison that leads to the most uniform partition of the set of all outcomes correctly identifies the comparison 1 : 2 as the root of the best decision tree for the restricted four-coins problem. Since most decision tree problems admit no technique for finding optimal solutions other than exhaustive search, this simple heuristic principle is worth remembering.

This section has illustrated the benefits that can be obtained from a precise definition of a class of algorithms. From the single idea of observing the flow of outcomes along the branches of a decision tree, we were able to obtain bounds on the number of comparisons needed, an efficient algorithm for constructing optimal trees in a special case, and a valuable heuristic principle that applies to any decision tree problem.

1.6 ANALYSIS OF ALGORITHMS

In the process of designing and implementing an algorithm, one naturally discovers some properties of the algorithm involved: for example, it was obvious from the description of Algorithm 1.3 of Section 1.1 that it goes through $\lceil \lg n \rceil$ iterations for a bit string of length n. As algorithms become more and more complex, it is less and less likely that their important properties will become apparent during the design and implementation stages. As a rule, some important aspects of an algorithm's behavior are difficult to determine, such as whether the algorithm is correct and how many operations or how much memory is needed. Hence a new algorithm generally requires a lengthy analysis stage before it is well understood.

Because the analysis of algorithms can be difficult, it is frequently neglected; instead the program is executed to see what happens (e.g., to measure the running time). This approach may be satisfactory if there is reason to believe that the algorithm's performance on the test cases is indicative of its performance on other, yet untested, cases; otherwise it yields little information of value. Even in those cases in which testing provides a fair indication of an algorithm's performance, it never answers the nagging question of whether better algorithms might exist for solving the same problem. Only analysis can answer the questions of optimality.

There are two fundamental questions in the analysis of algorithms:

1.   What are the properties of a given algorithm?

2.   What properties must *any* algorithm have if it solves a given problem?

The fundamental difference between the two questions is in the approach taken to answer them. In the first case, an algorithm is given and conclusions are drawn by studying properties inherent in the algorithm. In the second case, a problem is given and a framework for algorithms is specified; conclusions are drawn by studying the nature of the problem vis-à-vis the class of algorithms.

The discussion of the false-coin problem in Section 1.5 provided examples of both types of questions, in a setting where they could be answered with relative ease. In this section we resume the discussion of difference-preserving codes in order to provide nontrivial examples of the analysis of an individual algorithm and of the class of all algorithms for solving a given problem.

Analysis of a Code-Construction Algorithm.   Consider the algorithm that constructs a sequence $\mathcal{D}_0, \mathcal{D}_1, \mathcal{D}_2, \ldots$ of difference-preserving codes by starting with an arbitrary code $\mathcal{D}_0$ and constructing $\mathcal{D}_i$ by composing $\mathcal{D}_{i-1}$ with itself according to the range-extension technique of Section 1.3. Let us compute, as a function of the codeword length, the range of the codes $\mathcal{D}_i$ constructed by this algorithm. Let $R(n)$ be the range of an n-bit code $\mathcal{D}_i$ and, accordingly, let $R(2n)$ be the range of the code $\mathcal{D}_{i+1}$ obtained by composing $\mathcal{D}_i$ with itself. Then we know from equation (1.1) of Section 1.3 that

$$R(2n) = \left\lceil \frac{R(n)}{t+1} \right\rceil R(n) + t \left\lfloor \frac{R(n)}{t+1} \right\rfloor.$$

In order to determine what kind of function satisfies this equation and, in particular, how fast the solutions to this equation grow, let us simplify the right-hand side (without changing it too much) and study instead the functional equation

$$r(2n) = \frac{r^2(n)}{t+1}, \tag{1.2}$$

hoping that $r(n)$ is not too different from $R(n)$. In Exercise 11 we pursue the issue of how good an approximation $r(n)$ is to $R(n)$.

Systematic techniques for solving such recursive equations are discussed in more detail in Chapter 3, but in this case trial and error is sufficient. After some experimentation, we are led to an exponential solution of the form

$$r(n) = c2^{an},$$

where c and a are constants to be determined. Substituting this attempted solution into the equation (1.2) yields

$$c2^{2an} \stackrel{?}{=} \frac{c^2 2^{2an}}{t+1},$$

which is satisfied if $c = t + 1$, regardless of the value of a. Thus the range-extension algorithm constructs codes with ranges that are approximately of the form

$$r(n) = (t+1)2^{an}. \tag{1.3}$$

The recursive equation imposed no constraint on the choice of a, but there is another constraint to be satisfied—namely, the initial condition: the range and the length of the code $\mathcal{D}_0$ with which the composition scheme was started. If we choose as $\mathcal{D}_0$ the optimal 4-bit DP-1 code with range 8, we obtain the initial condition

$$r(4) = 8 = 2 \cdot 2^{4a},$$

forcing $a = \frac{1}{2}$, and hence

$$r(n) = 2 \cdot 2^{n/2}.$$

We conclude that the range-extension algorithm starting with this particular code $\mathcal{D}_0$ constructs n-bit codes having about $2^{n/2+1}$ codewords out of the 2^n possible bit strings of length n. One way to state this result is as follows. Only about half the bits in this code can be utilized for information; the other half are determined partly by the problem (the face that the code must be difference-preserving) and partly by the framework imposed by the range-extension technique.

Can we do better? Is it possible to find algorithms that construct n-bit codes whose ranges are closer to 2^n? This question can be approached in two different ways. The obvious one is to try to improve the given algorithm. A more sophisticated approach is first to see whether there is room for improvement. Before discussing the more sophisticated approach, let us briefly examine the obvious one.

A minor improvement on our example can be obtained by starting the range-extension technique with a more efficient code $\mathcal{D}_0$. For example, the optimal 5-bit DP-1 code with range 14 yields $r(5) = 14 = 2 \cdot 2^{5a}$, or $a = \frac{1}{5} \lg 7 \approx 0.56$, a slight improvement. However, equation (1.3), together with the fact that $a < 1$ (the range of a code of length n obviously cannot exceed 2^n), tells us that, in any code constructed by the range-extension algorithm, a fraction $(1 - a)$ of the n bits of a codeword cannot be used for information. In order to obtain a more significant improvement, we must drop the range-extension technique described and search for a different code-construction principle. The result of the following analysis provides the motivation for embarking on such an extensive investigation.

An Upper Bound on the Range of (n, t)-Codes.   The two properties required of difference-preserving codes

1. $|i - j| \leq t$ implies $H(D_i, D_j) = |i - j|$ and
2. $|i - j| > t$ implies $H(D_i, D_j) > t$

force the codewords D_i to have certain minimal distances from each other, which, in turn, prevents the use of all vertices of an n-cube as codewords of an n-bit code. Figure 1.2 illustrates this phenomenon. In this section we derive an upper bound on the range of an n-bit DP-t code, assuming nothing but the definition of such codes. Hence this bound, which is inherent in the problem statement, must hold for any code-construction algorithm.

The upper bound that we will obtain for the range N of an n-bit DP-t code is of the form

$$N \leq c\,2^{n}n^{-\lfloor t/2 \rfloor}, \tag{1.4}$$

where c depends on t but not on n. The right-hand side of this inequality grows much faster than the estimate $r(n) = (t + 1)2^{an}$, $a < 1$, that we obtained in equation (1.3) for codes constructed according to the range-extension technique. The discrepancy between this upper bound and the lower bound provided by equation (1.3) leaves a great deal of uncertainty about what ranges are achievable: is the upper bound loose or is the range-extension technique inefficient? This uncertainty provides the typical setting for a research problem. Our particular problem is too specialized to justify further discussion here, and so we will merely assert that the upper bound is reasonably tight and that codes can be constructed whose ranges come close to it. The interested reader is referred to the references in Section 1.7. In this section we want only to outline the problems that must be overcome in analyzing nontrivial algorithms. Let us sketch the derivation of the upper bound in inequality (1.4). The details are left for the reader to fill in.

Proof of Inequality 1.4:   Let $\mathcal{D} = \{D_1, D_2, \ldots, D_N\}$ be an n-bit DP-t code and identify the codewords D_i with vertices in an n-dimensional code, as shown in Figure 1.2. To every vertex V of the n-cube, assign a weight $W_r(V)$, equal to the number of codewords that are at Hamming distance r from V, where $r = \lfloor t/2 \rfloor + 1$:

$$W_r(V) = (\text{number of } D_i \text{ such that } H(V, D_i) = r).$$

First, we show that

$$W_r(V) \leq n \tag{1.5}$$

or, equivalently, that among the $\binom{n}{r}$ vertices at distance r from V there can be no more than n codewords of $\mathcal{D}$. If V is a codeword in $\mathcal{D}$, then $W_r(V) \leq 2 \leq n$ (why?). To settle the more interesting case when V is not in $\mathcal{D}$, assume, without loss of generality, that V is the vertex whose n coordinates are all zero. Then the vertices at distance r from V have exactly r ones and $n - r$ zeros. Among them, let $C_1, C_2, \ldots, C_s$ be all the vertices that are codewords in $\mathcal{D}$. In order to show that $s \leq n$, write the coordinates of $C_1, \ldots, C_s$ as rows in a matrix of s by n bits.

$$
\begin{array}{c|ccccccc}
 & \multicolumn{7}{c}{n} \\
\hline
C_1 & 0 & 0 & 1 & 1 & 0 & \cdots \\
C_2 & 1 & 0 & 1 & 0 & 1 & \cdots \\
 & \cdot & \cdot & \cdot & \cdot & \cdot & \cdots \\
C_s & 1 & 1 & 0 & 1 & 1 & \cdots \\
\end{array}
$$

and verify that

1.   Each row has exactly r ones, and hence the total number of ones in the matrix is sr.

2.   In each column there are at most r ones (the verification of this point is the subject of Exercise 12), and hence the total number of ones in the matrix is at most nr.

Combining these, we obtain $sr \leqq nr$, or $s \leqq n$. (The simple trick of counting the elements of a matrix by rows and by columns and equating the result of the two counts goes amazingly far, as Section 3.4.2 will show).

Having now established that $W_r(V) \leqq n$, observe the following facts about the weight-distribution in the n cube:

1.   The total weight in the cube is $N\binom{n}{r}$, since each of the N codewords contributes a weight of one to each of the $\binom{n}{r}$ vertices at distance r from it.

2.   The total weight of all codewords D_i is $2(N - r)$ (Exercise 13), and hence the total weight of all vertices outside $\mathscr{D}$ is $N\binom{n}{r} - 2(N - r)$.

3.   Since there are $2^n - N$ vertices V outside $\mathscr{D}$ and each of them has weight $Wr(V) \leqq n$, the total weight of all vertices outside $\mathscr{D}$ is at most $(2^n - N)n$.

Combining (2) and (3), we obtain the inequality

$$N\binom{n}{r} - 2(N - r) \leqq (2^n - N)n$$

or

$$N\left[\binom{n}{r} + n - 2\right] \leqq n\,2^n - 2r.$$

For fixed r and increasing n, $\binom{n}{r}$ grows approximately as bn^r, where b depends on r but not on n. With this information, we can show that

$$N \leqq c\frac{n\,2^n}{n^r} = c\frac{2^n}{n^{\lfloor t/2 \rfloor}},$$

which is inequality (1.4).

The reader who has worked through this chapter diligently, starting with the trivial bit sum algorithm that scans each bit in turn and ending with the demanding exercise of filling in the details of the derivation of the upper bound (1.4), has had a bird's-eye view of the entire field of combinatorial algorithms. In particular, every topic encountered in this book will be a reminder of the principles of algorithm design illustrated in Sections 1.3 and 1.4, the consideration of details of implementation as discussed in Sections 1.1 and 1.2, and the mathematical techniques of algorithm analysis used in Sections 1.5 and 1.6.

1.7 REMARKS AND REFERENCES

Although combinatorial mathematics is an old field (it was given its name by Leibnitz in 1666 in his *Dissertatio de Arte Combinatoria*), combinatorial algorithms with its emphasis on the design, analysis, and implementation of practical algorithms, is a product of the computer age. Reference to pioneering papers dealing with special aspects of this field will be given in later chapters, but more general references appear below.

LEHMER, D. H., "Combinatorial Problems with Digital Computers," *Proc. Fourth Canadian Math. Congress, 1957,* University of Toronto Press, 1960, 160–173.

LEHMER, D. H., "Teaching Combinatorial Tricks to a Computer," Chapter 15 in *Combinatorial Analysis*, R. Bellman and M. Hall (eds.), *Proc. Symp. Applied Math., Vol.* 10, American Math. Society, Providence, R.I., 1960, 179–193.

LEHMER, D. H., "The Machine Tools of Combinatorics," Chapter 1 in *Applied Combinatorial Mathematics*, E. F. Beckenbach (Ed.), Wiley, New York, 1964, 5–31.

HALL, M., and D. E. KNUTH, "Combinatorial Analysis and Computers," *American Math. Monthly*, **72**, Pt. II (1965), 21–28.

More recent textbooks include

WELLS, M. B., *Elements of Combinatorial Computing*, Pergamon Press, Oxford, 1971,

EVEN, S., *Algorithmic Combinatorics*, Macmillan, New York, 1973,

AHO, A. V., J. E. HOPCROFT, and J. D. ULLMAN, *The Design and Analysis of Computer Algorithms*, Addison-Wesley, Reading, Mass., 1974,

and

NIJENHUIS, A., and H. S. WILF, *Combinatorial Algorithms*, Academic Press, New York, 1975.

A wealth of knowledge on combinatorial algorithms can be found in

KNUTH, D. E., *The Art of Computer Programming*, Vol. 1, Fundamental Algorithms (1968), Vol. 2, Seminumerical Algorithms (1969), Vol. 3, Sorting and Searching (1973), Vol. 4, Combinatorial Algorithms (to appear), Addison-Wesley, Reading, Mass.

The logarithmic bit sum algorithm of Section 1.1 has probably been invented independently by several people. One of the first implementations is due to David Muller of the University of Illinois, who programmed it for the Illiac I computer in 1954.

The difference-preserving codes of Section 1.2, or minor variations of them, have been known under various names (path codes, circuit codes, snake-in-a-box codes) since the special case $t = 1$ was introduced in

KAUTZ, W. H., "Unit-Distance Error-Checking Codes," *IRE Trans. Electronic Computers,* **EC-7** (1958), 179–180.

This paper, as well as several subsequent ones, is concerned with the application of such codes to error-detection and error-correction problems, particularly in analog-to-digital conversion. A bibliography and a discussion of the application of difference-preserving codes to pattern recognition are given in

PREPARATA, F. P., and J. NIEVERGELT, "Difference-Preserving Codes," *IEEE Trans. Information Theory* **IT-20**, (1974), 643–649.

An analysis of a more general version of the false-coin problem of Section 1.6 is presented in

SMITH, C. A. B., "The Counterfeit-Coin Pr blem," *The Mathematical Gazette,* **21** (1947), 31-39.

Analysis of combinatorial algorithms is a topic that has recently received much attention. Two survey articles are

KNUTH, D. E., "Mathematical Analysis of Algorithms," *Information Processing 71* (Proceedings of the 1971 IFIP Congress), North-Holland Publishing Co., Amsterdam, 1972, 19–27,

and

FRAZER, W. D., "Analysis of Combinatory Algorithms—A Sample of Current Methodology," *Spring Joint Computer Conference, AFIPS Conference Proceedings*, Vol. 40 (1972), 483–491.

1.8 EXERCISES

1. Algorithm 1.3 (the fast bit sum algorithm of Section 1.1) can be modified slightly so that instead of computing the bit sum $S(B)$ of a bit string B, it computes certain other functions of bit strings. State necessary and/or sufficient conditions for a function to be computable by an algorithm that differs from the fast bit sum algorithm only in that addition is replaced by some other operation.

2. The well-known Gray codes are partially characterized by the requirement that G_i and G_{i+1} differ in exactly one bit position, where G_i denotes the bit string that encodes the integer i. Prove that, for all n, there exist Gray codes of codeword length n for the integers $1, 2, \ldots, 2^n$. Show that the additional requirement that G_{2^n} and G_1 differ in exactly one bit position can be satisfied. (*Hint*: Try the composition principle approach. Assume that you have an n-bit Gray code for the integers $1, 2, \ldots, 2^n$ satisfying all the requirements. How would you construct an $(n + 1)$-bit Gray code for the integers $1, 2, \ldots, 2^{n+1}$?)

3. The *Tower of Hanoi problem* is illustrated in the following figure.

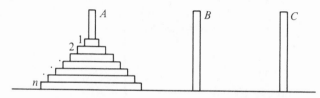

There are three poles labeled A, B, C. On pole A, n discs of radius $1, 2, \ldots, n$ (each with a hole in the middle) are piled up in such a way that the disc of radius i is the ith from the top. The problem is to move all n discs to pole C so that again the disc of radius i is the ith from the top. Only one disc may be moved at a time, from any pole to any other one, subject to the restriction that no disc may ever rest above a smaller disc on the same pole. Show how the decomposition principle of Section 1.4 leads to a solution of the problem of the Tower of Hanoi. (*Hint*: Solve Exercise 2 first.)

4. Show that any n-bit DP-1 code with range N can be used to produce an $(n + 1)$-bit DP-1 code with range $\lceil 3N/2 \rceil$.

5. Show that any strategy for splitting B (in Section 1.4) yields an algorithm that requires $n - 1$ additions.

6. Design a decision tree to sort four coins of arbitrary weight.

7. Complete the proof of Theorem 1.4 by showing how the n-coins problem can be solved in l comparisons when $n < K_l$.

8. Prove a result analogous to Theorem 1.4 for the restricted n-coins problem. When no additional coin known to be true is available, then the n-coins problem can be solved in l comparisons if $n \leq \frac{1}{2}(3^l - 1) - 1$.

★9. Estimate the number of essentially different (i.e., nonisomorphic) decision trees that solve the n-coins problem.

10. Analyze the threshold addition technique of Section 1.3 in a way similar to the analysis (in Section 1.6) of the range-extension technique for constructing difference-preserving codes.

11. This problem is concerned with estimating the difference between the approximation $r(n)$ and the function $R(n)$ in Section 1.6.
    (a) Show that, for any integer $t \geq 1$ and any initial condition $R(1) = c \geq 1$, the function $R(n)$ grows monotonically.
    (b) Show that under the same initial conditions $R(1) = r(1) > 1$ we have $R(n) \geq r(n)$ for all $n > 1$.
    (c) Consider the recursion formula

    $$\hat{r}(2n) = \frac{\hat{r}^2(n)}{t+1} + 2\hat{r}(n).$$

    Show that under the same initial condition $R(1) = \hat{r}(1) \geq 1$ we have $R(n) < \hat{r}(n)$ for all $n > 1$.
    (d) Find a closed-form expression for the solutions $\hat{r}(n)$ of the recursion formula in part (c).
    (e) Since $r(n)$ and $\hat{r}(n)$ are lower and upper bounds for $R(n)$, respectively, $[\hat{r}(n) - r(n)]/r(n)$ is a bound for the *relative* error of the approximation $r(n)$ to the function $R(n)$. Determine the growth rate of this bound for the relative error.

12. Complete step (2) of the proof of inequality (1.5) by proving each of the following statements.

    (a) $H(C_i, C_j)$ is even for $1 \leq i, j \leq s$.
    (b) $H(C_i, C_j) \leq 2r \leq t + 2$.
    (c) If C_i, C_j agree in any one component, then $H(C_i, C_j) \leq 2r - 2 \leq t$.

(d) Among all those C_i that have a 1 in a given column, there can be at most r at pairwise even distance.

13. Show that for a DP-t code at range N and for $r \leqq t$,

(a) $N - 2r$ codewords have exactly two other codewords at distance r.

(b) $2r$ codewords have exactly one other codeword at distance r.

chapter **2**

Representation of Combinatorial Objects

Most computers allow only bit strings, integers, and characters as basic data types; therefore more complex objects must be encoded as bit strings, integers, or characters before they can be manipulated. For example, floating-point numbers are encoded as an integer pair—the mantissa and the exponent—but this encoding is usually invisible to the user. In contrast, the encoding of objects discussed in this chapter (sets, sequences, and trees) is almost always left to the user.

Any given type of object can have several possible representations, and the best choice depends critically on the intended use of the object and the type of manipulations that it will undergo. We are thus impelled to discuss not only the properties of the representations themselves but also some applications.

2.1 INTEGERS

The integers are basic to combinatorial computing, and although readily available on all computers, it is sometimes convenient to represent them in a form other than the one directly available. In various computational number-theoretic investigations, it is the integers themselves that are to be studied, but our use of them will be primarily for counting and indexing. It is in the latter role that alternative representations are useful. In this section we will discuss the important and general class of radix (i.e., positional) representations, leaving other more specialized ones to Exercises 4, 5, and 6.

We will consider only the nonnegative integers $0, 1, 2, \ldots$, since negative counts and indices are not generally needed. Furthermore, it is easy to adjoin a single sign-bit to any representation of the nonnegative integers.

Radix systems for representing integers are extremely well known, for they occur in so many different subjects ranging from "the new math" to advanced courses in number theory. In the *base r radix system* every positive integer has a unique representation as a finite sequence

$$(d_0, d_1, d_2, d_3, \ldots, d_k) \tag{2.1}$$

in which each d_i is an integer satisfying $0 \leq d_i < r$ and $d_k \neq 0$. Zero is represented by the sequence (0). $r > 1$ is the *radix* or *base* of the system. The integer that corresponds to the sequence (2.1) is

$$N = d_0 + d_1 r + d_2 r^2 + d_3 r^3 + \ldots + d_k r^k,$$

and it is common to express this fact as

$$N = (d_k d_{k-1} \ldots d_1 d_0)_r.$$

Various values of r have been used throughout history. For example, the ancient Babylonians used $r = 60$ and the ancient Mayas $r = 20$. The most well known today are $r = 10$, the decimal system which we inherited from the Arabs, and $r = 2$, the binary system which is the basis of modern computers. Actually, it is only at the bottommost hardware level that the binary system is used; a computer's architecture and basic language are more likely to use $r = 8$ or $r = 16$.

The uniqueness of this representation follows from a proof by contradiction. $N = 0$ and $N = 1$ obviously have unique representations. Suppose that the representation is not unique and let $N > 1$ be the least integer having two different representations

$$N = (d_k d_{k-1} \ldots d_0)_r$$

$$= (e_l e_{l-1} \ldots e_0)_r.$$

If $k \neq l$, without loss of generality, suppose that $k > l$. Then since

$$\sum_{i=0}^{l} (r - 1) r^i = r^{l+1} - 1 < r^{l+1} \leq r^k$$

and since $d_k \neq 0$, we conclude that

$$(d_k d_{k-1} \ldots d_0)_r > (e_l e_{l-1} \ldots e_0)_r, \tag{2.2}$$

a contradiction. Thus we must have $k = l$. Similarly, if $d_k > e_k$, we would again have inequality (2.2), and so we must have $d_k = e_k$. Consequently,

$$N - d_k r^k = (d_{k-1} d_{k-2} \ldots d_0)_r$$

$$= (e_{k-1} e_{k-2} \ldots e_0)_r$$

has two different representations, contradicting the supposition that N was the least such integer.

To prove that every positive integer has a base r representation, it suffices to give an algorithm for constructing the (necessarily unique) representation of a given integer N. Algorithm 2.1 does so as follows. It constructs the sequence $d_0, d_1, d_2, \ldots, d_k$ by repeatedly dividing by r and recording the remainder. First N is divided by r, the remainder being d_0. The quotient of one division becomes the dividend of the next, and the sequence of remainders is the desired base r representation.

$$d_0 \leftarrow 0$$

$$q \;\leftarrow N$$

$$k \;\leftarrow 0$$

$$\textbf{while } q \neq 0 \textbf{ do} \begin{cases} d_k \leftarrow q \bmod r \\[2mm] q \;\;\leftarrow \left\lfloor \dfrac{q}{r} \right\rfloor \\[2mm] k \;\;\leftarrow k + 1 \end{cases}$$

$$\textbf{if } k \neq 0 \textbf{ then } k \leftarrow k - 1$$

Algorithm 2.1 Conversion of N to its base r representation $(d_k d_{k-1} \ldots d_1 d_0)_r$.

It is obvious that the sequence $d_0, d_1, \ldots, d_k$ computed by Algorithm 2.1 satisfies $0 \leqq d_i < r$ and $d_k \neq 0$ except for $N = 0$. A simple proof by induction shows that $N = \sum_{i=0}^{k} d_i r^i$.

An important generalization of radix systems is *mixed radix systems*, in which we are given not a single radix r but rather a radix sequence $r_0, r_1, r_2, \ldots$ of integers and the sequence (2.1) corresponds to the integer

$$N = d_0 + d_1 r_0 + d_2 r_0 r_1 + d_3 r_0 r_1 r_2 + \cdots + d_k \prod_{i=0}^{k-1} r_i,$$

only now each d_i satisfies $0 \leqq d_i < r_i$ and $d_k \neq 0$ except for $N = 0$. The fact that each such sequence (2.1) corresponds to a unique integer and that each positive integer has a unique representation follow from an easy generalization of the results for ordinary radix systems, which correspond to the special case $r_i = r$, $i \geqq 0$.

Mixed radix systems may seem strange at first, but actually they are almost as common in everyday life as the decimal system. Consider, for example, our system for measuring time—seconds, minutes, hours, days, weeks, and years. It is precisely a mixed radix system with $r_0 = 60$, $r_1 = 60$, $r_2 = 24$, $r_3 = 7$, and $r_4 = 52$.

In Chapter 5 we will use certain mixed radix systems to establish correspondences between permutations and integers. In particular, we will use the *factorial number system* in which the sequence (2.1) corresponds to the integer

$$N = d_0 0! + d_1 1! + d_2 2! + d_3 3! + \cdots + d_k k!,$$

$0 \leq d_i < i + 1$, and $d_k \neq 0$ except for $N = 0$; notice that we always have $d_0 = 0$. This system is a mixed radix system in which $r_i = i + 1$. We will also use, for a fixed value of $n > k$, the *falling factorial number system* in which the sequence (2.1) corresponds to the integer

$$N = d_0 + d_1 n + d_2 n(n - 1) + \cdots + d_k \frac{n!}{(n - k)!},$$

$0 \leq d_i < n - i$ and $d_k \neq 0$ except for $N = 0$; in this case, we always have $d_{n-1} = 0$. This system is a mixed radix system in which $r_i = n - i, n > i \geq 0$.

Conversion of an integer N to the mixed radix system $(r_0, r_1, r_2, \ldots)$, Algorithm 2.2, is a simple generalization of Algorithm 2.1. Instead of always using r as the divisor to produce d_k, it uses r_k.

$$d_0 \leftarrow 0$$

$$q \leftarrow N$$

$$k \leftarrow 0$$

$$\textbf{while } q \neq 0 \textbf{ do } \begin{cases} d_k \leftarrow q \bmod r_k \\ q \leftarrow \left\lfloor \dfrac{q}{r_k} \right\rfloor \\ k \leftarrow k + 1 \end{cases}$$

$$\textbf{if } k \neq 0 \textbf{ then } k \leftarrow k - 1$$

Algorithm 2.2   Conversion of N to its representation $(d_0, d_1, \ldots, d_k)$ in the mixed radix system $(r_0, r_1, r_2, \ldots)$.

2.2   SEQUENCES

An *infinite sequence*

$$s_1, s_2, s_3, \ldots$$

is formally defined as function f whose domain is the set of positive integers; $f(i) = s_i, i \geq 1$. In many instances, it is more convenient to begin indexing the elements of the sequence at 0; then the domain of f is the set of nonnegative integers. We similarly define a *finite sequence* or *list*

$$s_1, s_2, \ldots, s_n$$

as a function whose domain is the set $\{1, 2, \ldots, n\}$. The prime numbers are an example of an infinite sequence

$$
\begin{array}{rllllllllll}
i: & 1 & 2 & 3 & 4 & 5 & 6 & 7 & 8 & 9 & 10 & \cdots \\
p_i: & 2 & 3 & 5 & 7 & 11 & 13 & 17 & 19 & 23 & 29 & \cdots
\end{array}
\tag{2.3}
$$

and a permutation

$$
\Pi(1, 2, 3, 4, 5, 6) = (6, 2, 5, 1, 3, 4)
$$

is an example of a finite sequence.

In combinatorial algorithms, we will frequently have to represent and manipulate finite sequences (or, equivalently, initial segments of infinite sequences). This section examines various techniques for such representations and manipulations.

2.2.1 Sequential Allocation

The simplest computer representation of a finite sequence is an explicit list of its members, arranged in order in adjacent memory locations. Thus s_1 is stored, say, in location l_1, s_2 is stored in location $l_2 = l_1 + d$, s_3 is stored in location $l_3 = l_1 + 2d$, and so on, where d is the number of locations required to store a single element of the sequence. This sequential representation is illustrated in Figure 2.1.

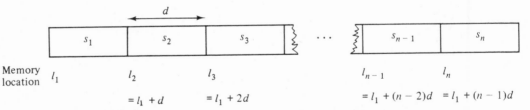

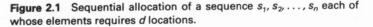

Figure 2.1 Sequential allocation of a sequence $s_1, s_2, \ldots, s_n$ each of whose elements requires d locations.

Such a representation of a sequence has a number of advantages. First, it is easy to implement and requires little overhead in terms of storage. Also, it is useful in many instances, since there is a simple, fixed relationship between the value of i and the location of s_i:

$$
l_i = l_1 + (i - 1)d.
$$

This relationship allows for direct, immediate access to any element of the sequence. Finally, this sequential representation is reasonably general and includes, as an important special case, the representation of multidimensional arrays (Exercise 7).

For example, to represent the $n \times m$ array

$$
\begin{pmatrix}
a_{11} & a_{12} & \cdots & a_{1m} \\
a_{21} & a_{22} & \cdots & a_{2m} \\
\vdots & \vdots & & \vdots \\
a_{n1} & a_{n2} & \cdots & a_{nm}
\end{pmatrix},
\tag{2.4}
$$

consider it as a sequence $s_1, s_2, \ldots, s_n$ in which each s_i is a sequence of m elements consisting of the ith row of the matrix. Thus d, the number of locations required to represent an element s_i, is $m\hat{d}$, where $\hat{d}$ is the number of locations required to represent an array element a_{ij}. Since the sequence s_i begins at location

$$
\begin{aligned}
l_i &= l_1 + (i - 1)d \\
&= l_1 + (i - 1)m\hat{d},
\end{aligned}
$$

the location of a_{ij} is therefore

$$
l_i + (j - 1)\hat{d} = l_1 + [(i - 1)m + (j - 1)]\hat{d}.
$$

This representation is known as *rowwise* storage of the matrix; the *columnwise* storage is obtained by considering the array (2.4) to be a sequence $t_1, t_2, \ldots, t_m$ in which each t_i is a sequence of n elements consisting of the ith column of the matrix.

As useful as sequential allocation of a sequence is, it does have some significant drawbacks. For example, if the sequence changes through the insertion and deletion of elements, then this representation is cumbersome; inserting a new element between s_i and s_{i+1} requires shifting $s_{i+1}, s_{i+2}, \ldots, s_n$ to the right by one position. Similarly, deleting s_i requires shifting those elements to the left by one position. Such movement of the elements can be expensive in terms of processing time, and in the case of such dynamic sequences it is generally better to use the techniques of linked allocation discussed in the next section.

Characteristic Vectors.   An important variant of sequential allocation is possible if the sequence to be represented is a subsequence of an underlying sequence $s_1, s_2, s_3, \ldots .$[†] In such a case, it may be advantageous to represent the sequence by means of its *characteristic vector*, a sequence of 0s and 1s in which the ith bit is 1 if s_i is in the sequence and 0 if it is not.

[†]The sequence $t_1, t_2, t_3, \ldots$ is a *subsequence* of the sequence $s_1, s_2, s_3, \ldots$, provided that $t_j = s_{i_j}$ for $1 \leq i_1 < i_2 < i_3 < \cdots$. Thus, for example, the primes that are congruent to 1 modulo 4—that is, 5, 13, 17, 29, . . . —form a subsequence of the primes 2, 3, 5, 7, 11, 13, 17, 19, 23, 29, ...

For example, the characteristic vector for an initial segment of the sequence (2.3) of primes is

$$s_i: \quad 1 \quad 2 \quad 3 \quad 4 \quad 5 \quad 6 \quad 7 \quad 8 \quad 9 \quad 10$$

characteristic
vector for primes: $\quad 0 \quad 1 \quad 1 \quad 0 \quad 1 \quad 0 \quad 1 \quad 0 \quad 0 \quad 0.$

Here the underlying sequence is the sequence of positive integers. On a computer with 32-bit words, storing the sequence of primes less than 10^6 would require $10^6/32 = 31250$ words. Furthermore, by noticing that $2i$ is not a prime for $i > 1$, we can save half that space by recording only the bits for $2i + 1$, $i \geqq 1$, and remembering that the prime 2 has been omitted. Thus the primes less than 10^6 can be recorded, in a sense, in only 15625 words. Since there are 78498 primes less than 10^6, the sequential representation described previously requires more than five times as much space.

Characteristic vectors are useful in a number of instances that are discussed in the sieve techniques of Section 4.2. In general, their utility stems from their compactness, the fact that there is a simple, fixed relationship between i and the location of the ith bit, and the fact that it is very easy to delete elements with this representation. Actually, the examples in Section 4.2 all start with an underlying sequence (e.g., the positive integers) and compute the desired subsequence (e.g., the prime numbers) by successively deleting elements discovered *not* to be in the subsequence.

The major disadvantages of characteristic vectors are that they are wasteful except for "dense" subsequences of sequences $s_1, s_2, s_3, \ldots$ and that they are difficult to use unless there is a simple relationship between i and s_i. If the relationship is complex, the result may be much overhead in processing time. If the subsequence is not sufficiently dense in $s_1, s_2, s_3, \ldots$, then the overhead in storage may be significant. In the case of the prime numbers, the simple relation between i and s_i is $s_i = i$ (or $s_i = 2_i + 1$ if only the odd numbers are considered). The Prime Number Theorem tells us that the number of primes less than n is approximately $n/\ln n$; thus the primes are relatively dense in the integers.

2.2.2   Linked Allocation

The inefficiency of inserting and deleting elements in a sequentially allocated sequence occurs because the order of the sequence elements is recorded implicitly by requiring that adjacent elements of the sequence be in adjacent memory locations. The result is that many elements of the sequence may have to be moved during an insertion or deletion. If this requirement is dropped, then we can perform these operations without having to move elements of the sequence. Of course, it is imperative that the information about the ordering of the sequence elements be retained, but that can be done explicitly instead of implicitly. In particular, *linked allocation* of a sequence (a *linked list*) means that there is a pointer P_i associated with each s_i that records the location at which s_{i+1} and P_{i+1} are stored. There is also a pointer P_0 that records the location of the *head* of the sequence—that is, the location of s_1 and P_1. This is illustrated in Figure 2.2.

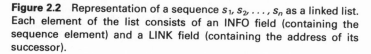

Figure 2.2 Representation of a sequence $s_1, s_2, \ldots, s_n$ as a linked list. Each element of the list consists of an INFO field (containing the sequence element) and a LINK field (containing the address of its successor).

Each node in Figure 2.2 consists of two fields, the sequence element itself in the INFO field and the pointer to its successor in the LINK field. Since there is no successor to $s_n = \text{INFO}(l_n)$, we use the notation $P_n = \text{LINK}(l_n) = \Lambda$, where Λ is the *empty* or *null* pointer. The exact locations $l_1, l_2, \ldots, l_n$ are unimportant, and so it is more common to depict the sequence of Figure 2.2 as shown in Figure 2.3.

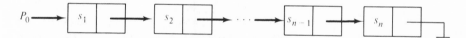

Figure 2.3 An alternate, more common way of depicting the list in Figure 2.2.

This linked representation facilitates the operations of insertion of a sequence element after s_i and deletion of the sequence element s_{i+1} if we know the location of s_i. All that is necessary is to change the values of some pointers. For example, to delete the element s_2 from the sequence in Figure 2.3, it is only necessary to set $\text{LINK}(l_1) = \text{LINK}(l_2)$ and the element s_2 is no longer in the sequence (see Figure 2.4). To insert a new element $s_{1.5}$ to the sequence in Figure 2.3, it is only necessary to create a new element at some location $l_{1.5}$ with

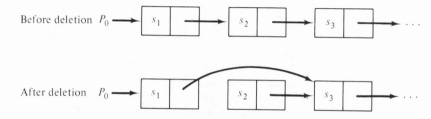

Figure 2.4 Deletion of an element from a linked list.

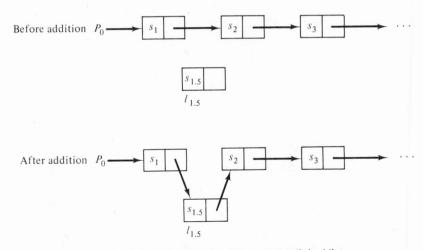

Figure 2.5   Insertion of an element to a linked list.

INFO $(l_{1.5}) = s_{1.5}$ and LINK $(l_{1.5}) =$ LINK (l_1) and to set LINK $(l_1) \leftarrow l_{1.5}$ (see Figure 2.5). (What is required to add an element $s_{0.5}$ to the sequence?) Concatenation of sequences and splitting a sequence into subsequences are also easy.

The use of linked allocation implies the existence of some mechanism for allocating new locations as they are needed and collecting old locations as they are released, but any discussion of this topic is beyond the scope of this book. We will assume, for the sake of presentation in various algorithms, that there is a primitive operation called **get-cell**, which, when used on the right side of an assignment statement, produces the address (location) of a new, unused memory cell. Thus to add the element $s_{1.5}$ as shown in Figure 2.5, we would actually have to use **get-cell** to find the value of $l_{1.5}$. We will completely ignore the problem of releasing unneeded memory locations, assuming that they are somehow collected for later use (this process is generally known as *garbage collection*).

We have lost quite a bit of information in order to gain this increased flexibility. The fixed relationship between i and the location of s_i in a sequential allocation allows us, for example, to have immediate and direct access to any element of the sequence. In a linked allocation there is no such relationship, and access to sequence elements other than the first is indirect and inefficient. For instance, given the length of a sequence, it is easy to find the middle element if the sequence is sequentially allocated, but relatively difficult if the sequence is represented as a linked list. Furthermore, a price is paid in terms of storage overhead for the pointers P_i.

In choosing between sequential and linked allocation for a particular application, we must examine the types of operations that will be performed on the sequence in order to make an intelligent decision. If the operations are largely accessing random elements, searching for specific elements (see Chapter 6), or sorting the elements into order (see Chapter 7), then sequential allocation is usually better. On the other hand, linked allocation is preferable if the operations are largely inserting and/or deleting elements and concatenation and/or splitting of sequences.

Variants of Linked Lists.   A trivial modification of the linked list in Figure 2.3 results in a slightly more flexible linked representation of a sequence: If we set P_n to point to s_1, as shown in Figure 2.6, we obtain what is known as a *circular list*. This change makes it possible to access (albeit indirectly) any element of the sequence from any other element of the sequence. Insertion and deletion of elements are the same as in noncircular linked lists (but what about insertion into an empty list?), whereas concatenation and splitting are slightly more complicated.

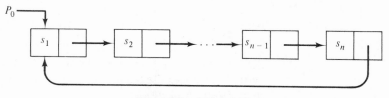

Figure 2.6   A circular list.

Even greater flexibility is achieved by *doubly linked lists* in which each sequence element s_i has two pointers associated with it instead of only one. These pointers record the locations of s_{i-1} and s_{i+1}, respectively, as shown in Figure 2.7. As a result, we have immediate, direct access to both the predecessor and the successor of a sequence element, thus facilitating such operations as the addition of a new element *before* s_i and deletion of the element s_i without already knowing its predecessor. If needed, doubly linked lists can be made circular in the obvious way.

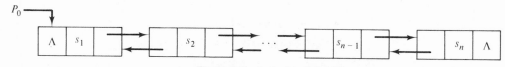

Figure 2.7   A doubly linked list.

2.2.3   Stacks and Queues

Two data structures based on dynamic sequences (i.e., sequences that change through insertions and deletions) are especially important in combinatorial algorithms. In each case, the insertions and deletions that the sequence undergoes are of a restricted type: they occur only at the ends of the sequence. A *push down stack*, or *stack* for short, is a sequence in which all insertions and deletions occur only at the right end, in this case, called the *top* of the stack (correspondingly, the left end of the sequence is called the *bottom*). Thus the elements enter and leave the stack in a first-in, last-out order. A *queue* is a sequence in which all insertions are made at the right end (the *rear* or *back*), whereas all deletions are made at the left end (the *front*). In contrast to a stack, a queue operates in a first-in, first-out order.

The importance of stacks and queues lies in their use as bookkeeping devices. In order to perform a certain task, we may first have to perform a number of subtasks. Each subtask may, in turn, lead to other subtasks that must be performed. Both stacks and queues provide a mechanism for keeping track of the subtasks that are yet to be performed, and the order in which they must be performed. In some cases, that order is first-in, last-out, and so a stack is appropriate; in other cases, the order is first-in, first-out and a queue is appropriate. We will see important examples of each case in Section 2.3.2.

Since the operations of insertion and deletion from stacks and queues occur frequently in the algorithms discussed in this book, we use the following notation:

$D \Leftarrow x$ Add x to D. If D is a stack, x is added at the top. If D is a queue, x is added at the rear.

$x \Leftarrow D$ Set the value of x to be the element at the top of D if D is a stack or the element at the front of D if D is a queue. *This element is then deleted from D.*

We will use these notations without regard for the technique used to implement the stack or queue.

Sequential Implementation. For stacks, this allocation is simple and extremely convenient. All that is needed is a variable, say t, to keep track of the top of stack S. Assuming that locations $S(1), S(2), \ldots, S(m)$ have been set aside for S, the empty stack corresponds to $t = 0$, and the operations of insertion and deletion are simply

$S \Leftarrow x$ $t \leftarrow t + 1$

                if $t > m$ **then overflow**

                     else $S(t) \leftarrow x$

$x \Leftarrow S$ **if** $t = 0$ **then underflow**

$$\textbf{else} \begin{cases} x \leftarrow S(t) \\ t \leftarrow t - 1 \end{cases}$$

underflow means that an attempt has been made to delete an element from an empty stack; it is generally a meaningful end condition in an algorithm. Conversely, **overflow** means that there is no room to add x to the stack, which usually means disaster.

The sequential allocation of a queue is more complicated because it grows at one end and shrinks at the other; if we are not careful, it can "inch" its way around and try to overrun the locations set aside for it. Thus we use the m locations $Q(0), Q(1), \ldots, Q(m - 1)$ allocated for the queue Q in a circular fashion, and we consider $Q(0)$ to follow $Q(m - 1)$. Using f as a pointer to the location just before the front of the queue and r as the pointer to the rear of the queue, the queue consists of the elements $Q(f + 1), Q(f + 2), \ldots, Q(r)$. With this definition, the empty queue corresponds to $r = f$. We have

$$Q \Leftarrow x \qquad r \leftarrow (r + 1) \bmod m$$

$$\textbf{if } r = f \textbf{ then overflow}$$

$$\textbf{else } Q(r) \leftarrow x$$

$$x \Leftarrow Q \qquad \textbf{if } r = f \textbf{ then underflow}$$

$$\textbf{else } \begin{cases} f \leftarrow (f + 1) \bmod m \\ x \leftarrow Q(f) \end{cases}$$

As in the case of a stack, **underflow** is generally a meaningful end condition and **overflow** is generally disaster. Notice that **overflow** occurs while attempting to add an mth element to a queue containing $m - 1$ elements; this means that one of the m allotted locations is wasted.

Linked Allocation.   The linked allocation of a stack is as easy as the sequential allocation. We maintain a pointer t whose value is the location of the top of the stack and use the LINK field of a stack element to point to the element below it on the stack. The bottom stack element has the null pointer Λ in its LINK field, and $t = \Lambda$ corresponds to the empty stack. We have

$$S \Leftarrow x \qquad l \leftarrow \textbf{get-cell}$$

$$\text{INFO}(l) \leftarrow x$$

$$\text{LINK}(l) \leftarrow t$$

$$t \leftarrow l$$

$$x \Leftarrow S \qquad \textbf{if } t = \Lambda \textbf{ then underflow}$$

$$\textbf{else } \begin{cases} x \leftarrow \text{INFO}(t) \\ t \leftarrow \text{LINK}(t) \end{cases}$$

In this case, the **overflow** condition occurs in the operation **get-cell** if no cells are available.

For queues, the linked allocation is essentially the same as for stacks [i.e., LINK(x) points to the element behind x in the queue] except that we use f instead of t as a pointer to the front, and, in addition, we use r to point to the rear. To add an element to the queue, we use

$$Q \Leftarrow x \qquad l \leftarrow \textbf{get-cell}$$

$$\text{INFO}\,(l) \leftarrow x$$

$$\text{LINK}\,(l) \leftarrow \Lambda$$

$$\text{LINK}\,(r) \leftarrow l$$

$$r \leftarrow l$$

As before, the **overflow** condition is hidden in the operation **get-cell**.

It is convenient to have the empty queue correspond to $f = \Lambda$, but we must ensure that the value of r for the empty queue is defined so that the insertion algorithm works properly when the first element is inserted. A little thought shows that we must store f in the LINK field of a node that is pointed to by r. Deletion is then accomplished by

$$x \Leftarrow Q \qquad \textbf{if } f = \Lambda \textbf{ then underflow}$$

$$\textbf{else} \begin{cases} x \leftarrow \text{INFO}\,(f) \\[4pt] f \leftarrow \text{LINK}\,(f) \\[4pt] \textbf{if } f = \Lambda \textbf{ then } r \leftarrow \text{ the location in which the} \\ \qquad\qquad\text{pointer } f \text{ is stored } (f \text{ is stored} \\ \qquad\qquad\text{in the LINK field of this loca-} \\ \qquad\qquad\text{tion)} \end{cases}$$

2.3   TREES

A finite *rooted tree* T is formally defined as a nonempty, finite set of labeled nodes such that there is one distinguished node, called the *root* of the tree, and the remaining nodes are partitioned into $m \geqq 0$ disjoint subtrees $T_1, T_2, \ldots, T_m$. (Except in Chapter 8, all trees in this book are rooted, and the term "tree" will generally mean a rooted tree.) Nodes having no subtrees are called *leaves*; the remaining nodes are called *internal* nodes. Figure 2.8 illustrates these concepts.

We used trees briefly in Chapter 1 to study the number of weighings necessary in the false-coin problem with n coins. It is not coincidental that trees occurred so early, for every chapter in the book uses the notion of trees in some important way. Because trees epitomize hierarchical organization, they are the most important nonlinear data structures in combinatorial algorithms.

In describing the relationships between nodes in a tree, we use the terminology of family trees. Thus all the nodes in a tree (or a subtree) are said to be *descendants* of its root; conversely, the root is an *ancestor* of all of its descendants. Furthermore, we refer to the root as the *father* of the roots of its subtrees; these nodes are, in turn, the *sons* of the root. Sons of the same node are called *brothers*. For example, in Figure 2.8, node *A* is the father of nodes *B*, *G*, and *I*; *J* and *K* are the sons of *I*; and *C*, *E*, and *F* are brothers.

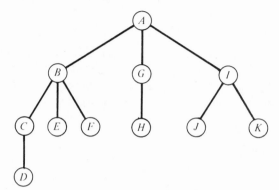

Figure 2.8 A tree with 11 nodes labeled *A* through *K*. The nodes labeled *D, E, F, H, J,* and *K* are leaves; the other nodes are internal nodes. The node labeled *A* is the root.

All the trees that we will consider will be *ordered*; that is, the relative order of the subtrees of each node is important. Thus

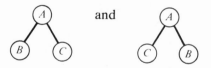

are different trees.

We define a *forest* to be an ordered set of trees, and so we can rephrase the definition of a tree: a tree is a nonempty set of nodes such that there is one distinguished node, called the root of the tree, and the remaining nodes are partitioned into a forest of $m \geq 0$ subtrees of the root.

An important variant of rooted trees is the class of binary trees. A *binary tree* T is either *empty* or consists of a distinguished node called the root and two binary subtrees T_l and T_r, the left and right subtrees, respectively. Binary trees are *not* a subset of trees in general but a a different structure entirely, since

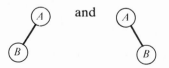

are not the same binary tree. As trees, however, they are both indistinguishable from

Therefore the differences between a tree and a binary tree are that a tree cannot be empty and each node of a tree can have an arbitrary number of subtrees; a binary tree can be empty, each of its nodes can have either 0, 1, or 2 subtrees, and a distinction is made between left and right subtrees.

2.3.1   Representations

Almost all computer representations of trees are based on linked allocation (sequentially allocated representations for certain types of binary trees will be introduced in Sections 6.3.1 and 7.1.3). Each node consists of an INFO field and some pointer fields. For example, a representation that we will find convenient in Section 2.4 has a single pointer field FATHER for each node, pointing to the father of the node. In this representation, the tree in Figure 2.8 would look as shown in Figure 2.9. Such a representation is useful if, as in Section 2.4, we need to move up the tree from descendants to ancestors. Unfortunately, this is rarely the case; it is far more common to need to move down the tree from ancestors to descendants.

Representing a tree (or forest) with pointers that go from ancestors to

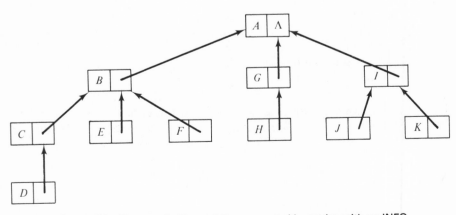

Figure 2.9   The tree in Figure 2.8 represented by nodes with an INFO field and a FATHER pointer.

descendants is complicated because, although a node has at most one father, it can have arbitrarily many sons. In other words, the nodes in the representation will need to vary in size, a definite inconvenience. However, one way around this problem is to define a correspondence between trees and binary trees, since binary trees are easily represented by nodes of fixed size. Each node has three fields: LEFT (pointing to the location of the root of the left subtree), INFO (the contents of the node), and RIGHT (pointing to the location of the root of the right subtree. This is illustrated in Figure 2.10.

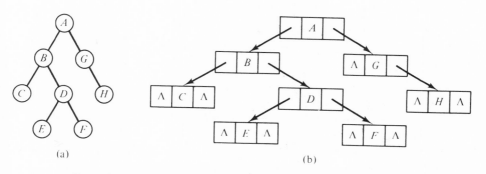

Figure 2.10 A binary tree (a), and its representation (b) by nodes with the tree fields respectively, LEFT, INFO, and RIGHT.

We can represent trees as binary trees (using nodes of fixed size) by representing every node in a forest as a node consisting of LEFT, INFO, and RIGHT fields and by using the LEFT field of a node to point to the leftmost son of that node and the RIGHT field of a node to point to the next brother of that node. For example, the forest shown in Figure 2.11(a) is transformed into the binary tree shown in Figure 2.11(b). Thus we are using the LEFT field of a node to point to a linked list of the sons of that node; that list is linked together by RIGHT fields. We call this the *natural correspondence* between forests and binary trees.

Other representations of binary trees and forests are discussed in Exercises 17 through 20.

2.3.2 Traversals

In many applications, it is necessary to traverse forests, visiting (i.e., processing) the nodes in some systematic order. The visit at each node might be as simple as printing its contents or as complicated as computing a function. The only assumption that we make about the visit is that it does not change the structure of the forest. The four basic forest traversals that we will find useful are depth-first order, bottom-up order, level order, and, for binary trees, symmetric order.

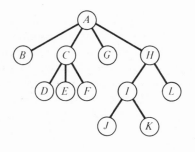
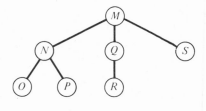

(a)

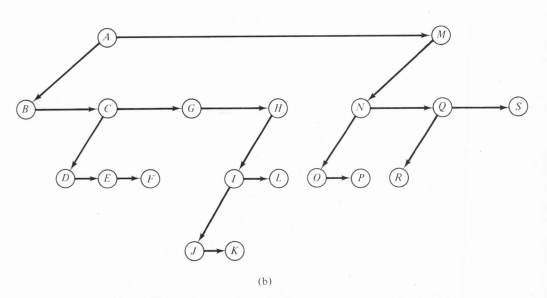

(b)

Figure 2.11   A forest (a) and its binary tree representation (b).

The *depth-first order*, sometimes known as *preorder*, visits the nodes of the forest as described by the following recursive procedure.

1. Visit the root of the first tree.
2. Traverse the subtrees, if any, of the first tree in depth-first order.
3. Traverse the remaining trees, if any, in depth-first order.

For instance, in the forest shown in Figure 2.11(a), the nodes would be visited in the order *A, B, C, D, E, F, G, H, I, J, K, L, M, N, O, P, Q, R, S*. The name "depth-first" refers to the fact that, after visiting a node, we continue by going deeper into the tree whenever possible. This order is extremely useful in various search procedures, as we will see in Chapters 4, 5, and 8.

For a binary tree, the recursive procedure simplifies to

1.   Visit the root of the binary tree.
2.   Traverse the left subtree in depth-first order.
3.   Traverse the right subtree in depth-first order.

In this case, the empty tree is traversed by doing nothing. Notice that traversing a forest in depth-first order is exactly the same as using depth-first order to traverse the binary tree arising by the natural correspondence. It is this fact that makes the correspondence "natural."

The *bottom-up order*, sometimes known as *postorder* or *endorder*, visits the nodes of the forest according to the following recursive procedure.

1.   Traverse the subtrees of the first tree, if any, in bottom-up order.
2.   Visit the root of the first tree.
3.   Traverse the remaining trees, if any, in bottom-up order.

The name "bottom-up" refers to the fact that at the time a node is visited, all its descendants have already been visited. This order is useful because it allows us to evaluate recursively defined functions on forests (see Exercises 22 and 23). In the forest in Figure 2.11(a), this order visits the nodes in the order $B, D, E, F, C, G, J,$ $K, I, L, H, A, O, P, N, R, Q, S, M.$ The recursive procedure for the bottom-up traversal applied to binary trees simplifies to

1.   Traverse the left subtree in bottom-up order.
2.   Traverse the right subtree in bottom-up order.
3.   Visit the root.

The *symmetric order* is defined recursively for binary trees as

1.   Traverse the left subtree in symmetric order.
2.   Visit the root.
3.   Traverse the right subtree in symmetric order.

This order is also known as *lexicographic order* (for reasons that will become clear in Chapter 6) or *inorder*. Notice that traversing a forest in bottom-up order is equivalent to traversing the binary tree corresponding to the forest (by the natural correspondence) in the symmetric order.

Comparing the recursive procedures for the depth-first, symmetric, and bottom-up binary tree traversals, we find considerable similarity.

depth-first	*symmetric*	*bottom-up*
1. visit root	1. left subtree	1. left subtree
2. left subtree	2. visit root	2. right subtree
3. right subtree	3. right subtree	3. visit root

This similarity allows us to construct a general nonrecursive algorithm that can be adapted to each of these orders for binary tree traversal. We use a stack S to store pairs consisting of a node in the binary tree and an integer i whose value tells which of the three operations (the first, second, or third) is to be performed when the pair reaches the top of the stack. This general algorithm is shown in Algorithm 2.3. The operations are

visit root   visit node p

left subtree   **if** LEFT $(p) \neq \Lambda$ **then** $S \Leftarrow$ (LEFT $(p), 1$)

right subtree   **if** RIGHT $(p) \neq \Lambda$ **then** $S \Leftarrow$ (RIGHT $(p), 1$)

So, for example, the straightforward specialization of Algorithm 2.3 to the depth-first binary tree traversal yields Algorithm 2.4(a). Notice, however, that

$S \leftarrow$ empty stack

$S \Leftarrow$ (root, 1)

$$\text{while } S \neq \text{empty do} \begin{cases} (p, i) \Leftarrow S \\ \text{case} \begin{cases} i = 1: \begin{cases} S \Leftarrow (p, 2) \\ \text{operation 1} \end{cases} \\ i = 2: \begin{cases} S \Leftarrow (p, 3) \\ \text{operation 2} \end{cases} \\ i = 3: \text{operation 3} \end{cases} \end{cases}$$

Algorithm 2.3   General binary tree traversal. The operations are described in the text.

after we visit node p, when $(p, 2)$ or $(p, 3)$ comes to the top of the stack, the *only* thing that happens is that (LEFT (p), 1) or (RIGHT (p), 1) is put on the stack. This step can be done earlier, when we first visit node p, and therefore we can simplify Algorithm 2.4(a). The result is Algorithm 2.4(b).

$S \leftarrow$ empty stack

$S \Leftarrow$ (root, 1)

$$\text{while } S \neq \text{empty do} \begin{cases} (p, i) \Leftarrow S \\ \text{case} \begin{cases} i = 1: \begin{cases} S \Leftarrow (p, 2) \\ \text{visit node } p \end{cases} \\ i = 2: \begin{cases} S \Leftarrow (p, 3) \\ \text{if LEFT } (p) \neq \Lambda \text{ then } S \Leftarrow (\text{LEFT } (p), 1) \end{cases} \\ i = 3: \quad \text{if RIGHT } (p) \neq \Lambda \text{ then } S \Leftarrow (\text{RIGHT}(p), 1) \end{cases} \end{cases}$$

Algorithm 2.4(a)   Straightforward specialization of Algorithm 2.3 to the depth-first order traversal of a binary tree.

$S \leftarrow$ empty stack

$S \Leftarrow$ root

while $S \neq$ empty do $\begin{cases} p \Leftarrow S \\ \text{visit node } p \\ \textbf{if } \text{RIGHT}(p) \neq \Lambda \textbf{ then } S \Leftarrow \text{RIGHT}(p) \\ \textbf{if } \text{LEFT}(p) \neq \Lambda \textbf{ then } S \Leftarrow \text{LEFT}(p) \end{cases}$

Algorithm 2.4(b)   Depth-first binary tree traversal, a simplified version of Algorithm 2.4(a).

Specializing Algorithm 2.3 for the symmetric order traversal of binary trees and simplifying the result as above, we get Algorithm 2.5. Specializing Algorithm 2.3 for the bottom-up order results in Algorithm 2.6.

$S \leftarrow$ empty stack

$S \Leftarrow$ (root, 1)

while $S \neq$ empty do $\begin{cases} (p, i) \Leftarrow S \\ \textbf{if } i = 1 \textbf{ then } \begin{cases} S \Leftarrow (p, 2) \\ \textbf{if } \text{LEFT}(p) \neq \Lambda \textbf{ then } S \Leftarrow (\text{LEFT}(p), 1) \end{cases} \\ \qquad\quad \textbf{else } \begin{cases} \text{visit node } p \\ \textbf{if } \text{RIGHT}(p) \neq \Lambda \textbf{ then } S \Leftarrow (\text{RIGHT}(p), 1) \end{cases} \end{cases}$

Algorithm 2.5   Symmetric order binary tree traversal, a simplified specialization of Algorithm 2.3.

$S \leftarrow$ empty stack

$S \Leftarrow$ (root, 1)

while $S \neq$ empty do $\begin{cases} (p, i) \Leftarrow S \\ \textbf{case } \begin{cases} i = 1 : \begin{cases} S \Leftarrow (p, 2) \\ \textbf{if } \text{LEFT}(p) \neq \Lambda \textbf{ then } S \Leftarrow (\text{LEFT}(p), 1) \end{cases} \\ i = 2 : \begin{cases} S \Leftarrow (p, 3) \\ \textbf{if } \text{LEFT}(p) \neq \Lambda \textbf{ then } S \Leftarrow (\text{RIGHT}(p), 1) \end{cases} \\ i = 3 : \text{visit node } p \end{cases} \end{cases}$

Algorithm 2.6   Bottom-up order binary tree traversal, a specialization of Algorithm 2.3.

The final forest traversal order, which we mention only briefly, is the *level order* or *breadth-first order*. This traversal visits the nodes of the forest from the left to right, level by level from the roots down. Thus the nodes in the forest in Figure 2.11(a) would be visited in the order $A, M, B, C, G, H, N, Q, S, D, E, F, I, L, O, P, R, J, K$. This traversal of a tree is useful in certain graph algorithms (see Chapter 8). In a binary tree it is accomplished by having S be a queue instead of a stack in Algorithm 2.4(b) and by interchanging the last two statements of the **while** loop. We leave the level order traversal of a forest to Exercise 24.

2.3.3 Path Length

In other parts of this book, trees are used not only as a data structure but also as a vehicle to analyze the behavior of certain algorithms. In this connection, we need quantitative measures of different aspects of trees and, in particular, of binary trees.

The most important quantitative aspects of trees involve the *levels* of the nodes. The level of p is defined recursively as 0 if p is the root of T; otherwise the level of p is $1 + \text{level}(\text{FATHER}(p))$. The notion of level gives us a simple way to define the *height* $h(T)$ of a tree T as

$$h(T) = \max_{p \in T} \text{level}(p).$$

In other words, the height of a tree is the maximum number of edges along a path from the root of the tree to a leaf of the tree. We can also define the height of a tree recursively as

$$h(\bigcirc) = 0$$

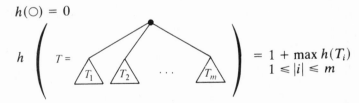

$$= 1 + \max_{1 \le |i| \le m} h(T_i)$$

As we have already seen in the false-coin problem of Section 1.5, the height of a ternary decision tree can correspond to the number of decisions required by an algorithm in the worst case (see Theorem 1.1). We will use this correspondence again in Chapters 6 and 7.

In Section 1.5 we also considered the average number of decisions required by an algorithm for the false-coin problem; this number corresponded to the sum of the levels of all leaves divided by the number of leaves in the ternary decision tree. In discussing this average distance to a leaf we will restrict our attention to binary trees, but the generalization to t-ary trees is obvious (a t-ary tree is a set of nodes that is either empty or that consists of a root and t ordered, disjoint t-ary trees).

Given a binary tree in which each of the nodes represents a binary decision, two averages are of interest: the average distance from the root to a node in the tree and the average distance from the root to a "hole" at the bottom of the tree. To be more precise, suppose that we are given a binary tree as shown in Figure 2.12(a). Then the average distance from the root to a node is of interest, and so is the average distance from the root to the external nodes (leaves) that have been added in Figure 2.12(b) to form an *extended binary tree*. In an extended binary tree with n internal nodes, there are always $n + 1$ external nodes (Exercise 25).

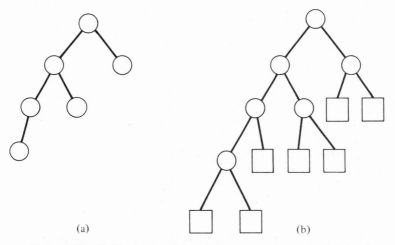

(a)   (b)

Figure 2.12 A binary tree (a) and the same tree extended by the addition of external nodes (b).

The *external path length* $E(T)$ of an extended binary tree T with n internal nodes is the sum of the levels of all the external nodes; the *internal path length* $I(T)$ is the sum of the levels of all the internal nodes. Therefore in Figure 2.12(b) the external path length is 21 and the internal path length is 9. These sums divided by the number of external nodes or the number of internal nodes respectively, give the desired averages.

$$\text{Average distance to an external node} = \frac{E(T)}{n + 1}.$$

$$\text{Average distance to an internal node} = \frac{I(T)}{n}.$$

It is easy to define both $E(T)$ and $I(T)$ recursively for an extended binary tree T with n internal nodes.

$$I(\square) = 0$$

$$I\left(T = \overset{\circ}{\diagup \diagdown} {}_{T_l} \quad {}_{T_r} \right) = I(T_l) + I(T_r) + n - 1 \tag{2.5}$$

$$E(\square) = 0$$

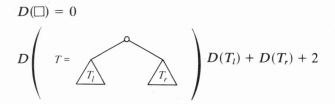

$$= E(T_l) + E(T_r) + n + 1 \qquad (2.6)$$

To understand the recursive part of the definition (2.5), notice that T_l and T_r contain between them $n - 1$ internal nodes and that adding an internal node above them as the root increases the level of each by 1. Similar reasoning explains the recursive part of the definition (2.6). We can relate $E(T)$ and $I(T)$ by considering the difference $D(T) = E(T) - I(T)$. Using (2.5) and (2.6), we find

$$D(\square) = 0$$

$$D(T_l) + D(T_r) + 2$$

which tells us (Exercise 26) that $D(T) = 2n$ for an extended binary tree T with n internal nodes. Thus

$$E(T) = I(T) + 2n, \qquad (2.7)$$

and hence we need only study one of $E(T)$ or $I(T)$ to determine the properties of both.

In later chapters we will be particularly interested in the range of values of $E(T)$ and $I(T)$. For instance, it is easy to see (Exercise 27) that, over all extended binary trees with n internal nodes, one with the maximum value of $I(T)$ is the tree

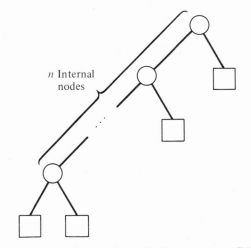

Figure 2.13   An extended binary tree with the largest possible internal and external path lengths.

shown in Figure 2.13. In this case,

$$I(T) = \sum_{i=0}^{n-1} i = \tfrac{1}{2}n(n-1)$$

and

$$E(T) = \tfrac{1}{2}n(n+3).$$

In Exercise 18 of Chapter 3 we will consider the average value of $I(T)$ when each of the possible extended binary trees with n internal nodes is equally probable. That average turns out to be approximately $n\sqrt{\pi n}$.

The remainder of this section is devoted to deriving the minimum value of $I(T)$. It is easier, however, to determine the minimum value of $E(T)$ and then use equation (2.7). We have

Lemma 2.1

An extended binary tree of n internal nodes with minimum external path length has all of its external nodes on levels l and $l + 1$, for some l. Such a tree is called a *completely balanced binary tree* of n internal nodes.

Proof: Let T be an extended binary tree with minimum external path length and let L and l be the maximum and minimum, respectively, of all the levels on which external nodes appear, $L \geq l$. Suppose that $L \geq l + 2$. Remove two external nodes that are brothers from level L and make their father an external node. Then put them below an external node on level l, which thus becomes an internal node. This process is shown in Figure 2.14. Such a modification of the tree preserves the number of internal and external nodes, but it decreases the external path length by $L - (l + 1)$, which is positive since $L \geq l + 2$. This contradicts the minimality of the external path length of T.

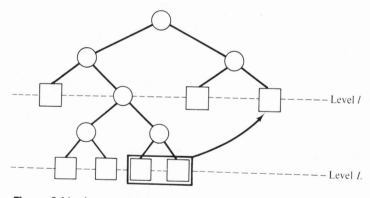

Figure 2.14 An example of how the external path length can be decreased if $L \geq l + 2$ in Lemma 2.1.

Lemma 2.2

If $l_1, l_2, \ldots, l_{n+1}$ are the levels of the $n + 1$ external nodes in an extended binary tree with n internal nodes, then $\sum_{i=1}^{n+1} 2^{-l_i} = 1$.

Proof: Exercise 28.

Now we are ready to compute the minimum external path length of an extended binary tree with n internal nodes and $n + 1$ external nodes. Using Lemma 2.1, let there be k external nodes on level l and $n + 1 - k$ on level $l + 1$, $1 \leq k \leq n + 1$ (i.e., all the external nodes may be on level l). Lemma 2.2 tells us that

$$k2^{-l} + (n + 1 - k)2^{-l-1} = 1$$

and hence

$$k = 2^{l+1} - n - 1. \tag{2.8}$$

Since $k \geq 1, 2^{l+1} > n + 1$ and since $k \leq n + 1$, we have $2^l \leq n + 1$; that is,

$$l = \lfloor \lg (n + 1) \rfloor. \tag{2.9}$$

Combining (2.8) and (2.9) gives

$$k = 2^{\lfloor \lg (n+1) \rfloor + 1} - n - 1,$$

and the minimum external path length is thus

$$l \cdot k + (l + 1)(n + 1 - k) = (n + 1)\lfloor \lg (n + 1) \rfloor + 2(n + 1) - 2^{\lfloor \lg (n+1) \rfloor + 1}.$$

Therefore we have proved

Theorem 2.1

The minimum external path length of an extended binary tree with n internal nodes is

$$(n + 1) \lg (n + 1) + (n + 1)(2 - \theta - 2^{1-\theta}), \tag{2.10}$$

where $\theta = \lg (n + 1) - \lfloor \lg (n + 1) \rfloor, 0 \leq \theta < 1$.

It is not difficult to compute similar formulas for extended t-ary trees with n internal nodes. In this case, the analog of equation (2.7) is

$$E(T) = (t - 1)I(T) + t \cdot n,$$

and the minimum external path length is

$$[n(t - 1) + 1]l - \frac{t^{l+1} - t}{t - 1} + t \cdot n, \qquad l = \lfloor \log_t [(t - 1)n + 1] \rfloor. \tag{2.11}$$

We can apply formula (2.11) to the false-coin problem of Chapter 1 by taking $t = 3$. In this case we find that the minimum external path length of an extended ternary tree with n internal nodes (and hence $2n + 1$ external nodes; see Exercise 25) is

$$(2n + 1) \log_3 (2n + 1) + \tfrac{1}{2}(2n + 1)(3 - 2\theta - 3^{1-\theta}),$$

where $0 \le \theta < 1$ is defined by $\theta = \log_3 (2n + 1) - \lfloor \log_3 (2n + 1) \rfloor$. So the minimum possible average distance from the root to an external node is

$$\log_3 (2n + 1) + \tfrac{1}{2}(3 - 2\theta - 3^{1-\theta}).$$

Since any solution to the false-coin problem with n coins corresponds to an extended ternary tree (the ternary decision tree) and since $3 - 2\theta - 3^{1-\theta} > 0$ for $0 < \theta < 1$, we conclude that no solution can use fewer than $\log_3 (2n + 1)$ weighings on the average and that that average is achievable if and only if $\theta = 0$, that is, if and only if $2n + 1$ is a power of 3. This proves Theorem 1.2.

2.4 SETS AND MULTISETS

There is no formal definition for the notion of a *set*; it is assumed to be too primitive to be defined. Thus we say that a set is a collection of distinct elements, but we leave the terms "collection" and "elements" undefined. A *multiset* is a collection of not necessarily distinct elements; it may be thought of as a set in which each element has an associated positive integer, its *multiplicity*.

We write a finite set S as

$$S = \{s_1, s_2, \ldots, s_n\},$$

where $s_1, s_2, \ldots, s_n$ are the (necessarily distinct) elements of S. The *size* or *cardinality* of S is written as $|S|$ and, in the example above, $|S| = n$. If S is a finite multiset, then we write

$$S = \{\underbrace{s_1, s_1, \ldots, s_1}_{m_1 \text{ times}}, \underbrace{s_2, s_2, \ldots, s_2}_{m_2 \text{ times}}, \ldots, \underbrace{s_n, s_n, \ldots, s_n}_{m_n \text{ times}}\}$$

$$= \quad \{m_1 \bullet s_1, m_2 \bullet s_2, \ldots, m_n \bullet s_n\},$$

where the s_i are distinct and m_i is the multiplicity of s_i. In this case, the cardinality is

$$|S| = \sum_{i=1}^{n} m_i.$$

The most common operations on sets and multisets are *union* and *intersection*. For sets, we write these operations as $\cup$ and $\cap$, respectively. For multisets, we write $\uplus$ and $\Cap$.

Both the sequential and linked representations for sequences (Section 2.2) can be used for sets and multisets in the obvious way. By inducing an artificial order of the set elements or by using a natural order if there is one, we can consider a set to be a sequence. Similarly, a multiset can be considered as a sequence or, to save space, as a sequence of pairs consisting of an element and its multiplicity. The hashing methods of Chapter 6 can be used to represent sets in which there is no natural order.

As with sequences, the best method for representing sets or multisets depends critically on the operations that will be performed on the sets. Suppose for instance, that we want to manipulate disjoint subsets of a set $S = \{s_1, s_2, \ldots, s_n\}$ and that the operations to be performed are merging two of the disjoint subsets and, given s_i, finding which of the subsets contains s_i. At any given time, we thus have a partition of S into nonempty disjoint subsets. The rest of this section will examine a set representation that allows extraordinarily efficient implementation of the union and find operations.

For identification, each of the disjoint subsets of S will have a *name*. The name is simply one of the elements of the subset and can be thought of as a subset representative. When we refer to the name of a subset, we are referring to its subset representative. Consider, for example,

$$S = \{1, 2, 3, 4, 5, 6, 7, 8, 9, 10, 11\}$$

partitioned into four disjoint subsets

$$\{1, 6, ⑦, 8, 11\} \quad \{②\} \quad \{③, 4, 5\} \quad \{9, ⑩\}; \tag{2.12}$$

in each case, the circled element is the name of the subset. If we ask to find the subset in which 8 is contained, the answer that we expect is 7, the name of the subset containing 8. If we ask to take union of the subsets named 2 and 10, we want the resulting partition of S to be

$$\{1, 6, ⑦, 8, 11\} \quad \{③, 4, 5\} \quad \{②\} \cup \{9, ⑩\}$$

in which the name of the set $\{②\} \cup \{9, ⑩\}$ can be chosen as either 2 or 10.

We assume that initially we have the partition of $S = \{s_1, s_2, \ldots, s_n\}$ into n singleton sets

$$\{ⓢ_1\} \quad \{ⓢ_2\} \quad \ldots \{ⓢ_n\}, \tag{2.13}$$

each of which set is named after its only element. This partition is modified by a sequence of union operations in which find operations are intermixed. This seemingly contrived problem is quite useful in certain combinatorial algorithms; we will see an instance of its usefulness in the "greedy" algorithm of Section 8.2.1.

We will give procedures *UNION* (x, y) and *FIND* (x) to implement the union and find operations. *UNION* (x, y) takes the *names* of two different subsets x and y and creates a new subset containing all the elements of x and y. *FIND* (x) returns as its value the name of the subset containing x. For example, if we want to cause the set containing a to be merged with the set containing b, we would use the sequence of instructions

$$x \leftarrow FIND\ (a)$$

$$y \leftarrow FIND\ (b)$$

if $x \neq y$ **then** *UNION* (x, y).

Suppose that we have a sequence of u union operations intermixed with f find operations and we start with $S = \{s_1, s_2, \ldots, s_n\}$ partitioned into the singleton sets of (2.13). We want a data structure to represent the disjoint subsets of S so that such a sequence of operations can be efficiently performed. The data structure that we will use is a forest representation based on father pointers, as illustrated in Figure 2.9. Each set element s_i will be a node in the forest, and the father of set element s_i will be another element in the same subset as s_i. If an element has no father (i.e., is a root), then *it* is the name of its subset. Thus the partition (2.12) might be represented as shown in Figure 2.15.

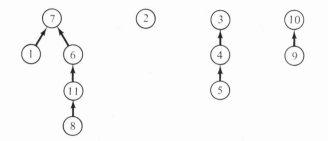

Figure 2.15   A forest representation of the partition (2.11).

With this representation, the operation *FIND* (x) consists of following father pointers up from x to the root (i.e., name) of its subset. The operation *UNION* (x, y) consists of somehow hooking together the trees rooted at x and y; for example, this could be done by making y the father of x.

After u union operations, the largest subset possible in the resulting partition of S contains $u + 1$ elements. Furthermore, since each union reduces the number of subsets by one, the sequence of operations can contain at most $n - 1$ unions; thus $u \leqq n - 1$. Since each union operation changes the name of the subset containing some of the elements, we can assume that each union is preceded by at least one find, and hence we assume that $f \geqq u$. The problem is to ascertain how efficiently we can perform a sequence of $u \leqq n - 1$ union operations intermixed with $f \geqq u$ find operations. The time required by the union operations is clearly proportional to u because only a fixed amount of work to rearrange some pointers is necessary for each union operation. We can therefore concentrate on the time required by the f find operations.

If the operation *UNION* (x, y) is performed by making x the father of y, then it is possible, after a sequence of u union operations, to produce the forest shown in Figure 2.16. In this case if the f find operations are done after all the union operations, and each find starts at the bottom of the chain of $u + 1$ set elements, it is clear that the time required by the find operations will be proportional to $f \cdot (u + 1)$. Obviously, it could not be worse than proportional to $f \cdot (u + 1)$.

By being more clever we can reduce this worst case considerably. If the operation *UNION* (x, y) keeps the trees in the forest "balanced" by making the

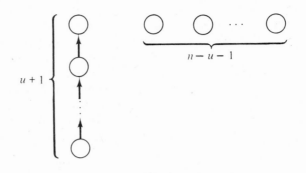

Figure 2.16   A possible forest resulting from u UNION operations.

root of the larger subset the father of the root of the smaller subset, then we pay a slight premium in storage, since each node in the forest must contain information about the size of the subtree beneath it, but we can prove

Lemma 2.3

For any $s_i \in S$, let $r(s_i)$, the *rank* of s_i, be the height of the subtree rooted at s_i and let $d(s_i)$ be the number of nodes in that subtree. If balancing is used in the union operation, then, for each $s_i \in S$, we have $d(s_i) \geq 2^{r(s_i)}$.

Proof: The proof is by induction on u, the number of union operations performed. For $u = 0$, the lemma is clearly true, since each element $s_i \in S$ is the root of a tree consisting of a single node, and so we have $r(s_i) = 0$ and $d(s_i) = 1$. Suppose that the lemma is true for $u = k$ and consider what happens on the $(k + 1)$st union operation, $UNION(x, y)$. By induction, we have (before that union operation) $d(x) \geq 2^{r(x)}$ and $d(y) \geq 2^{r(y)}$. Without loss of generality, let $d(x) \geq d(y)$. Then $UNION(x, y)$ causes x to become the father of y, and the rank of x after the union is $\max[r(x), r(y) + 1]$. After the union, we will have $d(x) \geq 2d(y) \geq 2^{r(y)+1}$; and both before and after the union, we have $d(x) \geq 2^{r(x)}$.

We conclude from Lemma 2.3 that after u union operations, each of the elements $s_i \in S$ satisfies

$$0 \leq r(s_i) \leq \lg(u + 1),$$

and thus f find operations will require time at most proportional to $f \lg(u + 1)$. Since $f \geq u$, the total time for the unions and finds is also at most proportional to $f \lg(u + 1)$. It is easy to see that a complete binary tree with $u + 1$ nodes can result from the unions, and therefore there is an example in which the finds can actually achieve the bound of $f \lg(u + 1)$.

We can improve the efficiency of the find operations by using *path compression*. After the operation $FIND(x)$, x and all the vertices on the path between x and the root are made sons of the root. For example, if we did a $FIND(8)$ on the forest in Figure 2.15, the value returned would be 7 as before, but in the meantime the forest would have been changed to the one shown in Figure 2.17. Path compression causes only a minor increase in the cost of a find operation, and, as we will see, its use has a significant effect on the time required by a sufficiently large number of find operations.

The union operation with balancing is given in Algorithm 2.7, and the find operation with path compression is given in Algorithm 2.8. In each algorithm we assume that the set element s_i is represented by a forest node at location l_i that consists of three fields: ELT $(l_i) = s_i$, FATHER (l_i) is the location of the node corresponding to the father of s_i, and SIZE (l_i) is $d(s_i)$, the number of elements in the subtree rooted at s_i.

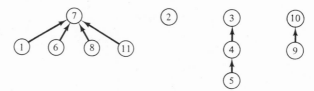

Figure 2.17   The result of FIND(8) on the forest in Figure 2.15 when path compression is used.

procedure *UNION* (x, y)

⟦x and y are assumed to be roots of trees in the forest stored at locations l_x and l_y, respectively⟧

if SIZE $(l_x) <$ SIZE (l_y) **then** $\begin{cases} \text{FATHER } (l_x) \leftarrow l_y \\ \text{SIZE } (l_y) \leftarrow \text{SIZE } (l_x) + \text{SIZE } (l_y) \end{cases}$

else $\begin{cases} \text{FATHER } (l_y) \leftarrow l_x \\ \text{SIZE } (l_x) \leftarrow \text{SIZE } (l_y) + \text{SIZE } (l_x) \end{cases}$

return

Algorithm 2.7   The union operation with balancing.

For the analysis of the effect of the simultaneous use of balancing and path compression, we introduce two important functions. We define

$$\lg^* k = \text{the number of times the } \lg \text{ function must be applied to } k \text{ to get a result less than or equal to 0}$$

and a function b that is essentially the functional inverse of $\lg^*$:

$$b(n) = \begin{cases} 0 & \text{if } n = 0, \\ 2^{b(n-1)} & \text{otherwise.} \end{cases}$$

procedure *FIND* (x)

⟦x is a node in the forest stored at location l_x⟧

$S \leftarrow$ empty stack

$t \leftarrow l_x$

while FATHER $(t) \neq \Lambda$ **do** $\begin{cases} S \Leftarrow t \\ t \leftarrow \text{FATHER}(t) \end{cases}$

while $S \neq$ empty **do** $\begin{cases} v \Leftarrow S \\ \text{FATHER}(v) \leftarrow t \\ \text{SIZE}(v) \leftarrow 1 \end{cases}$

$FIND \leftarrow t$

return

Algorithm 2.8   The find operation with path compression.

The function $\lg^* n$ grows extremely slowly and the function $b(n)$ extremely rapidly, as the following values indicate.

n	0	1	2	3	4	5	...	16	17	...	65536	65537	...	2^{65536}
$\lg^* n$	0	1	2	3	3	4	...	4	5	...	5	6	...	6

n	0	1	2	3	4	5	6
$b(n)$	0	1	2	4	16	65536	2^{65536}

Theorem 2.2

The time required by Algorithms 2.7 and 2.8 to perform a sequence of $u \leq n - 1$ union operations and a number f proportional to n intermixed find operations is at most proportional to $f \lg^* (u + 1)$.

Proof: It is useful to consider the problem as follows. Suppose that all u union operations are performed first, before any of the find operations. The result will be a forest in which the largest tree has at most $u + 1$ nodes. Each of the find operations is now "partial": to execute *FIND* (x), we follow the path in the forest from x to the highest ancestor of x corresponding to a union that appears before *FIND* (x) in the original sequence of operations. Thus the problem is reduced to a sequence of n union operations followed by f partial find operations.

Let F be the forest created by the sequence of u union operations. For any element $s_i \in S$, let $r(s_i)$ be the rank of s_i in F. $r(s_i)$ is defined *and fixed with respect to F*; it does not change even though the forest will change when the partial finds are performed. Notice that if x is the father of y before a partial find and $x' \neq x$ is the father of y after the partial find, then $r(x) < r(x')$. Consequently, even though the forest changes, the ranks are always strictly increasing along any path in the forest F traversed by a partial find.

We partition the elements $s_i \in S$ into $\lg^*(u + 1) + 1$ groups $G_0, G_1, \ldots, G_{\lg^*(u+1)}$ as follows.

$$G_i = \{x \in S | b(i) \leq r(x) < b(i + 1)\}.$$

Thus G_0 is the set of all elements with rank 0 (i.e., all leaves), G_1 is the set of all elements with rank 1, G_2 is the set of all elements with rank either 2 or 3, G_3 is the set of all elements with ranks $4, 5, \ldots, 15$, G_4 is the set of all elements with ranks $16, 17, \ldots, 65535$, and so on. We will use this partition of the elements of $s_i \in S$ to partition the links from sons to fathers followed in the f partial finds.

The time required to perform the f find operations is clearly proportional to the number of links from sons to fathers that are traversed in performing the f partial finds. We partition these links into classes

$L = \{$links from sons to fathers that are the *last* links in a partial find path$\}$,

$D = \{$links from sons to fathers such that the link is not the last link in its partial find path and the son and father are in *different* groups $G_i\}$,

and for $k = 0, 1, \ldots, \lg^*(u + 1)$,

$E_k = \{$links from sons to fathers such that the link is not the last link in its partial find path and both the son and father are in $G_k\}$.

Clearly, each partial find path contains exactly one last link, and hence

$$|L| = f. \tag{2.14}$$

Since there are at most $\lg^*(u + 1) + 1$ groups G_i, each partial find path can contain at most $\lg^*(u + 1)$ links going from an element in one group to one in another group. Thus

$$|D| \leq f \lg^*(u + 1). \tag{2.15}$$

If we can show that for some constant C

$$|E_k| \leq C \cdot n, \tag{2.16}$$

we are done, since we would have

$$\sum_{k=0}^{\lg^*(u+1)} |E_k| \leq Cn \lg^*(u + 1) \leq \hat{C}f \lg^*(u + 1)$$

because we assumed that f was proportional to n. Combining this result with (2.14) and (2.15), we would have that the total number of links traversed in the f partial finds, $|L| + |D| + \sum |E_k|$, is at most proportional to $f \lg^*(u + 1)$.

To prove (2.16), observe that for any element x in G_k, each link traversed in a partial find (except the last link in the partial find) from x to its father in G_k causes x to get a father of higher rank. Since there are at most $b(k + 1) - b(k)$ different ranks in G_k, it follows that there can be at most $b(k + 1) - b(k)$ such links traversed in the partial finds for any element in G_k. Thus for each of the $|G_k|$ elements in G_k, there can be at most $b(k + 1) - b(k)$ links in E_k. Now notice that by Lemma 2.3 there are at most $n/2^r$ elements of rank r, since each element of rank r is the root of a subtree containing at least 2^r elements. Therefore

$$|G_k| \leq \sum_{r=b(k)}^{b(k+1)-1} \frac{n}{2^r} < \frac{n}{2^{b(k)}}\left(1 + \frac{1}{2} + \frac{1}{4} + \frac{1}{8} + \cdots\right) = \frac{2n}{2^{b(k)}} = \frac{2n}{b(k + 1)},$$

and so

$$\begin{aligned}|E_k| &\leq |G_k|[b(k + 1) - b(k)] \\ &\leq 2n\frac{b(k + 1) - b(k)}{b(k + 1)} \\ &\leq 2n.\end{aligned}$$

Comparing this theorem with the previous results, we find that if both the number of union operations and the number of find operations performed on $\{s_1, s_2, \ldots s_n\}$ are proportional to n, then the total time required for the operations is proportional to

n^2 if neither balancing nor path compression is used.

$n \lg n$ if balancing is used but path compression is not.

$n \lg^* n$ if both balancing and path compression are used.

2.5   REMARKS AND REFERENCES

The material presented in Sections 2.1, 2.2, and 2.3 is basic not only to combinatorial algorithms but also to programming in general. The discussion in these sections is not meant to explore all the interesting and important facets of these subjects but only to establish certain terminology, notation, and basic results for later chapters. The interested reader should consult the much more detailed presentations given in

KNUTH, D. E., *The Art of Computer Programming*, Vols. 1 and 2, Addison-Wesley, Reading, Mass., 1968 and 1969.

Section 4.1 of Knuth's Vol. 2 contains a thorough introduction to positional number systems and their history. Chapter 2 of Knuth's Vol. 1 is a compendium of almost everything known about lists, stacks, queues, and trees.

The union/find algorithms and their analysis have an interesting history. The problem was first motivated by the processing of EQUIVALENCE statements in

FORTRAN, and the straightforward forest representation (without balancing or path compression) is presented in

GALLER, B. A., and M. J. FISCHER, "An Improved Equivalence Algorithm," *Comm. ACM,* **7** (1964), 301–303.

In

FISCHER, M. J., "Efficiency of Equivalence Algorithms," in *Complexity of Computer Computations,* R. E. Miller and J. W. Thatcher (Eds.), Plenum Press, New York, 1972,

Fischer showed that with both balancing and path compression the time required by proportional to n union operations and proportional to n find operations was at most proportional to $n \lg \lg n$. This bound was improved to $n \lg^* n$ in

HOPCROFT, J. E., and J. D. ULLMAN, "Set-Merging Algorithms," *SIAM J. Comput.,* **2** (1973), 294–303,

and it is essentially their proof that is presented in Section 2.4.

To understand the final result concerning the union/find problem, it is necessary to introduce some notation. Let the function $A(i, j)$ be defined on nonnegative integers i and j as follows:

$$A(0, j) = 2j, \quad j \geqq 0$$

$$A(i, 0) = 0, \quad i \geqq 1$$

$$A(i, 1) = 2, \quad i \geqq 1$$

and

$$A(i, j) = A(i - 1, A(i, j - 1)) \quad \text{for} \quad i \geqq 1, \quad j \geqq 2.$$

This function grows enormously fast; for example,

$$A(3, 4) = \left. 2^{2^{\cdot^{\cdot^{2}}}} \right\} \begin{array}{l} 65536 \\ \text{twos} \end{array}. \tag{2.17}$$

Let

$$\alpha(f, n) = \min \left\{ z \geqq 1 \,\middle|\, A\left(z, 4\left\lceil \frac{f}{n} \right\rceil\right) > \lg n \right\}. \tag{2.18}$$

Since A grows so rapidly, α grows *extremely* slowly. In

TARJAN, R. E., "Efficiency of a Good but Not Linear Set Union Algorithm," *J. ACM,* **22** (1975), 215–225

it is shown that the time required for $f \geqq n$ finds and $n - 1$ unions, using both balancing and path compression is at most proportional to $f\alpha(f, n)$. From (2.17) and (2.18), we see that $\alpha(f, n) \leqq 3$ for

$$n \leqq \left. 2^{2^{\cdot^{\cdot^{2}}}} \right\} \begin{array}{l} 65537 \\ \text{twos} \end{array}.$$

Moreover, Tarjan shows that there is a sequence of unions and finds for which the time required is proportional to $f\alpha(f, n)$.

There has been little analysis of the average performance of algorithms for the union/find problem. Some preliminary results are found in

YAO, A. C.-C., "ON the Average Behavior of Set Merging Algorithms (extended abstract)," *Proceedings of Seventh Annual ACM Symposium on Theory of Computing* (1976), 192–195.

In addition to the "greedy" algorithm of Section 8.2.1 and the processing of EQUIVALENCE statements, examples of the usefulness of the union/find problem can be found in

AHO, A. V., J. E. HOPCROFT, and J. D. ULLMAN, *The Design and Analysis of Computer Algorithms*, Addison-Wesley, Reading, Mass., 1974

and in

TARJAN, R. E., "Applications of Path Compression on Balanced Trees," *J. ACM*, to appear.

2.6 EXERCISES

1. Explore base r radix systems for negative values of r. What are their advantages and disadvantages?

2. Explore the properties of the *balanced ternary* system in which the radix $r = 3$ but, instead of the digits being 0, 1, and 2, they are 0, 1, and -1.

3. Algorithms 2.1 and 2.2 produce the representation of an integer least significant digits first (i.e., from right to left). Devise algorithms that produce the most significant digits first (i.e., from left to right).

4. What are the advantages and disadvantages of representing a number by its prime factorization? Which operations are made easier and which are made more difficult?

5. The *Fibonacci number system* is based on the sequence of integers (to be examined in more detail in Chapter 3)

$$F_0 = 0, \qquad F_1 = 1$$

$$F_{n+1} = F_n + F_{n-1}.$$

Prove that every positive integer can be represented uniquely as the sum of Fibonacci numbers

$$F_{i_1} + F_{i_2} + \cdots + F_{i_t},$$

where $i_1 \geqq i_2 + 2, i_2 \geqq i_3 + 2, \ldots, i_{t-1} \geqq i_t + 2, i_t \geqq 2$.

6. Let $m_1, m_2, \ldots, m_n$ be pairwise relatively prime positive integers. Prove that any integer N in the range $0 \leqq N < \prod_{i=1}^{n} m_i$ has a unique representation $(r_1, r_2, \ldots, r_n)$ in which $N \equiv r_i \pmod{m_i}$ and $0 \leqq r_i < m_i$ for $1 \leqq i \leqq n$. Devise algorithms to convert a number from this *residue representation* to its decimal representation.

7. Show how to allocate an $n_1 \times n_2 \times \cdots \times n_t$ multidimensional array sequentially. What is the connection between the allocation and the mixed radix number systems of Section 2.1?

8. A *lower triangular matrix* is a matrix $A = (a_{ij})$ in which $a_{ij} = 0$ for $i < j$, and thus it is written

$$A = \begin{pmatrix} a_{11} & & & & \\ a_{21} & a_{22} & & & \\ a_{31} & a_{32} & a_{33} & & \\ \vdots & \vdots & \vdots & \ddots & \\ a_{n1} & a_{n2} & a_{n3} & \cdots & a_{nn} \end{pmatrix}.$$

Design a sequential allocation for such matrices. The location of a_{ij} should be a simple function of i, j and the location of a_{11}.

9. A *tridiagonal matrix* is a matrix $A = (a_{ij})$ in which $a_{ij} = 0$ if $|i - j| > 1$, and thus it is written

$$A = \begin{pmatrix} a_{11} & a_{12} & & & & \\ a_{21} & a_{22} & a_{23} & & & \\ & a_{32} & a_{33} & a_{34} & & \\ & & \ddots & \ddots & \ddots & \\ & & & \ddots & \ddots & a_{n-1,n} \\ & & & & a_{n,n-1} & a_{n,n} \end{pmatrix}.$$

Design a sequential allocation for such matrices. The location of a_{ij} should be a simple function of i, j and the location of a_{11}.

10. A generalization of the Prime Number Theorem says that the number of primes less than n of the form $ai + b$, a and b relatively prime, is approximately $n/[\varphi(a) \ln n]$, where $\varphi(x)$ is the number of positive integers less than x and relatively prime to x. Use this result to determine the values of a for which it is more economical to use a sequentially allocated list instead of a characteristic vector to represent the primes less than n of the form $ai + b$.

11. What is the sequentially allocated analog of a circular list?

12. Design an algorithm to copy a circular list.

13. A matrix is called "sparse" if most of its elements are zero. Devise a linked representation in which only nonzero entries appear for sparse matrices. Design an algorithm to make a copy of such a structure.

14. Let x and y be the locations of the first elements of two linked lists. What value is returned by the following procedure?

$$\textbf{procedure } \text{MYSTERY } (x, y)$$

$$\textbf{if } x = \Lambda \textbf{ then } \text{MYSTERY} \leftarrow y$$

$$\textbf{else } \begin{cases} l \leftarrow \textbf{get-cell} \\ \text{INFO } (l) \leftarrow \text{INFO } (x) \\ \text{LINK } (l) \leftarrow \text{MYSTERY } (\text{LINK } (x), y) \\ \text{MYSTERY} \leftarrow l \end{cases}$$

return

15. A *deque* (*d*ouble *e*nded *que*ue) is a linear list in which all insertions and deletions are made at the ends of the list. Describe sequential and linked allocations for a deque. Give insertion and deletion algorithms for each case.

16. A *priority queue* is a linear list that operates on a "first-in, highest-priority-out" manner; in other words, with each element on the queue there is an associated priority number. Insertions are made at the back of the queue, but deletions are made anywhere in the queue, since the element deleted is always the element with the highest priority. Describe representations and insertion/deletion algorithms for priority queues.

17. Prove that a forest can be uniquely represented by the sequence of nodes in bottom-up order along with the degrees (i.e., number of sons) of each node; thus the forest in Figure 2.11(a) would be represented as

NODE	B	D	E	F	C	G	J	K	I	L	H	A	O	P	N	R	Q	S	M
DEGREE	0	0	0	0	3	0	0	0	2	0	2	4	0	0	2	0	1	0	3

18. Design a representation of a forest analogous to the one in Exercise 17 but based on depth-first order.

19. Prove that a binary tree is uniquely determined by the combination of the depth-first and symmetric orders of its nodes.

20. In a binary tree with n nodes, there will always be $n + 1$ Λ pointers at the bottom of the tree (see Exercise 25). To make use of this space, it is possible to use the empty pointers as follows. A *threaded tree* replaces these Λs with pointers ("threads") to the symmetric order predecessor (if the thread is in a LEFT link) or the symmetric order successor (if the thread is in a RIGHT link). The threads are distinguished from regular links by adjoining a tag bit to both the LEFT and RIGHT fields; this bit is set one way if the pointer is a tree edge and the other way if the pointer is a thread. Thus the binary tree in Figure 2.10 would be made threaded by adding the dotted threads as shown at the top of p. 69.

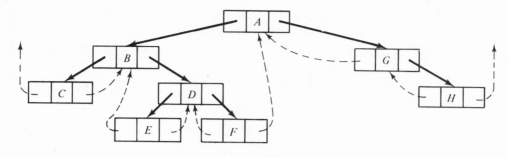

(a) What is the meaning of such threads in the binary tree that corresponds to a tree under the natural correspondence?

(b) Describe a very simple algorithm to traverse a threaded tree in symmetric order.

(c) Describe a simple algorithm to find the depth-first order successor of a node in a threaded tree.

21. Which of the tree traversal orders is suited to an algorithm to copy a tree? Design such an algorithm.

22. Define the *Strahler numbering* of the nodes of a binary tree as follows. The empty tree has the Strahler number 0. If the binary tree T has subtress T_l and T_r, the Strahler number $S(T)$ of T is defined to be

$$S(T) = \begin{cases} \max [S(T_l), S(T_r)] & \text{if } S(T_l) \neq S(T_r), \\ S(T_l) + 1 & \text{otherwise.} \end{cases}$$

Design an algorithm based on bottom-up traversal to compute $S(T)$ for a binary tree T.

23. Arithmetic expressions over the operations $+$, $-$, $*$, and $/$ can be represented as extended binary trees in an obvious way by having the internal nodes be operators and the leaves be operands. Thus, for example,

$\dfrac{A + B}{C - E}$ corresponds to

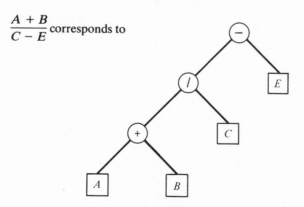

Design an algorithm to evaluate such expressions represented as extended binary trees. What is the significance of the depth-first, symmetric, and bottom-up traversal orders for such trees?

24. Design an algorithm for the level order traversal of a forest that is represented, via the natural correspondence, as a binary tree.

25. Prove that in an extended t-ary tree with n internal nodes, there are $(t - 1)n + 1$ external nodes.

26. Prove by induction on the height of binary trees that the solution to the recurrence relation

$$D(\square) = 0$$

$$D\left(T = \begin{array}{c} \bullet \\ \diagdown \end{array} \begin{array}{c} T_l \quad T_r \end{array} \right) = D(T_l) + D(T_r) + 2$$

is $D(T) = 2n$ for an extended binary tree with n internal nodes.

27. Prove by induction that the maximum value of $I(T)$ for an extended binary tree with n internal nodes is achieved by the tree shown in Figure 2.13.

28. Prove Lemma 2.2 by induction on the height of extended binary trees.

29. Explain why $E_0 = E_1 = \varnothing$ in the proof of Theorem 2.2.

chapter 3

Counting
and Estimating

The entire field of combinatorial algorithms is permeated by problems that require counting or estimating the number of elements in a finite set or that require listing all these elements, without duplication, in some specified order. Consequently, standard techniques for counting and estimating are necessary tools for anyone involved in combinatorial algorithms. The most generally applicable technique for counting and listing is backtrack (and its variants) because it makes the fewest assumptions about the underlying set; it is discussed in detail in Section 4.1. In this chapter we discuss techniques for counting that are applicable when the underlying set (or family of sets) has a well-understood structure. By exploiting this structure, these techniques frequently provide analytical techniques as opposed to the computational approach of backtrack.

There are two kinds of counting problems. In the simpler kind, one particular set is given, and the number of elements in it must be determined exactly. The exhaustive search techniques of Chapter 4 and Polya's counting theory of Section 3.4 apply to this type of problem. More common, however, is the type in which there is a family of sets indexed by a parameter, and we are interested in the size of the sets as a function of the parameter. In this case, exact numbers are rarely needed; more often an order-of-magnitude estimate suffices, and sometimes only an estimate of the growth rate of the function is needed. For example, the knowledge that the size of some set to be examined grows exponentially with some parameter may be sufficient reason to discard a proposed approach to a problem without investigating its details. The techniques of asymptotic expansion, recurrence relations, and generating functions (Sections 3.1, 3.2, and 3.3, respectively) all apply to this second kind of problem.

Because the techniques discussed here involve a fair amount of mathematics, we present only the main ideas and refer the interested reader to more advanced literature (see Section 3.5). Fortunately, the basic ideas of these counting techniques are simple, as will be shown. It is only their application to sets with a complex structure that can require advanced mathematical techniques.

3.1 ASYMPTOTICS

Asymptotics is the art of estimating and comparing the growth rates of functions. We say that the function $(1 + x^2)/x$ "behaves like x," or "grows at the same rate as x," as $x \to \infty$ and "behaves like $1/x$" as $x \to 0$; we say that $\log x$ grows less quickly than x^ε as $x \to \infty$, for any $\varepsilon > 0$, and that $\sum_{i=1}^{n} \log i$ grows no faster than $n \log n$ as $n \to \infty$. Such imprecise but intuitively meaningful statements are as useful for comparing functions as the relations $<$, $\leq$, and $=$ are for comparing numbers. This section shows how precise meanings can be attached to such statements.

Let us now define the three basic asymptotic relationships.

Definition

$f(x) \sim g(x)$ as $x \to x_0$ if and only if

$$\lim_{x \to x_0} \frac{f(x)}{g(x)} = 1.$$

We say that $f(x)$ is asymptotically equal to $g(x)$, or $f(x)$ grows at the same rate as $g(x)$, as $x \to x_0$.

Definition

$f(x) = o(g(x))$ as $x \to x_0$ if and only if

$$\lim_{x \to x_0} \left| \frac{f(x)}{g(x)} \right| = 0.$$

We say that $f(x)$ grows slower than $g(x)$, or that $f(x)$ is "little oh of" $g(x)$, as $x \to x_0$.

Definition

$f(x) = O(g(x))$ as $x \to x_0$ if and only if there exists a constant c such that

$$\limsup_{x \to x_0} \left| \frac{f(x)}{g(x)} \right| = c.$$

We say that $f(x)$ grows no faster than $g(x)$, or that $f(x)$ is "big oh of" $g(x)$, as $x \to x_0$.

The qualification "as $x \to x_0$" is omitted when the value of x_0 is understood; since we will be interested primarily in asymptotic growth rates for large values of the independent variable, we assume $x_0 = \infty$ throughout the book unless otherwise stated.

The equation $f(x) = g(x) + o(h(x))$ is shorthand for $f(x) - g(x) = o(h(x))$. Similarly, $f(x) = g(x) + O(h(x))$ means $f(x) - g(x) = O(h(x))$. The terms $O(\ )$ and $o(\ )$ can also be used in inequalities. For example,

$$x + o(x) < 2x \text{ as } x \to \infty$$

means that, for any function $f(x)$ such that $f(x) = o(x)$ as $x \to \infty$, it is true that $x + f(x) < 2x$ for all sufficiently large values of x.

It is important to remember that, in an equality containing instances of "little oh" or "big oh," the *two sides of the equality do not play a symmetric role*: the right-hand side always contains less information than the left-hand side and thus cannot be used to replace the left-hand side in every context. For example, the two (correct) asymptotic equations $x = O(x^2)$ and $x^2 = O(x^2)$ (as $x \to \infty$) do not imply the (false) statement that $x = x^2$. No difficulties of this nature arise if terms $O(\ )$ and $o(\ )$ appear only on the right-hand sides of equations. However, it is possible (and occasionally convenient) to use them on the left-hand sides of equations as long as it is understood that information is dropped in going from left to right. We can write, for example,

$$\frac{x}{x-1} = 1 + \frac{1}{x} + O\left(\frac{1}{x^2}\right) = 1 + \frac{1}{x} + o\left(\frac{1}{x}\right) = 1 + o(1) \text{ as } x \to \infty.$$

We now present some useful asymptotic equalities. By studying them carefully, the reader can test his understanding of the preceding definitions.

A polynomial is asymptotically equivalent to its leading term, and we have

$$\sum_{i=0}^{k} a_i x^i = O(x^k) \text{ as } x \to \infty; \tag{3.1}$$

$$\sum_{i=0}^{k} a_i x^i = o(x^{k+1}) \text{ as } x \to \infty; \tag{3.2}$$

$$\sum_{i=0}^{k} a_i x^i \sim a_k x^k \text{ as } x \to \infty, \quad \text{if } a_k \neq 0. \tag{3.3}$$

Sums of powers of the integers satisfy

$$\sum_{i=1}^{n} i^k \sim \frac{1}{k+1} n^{k+1} \text{ as } n \to \infty \tag{3.4}$$

(which has

$$\sum_{i=1}^{n} i \sim \frac{n^2}{2} \text{ as } n \to \infty$$

and

$$\sum_{i=1}^{n} i^2 \sim \frac{n^3}{3} \text{ as } n \to \infty$$

as special cases);

$$\sum_{i=1}^{n} i^k = \frac{1}{k+1} n^{k+1} + o(n^{k+1}) \text{ as } n \to \infty, \quad \text{for any integer } k \geq 0; \quad (3.5)$$

$$\sum_{i=1}^{n} i^k = \frac{1}{k+1} n^{k+1} + O(n^k) \text{ as } n \to \infty, \quad \text{for any } k \geq 0. \quad (3.6)$$

Notice that (3.4) and (3.5) contain exactly the same information but that (3.6) is a somewhat stronger statement than either (3.4) or (3.5). The equation

$$\sum_{i=1}^{n} i^k = \frac{1}{k+1} n^{k+1} + \frac{1}{2} n^k + O(n^{k-1}) \text{ as } n \to \infty, \quad \text{for any } k \geq 1, \quad (3.7)$$

is even stronger, since it contains a second term in the asymptotic expansion of $\sum_{i=1}^{n} i^k$ as $n \to \infty$.

It is worth dwelling on this example to exhibit the principle involved in obtaining an asymptotic expansion for a given function. Equation (3.4) says that $\sum_{i=1}^{n} i^k$ grows as $n^{k+1}/(k+1)$ as $n \to \infty$. Suppose that we would like more information about the difference between these two functions [all that follows immediately from the definition is that this difference is $o(n^{k+1})$, as equation (3.5) states]. We consider

$$\sum_{i=1}^{n} i^k - \frac{1}{k+1} n^{k+1} \text{ as } n \to \infty$$

and ask how fast this function grows. A reasonable guess is that it grows as n^k. A proof by induction on k (Exercise 2) shows that

$$\sum_{i=1}^{n} i^k - \frac{1}{k+1} n^{k+1} = \frac{1}{2} n^k + O(n^{k-1}) \text{ as } n \to \infty.$$

We could now continue and ask about the difference between $\sum_{i=1}^{n} i^k$ and $n^{k+1}/(k+1) + \frac{1}{2} n^k$, and so on, each time finding an additional term of an asymptotic series for $\sum_{i=1}^{n} i^k$.

In general,

$$c_1 g_1(x) + c_2 g_2(x) + c_3 g_3(x) + \cdots$$

is an *asymptotic series* for $f(x)$, or an *asymptotic expansion* of $f(x)$, as $x \to x_0$, denoted by

$$f(x) \sim \sum_{i=1}^{\infty} c_i g_i(x) \text{ as } x \to x_0,$$

if and only if

1. $g_{i+1}(x) = o(g_i(x))$ as $x \to x_0$, for $i \geq 1$, and

2. for each n,

$$f(x) = \sum_{i=1}^{n} c_i g_i(x) + o(g_n(x)) \text{ as } x \to x_0.$$

Asymptotic series need not converge. For example, consider

$$\frac{1}{x} + \frac{1}{x^2} + \frac{2!}{x^3} + \frac{3!}{x^4} + \cdots \text{ as } x \to \infty,$$

which is an asymptotic expansion of

$$f(x) = e^{-x} \int_1^x \frac{e^t}{t}\, dt$$

(see Exercise 4) but which converges for no value of x. The difference between a convergent series and an asymptotic series is that, in convergence, we are concerned with the behavior of the partial sums s_n for a fixed value of x as $n \to \infty$, whereas, in asymptotics, we are concerned with the behavior of the partial sums s_n for each fixed value of n as $x \to x_0$.

A nonconvergent asymptotic series $f(x) \sim \sum c_i g_i(x)$ as $x \to x_0$ can usually be used to compute good approximations of $f(x)$, provided that the argument x is sufficiently close to x_0 and that the summation is stopped at the correct place. Typically, for any fixed value of x, the terms initially decrease and then grow without bound. If the summation is carried too far, into the range where terms begin to increase, then the values of the partial sums become increasingly unrelated to the values of $f(x)$. A good place to stop the summation is either the first value of i for which $c_i g_i(x)$ is smaller than the allowable error or the value of i for which $c_i g_i(x)$ is minimal, whichever is smaller. Thus for

$$f(x) = e^{-x} \int_1^x \frac{e^t}{t}\, dt,$$

if we aim at an error bound of 10^{-3}, then $x = 10$ can be considered sufficiently close to $x_0 = \infty$.

$$\frac{1}{10} + \frac{1}{100} + \frac{2}{1000} + \frac{6}{10000} = 0.1126$$

gives us a reasonable approximation to $f(10) \approx 0.11306$.

We now present some useful asymptotic formulas whose derivations demonstrate a technique that is often useful for deriving asymptotic equations for sums: comparing the sum to an integral that can be evaluated in closed form. From

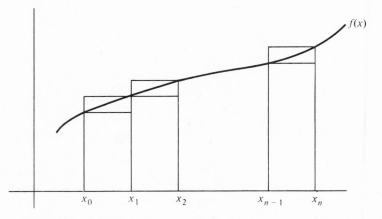

Figure 3.1   Lower and upper sums as bounds on an integral.

elementary calculus we know that if $f(x)$ is a monotonically increasing, integrable function, then $\int_a^b f(x)\, dx$ is bounded below and above by lower and upper sums, respectively, as shown in Figure 3.1:

$$\sum_{k=1}^{n} f(x_{k-1})(x_k - x_{k-1}) \leq \int_{x_0}^{x_n} f(x)\, dx \leq \sum_{k=1}^{n} f(x_k)(x_k - x_{k-1}).$$

Letting $x_i = i + 1$, this inequality becomes

$$\sum_{k=1}^{n} f(k) \leq \int_1^{n+1} f(x)\, dx \leq \sum_{k=1}^{n} f(k + 1),$$

and so

$$\int_1^{n+1} f(x)\, dx - f(n + 1) + f(1) \leq \sum_{k=1}^{n} f(k) \leq \int_1^{n+1} f(x)\, dx.$$

For example, for sums of terms involving logarithms, we can take $f(x) = \ln x$ so that $\int \ln x\, dx = x \ln x - x$ and

$$(n + 1) \ln (n + 1) - n - \ln (n + 1) \leq \sum_{k=1}^{n} \ln k \leq (n + 1) \ln (n + 1) - n.$$

This can be rewritten in a weaker form as

$$\sum_{k=1}^{n} \ln k = (n + 1) \ln (n + 1) - n + O(\log n). \tag{3.8}$$

Furthermore, dividing (3.8) by ln 2 yields

$$\sum_{k=1}^{n} \lg k = (n + 1) \lg (n + 1) - \frac{n}{\ln 2} + O(\log n). \tag{3.9}$$

Notice how this constant gets absorbed by the "big oh" notation. In a similar way (Exercise 5), we find that

$$\sum_{k=1}^{n} k \ln k = \frac{(n + 1)^2}{2} \ln (n + 1) - \frac{(n + 1)^2}{4} + O(n \log n), \tag{3.10}$$

and dividing by ln 2 transforms this to

$$\sum_{k=1}^{n} k \lg k = \frac{(n + 1)^2}{2} \lg (n + 1) - \frac{(n + 1)^2}{4 \ln 2} + O(n \log n). \tag{3.11}$$

By using more careful approximations to $\int_{x_0}^{x_n} f(x)\, dx$ as upper and lower bounds, we can improve on equation (3.8) and derive the well-known *Stirling's approximation* for factorial numbers (see Exercises 6 and 7):

$$n! \sim \sqrt{2\pi n}\, n^n e^{-n} \text{ as } n \to \infty. \tag{3.12}$$

Finally, this technique suffices to determine the growth rate of the *harmonic numbers*,

$$H_n = \sum_{k=1}^{n} \frac{1}{k} = \ln n + O(1). \tag{3.13}$$

Equation (3.13) can be sharpened to

$$H_n = \ln n + \gamma + O\left(\frac{1}{n}\right), \tag{3.14}$$

where $\gamma \approx 0.577$ is called *Euler's constant*.

3.2 RECURRENCE RELATIONS

Let us illustrate the concept of recurrence relations by using a classical problem, posed and studied around the year 1200 by Leonardo da Pisa, known as Fibonacci. The surprising importance of Fibonacci numbers in the analysis of combinatorial algorithms makes this example quite appropriate.

Fibonacci presented his problem with a story about the rate of growth of a population of rabbits, under the following assumptions. Start with one pair of rabbits. Each pair of rabbits becomes fertile after one month, after which it produces as offspring one other pair of rabbits every month. Rabbits never die and never stop reproducing.

Let F_n be the number of pairs of rabbits in the population after n months and let this population be composed of N_n newborn pairs and O_n old pairs—that is, $F_n = N_n + O_n$. The rules tell us that during the next month the following events happen:

$O_{n+1} = O_n + N_n = F_n$ The old population at time $n + 1$ is increased by those born at time n.

$N_{n+1} = O_n$ Every old pair at time n produced a newborn pair at time $n + 1$.

During the following month this pattern repeats:

$$O_{n+2} = O_{n+1} + N_{n+1} = F_{n+1}$$
$$N_{n+2} = O_{n+1}$$

Combining these equations, we obtain the recurrence relation

$$F_{n+2} = O_{n+2} + N_{n+2} = F_{n+1} + O_{n+1}$$

or, in other words,

$$F_{n+2} = F_{n+1} + F_n. \tag{3.15}$$

The initial conditions chosen for the sequence of Fibonacci numbers are unimportant; the essential properties of this sequence are determined by the recurrence relation. We will assume $F_0 = 0, F_1 = 1$ (another pair of conventional starting values is $F_0 = F_1 = 1$).

3.2.1   *Linear Recurrence Relations with Constant Coefficients*

The Fibonacci recurrence is a special case of *homogeneous linear recurrence relations with constant coefficients*, which have the form

$$x_n = a_1 x_{n-1} + a_2 x_{n-2} + \cdots + a_k x_{n-k}, \tag{3.16}$$

where the coefficients a_i are independent of n and $x_1, x_2, \ldots, x_k$ are specified. There is a general technique for solving linear recurrence relations with constant coefficients—that is, for determining x_n as a function of n. We will first use it for the Fibonacci sequence and then discuss it in general.

We seek a solution of the form

$$F_n = cr^n$$

with constants c and r. Substituting this into the Fibonacci recurrence relation yields

$$cr^{n+2} = cr^{n+1} + cr^n$$

or
$$cr^n(r^2 - r - 1) = 0.$$

This means that $F_n = cr^n$ is a solution if either $c = 0$ or $r = 0$ (and hence $F_n = 0$ for all n), or, more interestingly, if $r^2 - r - 1 = 0$. That is,

$$r = \frac{1}{2}(1 + \sqrt{5}) \quad \text{or} \quad r = \frac{1}{2}(1 - \sqrt{5})$$

with arbitrary constant c. The number $\frac{1}{2}(1 + \sqrt{5}) \approx 1.618$ is known as the *golden ratio*, because a rectangle with sides 1 and $\frac{1}{2}(1 + \sqrt{5})$ has been considered since antiquity to have the most pleasing proportions.

The sum of two solutions of a homogeneous linear recurrence relation is obviously also a solution, and, in fact, it is possible to show that the most general solution of the Fibonacci recurrence is

$$F_n = c\left(\frac{1 + \sqrt{5}}{2}\right)^n + c'\left(\frac{1 - \sqrt{5}}{2}\right)^n, \tag{3.17}$$

where the constants c and c' are determined by initial conditions. Assuming $F_0 = 0, F_1 = 1$, we obtain the linear equations

$$F_0 = c + c' = 0,$$

$$F_1 = c\frac{1}{2}(1 + \sqrt{5}) + c'\frac{1}{2}(1 - \sqrt{5}) = 1,$$

which yield

$$c = -c' = \frac{1}{\sqrt{5}}. \tag{3.18}$$

For all initial conditions that give rise to $c \neq 0$, sequences satisfying the Fibonacci recurrence grow approximately as $c\left(\frac{1 + \sqrt{5}}{2}\right)^n$, the terms of the form $c'\left(\frac{1 - \sqrt{5}}{2}\right)^n$ contributing very little (on an absolute as well as a relative basis), since

$$\left|\frac{1}{2}(1 - \sqrt{5})\right| < 1 < \frac{1}{2}(1 + \sqrt{5}).$$

Thus, for the initial conditions $F_0 = 0, F_1 = 1$, Fibonacci numbers grow asymptotically for large n as

$$F_n \sim \frac{1}{\sqrt{5}} \left(\frac{1 + \sqrt{5}}{2} \right)^n.$$

Let us now describe, without proof, how the solution technique just described is applied to determine the asymptotic behavior of solutions to general linear recurrence relations with constant coefficients. We begin with homogeneous relations and then discuss *inhomogeneous* relations, where a constant term is added to the right-hand side. Substituting $x_n = r^n$ leads to the polynomial equation

$$r^k - a_1 r^{k-1} - a_2 r^{k-2} - \cdots - a_k = 0$$

whose k roots $r_1, r_2, \ldots, r_k$ give rise to solutions of the form $x_n = r_i^n$ for $i = 1, 2, \ldots, k$. If all these roots are distinct, the k solutions $x_n = r_i^n$ form a basis for the space of all solutions: they are independent (i.e., no solution r_i^n can be obtained as a linear combination of the others) and span the space [i.e., any solution of equation (3.16) can be expressed as a linear combination of these solutions]. So, for the case of distinct roots, the most general solution of the recurrence relation (3.16) has the form

$$x_n = \sum_{i=1}^{k} c_i r_i^n.$$

If we are interested only in the asymptotic behavior of the sequence $x_1, x_2, \ldots$, we need only consider the terms $c_i r_i^n$ for r_i having the maximum absolute value among those whose corresponding coefficients c_i are not zero.

Initial conditions that give rise to some zero coefficients c_i are in a certain sense exceptional (in the space of all possible initial conditions), and, in the same sense, recurrence relations that give rise to roots of equal absolute values are exceptional in the space of all recurrence relations. Therefore the most frequent case, and the one worth remembering, is when one root dominates in absolute value; that is,

$$|r_1| > |r_i| \qquad \text{for } i = 2, 3, \ldots, k.$$

In this case, the asymptotic behavior of x_n is given by

$$x_n \sim c_1 r_1^n,$$

and this formula frequently provides good approximations of x_n for even moderate values of n. When $|r_i| < 1, i = 2, 3, \ldots, k$, the *absolute error* of this approximation, that is,

$$x_n - c_1 r_1^n = \sum_{i=2}^{k} c_i r_i^n,$$

tends to zero as $n \to \infty$. If $|r_i| > 1$ for some of $i = 2, 3, \ldots, k$, then only the *relative error* of the approximation $x_n \sim c_1 r_1^n$, that is, $(x_n - c_1 r_1^n)/x_n$, tends to zero as n increases.

When some of the roots $r_1, r_2, \ldots, r_k$ are identical, say $r_1 = r_2 = r_3$, then the solutions $x_n = c_i r_i^n$ are not independent and do not form a basis for the space of all solutions. It can be shown that the most general solution in this particular case has the form

$$x_n = c_1 r_1^n + c_2 n r_2^n + c_3 n^2 r_3^n + \sum_{i=4}^{k} c_i r_i^n.$$

Generalization of this formula to equations with an arbitrary number of multiple roots is straightforward.

An *inhomogeneous* linear recurrence relation with constant coefficients has the form

$$x_n = a_1 x_{n-1} + a_2 x_{n-2} + \cdots + a_k x_{n-k} + b. \tag{3.19}$$

Its most general solution is obtained by adding the *particular solution* of relation (3.19) to the most general solution of its *associated homogeneous relation*

$$y_n = a_1 y_{n-1} + a_2 y_{n-2} + \cdots + a_n y_{n-k},$$

which is obtained by the techniques already discussed. This leads to a solution with k parameters that are used to satisfy k initial conditions of equation (3.19).

A particular solution to (3.19) is found as follows. If $\sum_{i=1}^{k} a_i \neq 1$, then there is a particular solution $x_n = c$, where c is determined by the equation

$$c = a_1 c + a_2 c + \cdots + a_k c + b;$$

that is,

$$c = \frac{b}{1 - \sum_{i=1}^{k} a_i}.$$

If $\sum_{i=1}^{k} a_i = 1$, then there is a particular solution $x_n = cn$, where c is determined by the equation

$$cn = a_1 c(n-1) + a_2 c(n-2) + \cdots + a_k c(n-k) + b;$$

that is,

$$c = \frac{b}{\sum_{i=1}^{k} i a_i}.$$

$cn[1 - a_1 - a_2 - \ldots - a_k] + c[a_1 + 2a_2 + \ldots + ka_k]$
$= b$
$c[n - (n-1)a_1 - (n-2)a_2 - \ldots - (n-k)a_k] = b$

A word of caution about recurrence relations is appropriate. In combinatorial applications, it is generally necessary only to analyze properties of the solution, not to calculate a sequence from the recurrence relation. For this purpose, the preceding analysis is sufficient. If it is necessary to compute sequences numerically according to recurrence relations, and if this calculation cannot be carried out in the domain of integers (or with infinite precision), care should be taken to avoid pitfalls in numerical computation (instability) that may render useless any computation based on a simple-minded approach. A discussion of the stability problems that occur in the numerical computation of linear recurrence relations can be found in any textbook on the numerical solution of differential equations.

3.2.2   General Recurrence Relations

Recurrence relations that are not linear with constant coefficients have no general solution techniques comparable to the one discussed in the previous section. General recurrence relations are solved (or their solutions are approximated or bounded) by trial-and-error techniques. If the trial-and-error is guided by the general ideas discussed in this section, it will at least yield a good estimate of the asymptotic behavior of the solution of most recurrence relations.

Reducing the Number of Terms in the Relation.   Consider the recurrence relation

$$x_{n+1} = \sum_{i=0}^{n} x_i$$

with x_0 given as the initial condition. It is a linear recurrence relation but not of the type discussed in the previous section. It can be transformed into a linear recurrence relation with constant coefficients, however, by observing

$$x_{n+1} - x_n = \sum_{i=0}^{n} x_i - \sum_{i=0}^{n-1} x_i = x_n,$$

or

$$x_{n+1} = 2x_n,$$

whose solution is immediately seen to be

$$x_n = x_0 2^n \qquad \text{for } n \geq 0.$$

Similar observations often succeed in simplifying recurrence relations that involve a weighted sum (with constant weights) of all preceding terms, for example

$$x_{n+1} = a_{n+1} + \sum_{i=0}^{n} b_i x_i$$

leads to

$$x_{n+1} - x_n = a_{n+1} - a_n + b_n x_n$$

or

$$x_{n+1} = (a_{n+1} - a_n) + (1 + b_n)x_n.$$

The solution can thus be expressed in closed form as a sum of products, but unless the coefficients $a_{n+1} - a_n$ and $1 + b_n$ have a simple form, nothing will be gained by doing so.

Determining the Growth Rate of the Solution.   In analyzing the amount of work involved in an algorithm of the decomposition type (see Section 1.4), a recurrence relation of the form

$$x_n = \frac{2}{n} \sum_{k=0}^{n-1} x_k + an + b \qquad \text{for } n \geq 1 \qquad (3.20)$$

with x_0 given as an initial condition is sometimes encountered. Let us see first how this relation occurs and, then, how to solve it.

Imagine a class of problems P_n, $n = 1, 2, \ldots$, whose size, or complexity, is measured in terms of an integer parameter n. Let x_n denote the expected amount of work required to solve P_n. For example, assume that P_n requires imposing a binary tree structure over a sequence of n elements, $e_1, e_2, \ldots, e_n$, in the following way. First the sequence is scanned, say from left to right, and afterward a particular element e_k is chosen as "the root of the tree." The amount of work involved in the scan is proportional to n, which accounts for the term an in equation (3.20). As a result of this interpretation, we assume $a > 0$. Next, some work is carried out on element e_k, and we have denoted this by the term b.

The element e_k splits the sequence $e_1, \ldots, e_n$ into a left part, $e_1, \ldots, e_{k-1}$, and a right part, $e_{k+1}, \ldots, e_n$ (either or both of which can be empty). We have now decomposed P_n into two smaller problems, P_{k-1} and P_{n-k}. Both must be solved, and they require an amount of work x_{k-1} and x_{n-k}, respectively. For a fixed value of k, this yields

$$x_n = x_{k-1} + x_{n-k} + an + b.$$

Assume that, as a result of scanning the sequence, any element $e_1, e_2, \ldots, e_n$ is equally likely to be chosen as e_k. Thus k can assume any of the values from 1 to n with equal probability $1/n$; and the expected amount of work x_n required to solve P_n satisfies the equation

$$x_n = \frac{1}{n} \sum_{k=1}^{n} (x_{k-1} + x_{n-k}) + an + b,$$

which is easily transformed into (3.20).

Let us now present a systematic trial-and-error process whereby the asymptotic behavior of the solution of the recurrence relation (3.20) can be determined. We begin by trying the simplest possible solution, a constant. Substituting $x_n = c$ into (3.20) leads to

$$c \stackrel{?}{=} 2c + an + b,$$

which tells us that $x_n = c$ is not a solution. But this attempt tells us a great deal more, since, with our assumption that $a > 0$, the right-hand side is larger than the left-hand side. In other words, the attempt $x_n = c$ leads to the inequality

$$x_n < \frac{2}{n} \sum_{k=0}^{n-1} x_k + an + b.$$

So, in order to turn it into an equality, x_n must be larger in relation to $x_1, \ldots, x_{n-1}$; that is, the solution to the recurrence relation must grow faster as function of n than $x_n = c$ does.

Therefore, as our next attempt, let us try a linear function $x_n = cn$. Substituting it into the recurrence relation leads to

$$cn \overset{?}{=} \frac{2}{n}\sum_{k=0}^{n-1} ck + an + b = c(n-1) + an + b = (c+a)n - c + b.$$

There are several differences between the left and the right sides of this proposed equality, but at this stage of the analysis it is necessary to focus only on the *leading terms* of each side: cn on the left, $(c+a)n$ on the right. This discrepancy tells us that a linear function cannot be the solution; moreover, the fact that the right side is larger for sufficiently large values of n tells us that a linear function also grows too slowly to be a solution of our recurrence relation.

For our third attempt we choose a function that grows faster than linearly: $x_n = cn^2$ leads to

$$cn^2 \overset{?}{=} \frac{2}{n}\sum_{k=0}^{n-1} ck^2 + an + b = \frac{2}{n}\left[c\frac{n^3}{3} + O(n^2)\right] + an + b = \frac{2}{3}cn^2 + O(n).$$

For sufficiently large values of n, the right-hand side is smaller than the left-hand side, which tells us that a quadratic function grows too quickly to be a solution.

Having bounded the growth rate of the solution from below and from above, our next attempt should be a function whose growth rate lies between that of linear and quadratic functions. A power of n, say $x_n = n^{3/2}$, would still lead to a mismatch of the coefficients of the leading terms. In fact, for any $\varepsilon > 0$, $x_n = n^{1+\varepsilon}$ grows too quickly. This suggests $x_n = cn \lg n$, which gives

$$cn \lg n \overset{?}{=} \frac{2c}{n}\sum_{k=0}^{n-1} k \lg k + an + b.$$

Using a crude version of equation (3.11), we obtain

$$\sum_{k=0}^{n-1} k \lg k = \frac{1}{2}n^2 \lg n + o(n^2 \log n), \tag{3.21}$$

and our tentative equality becomes

$$cn \lg n \overset{?}{=} cn \lg n + o(n \log n) + an + b.$$

Thus $x_n = cn \lg n$ satisfies the first prerequisite of a solution—namely, that the leading terms on both sides match. We can now conclude that the solution's asymptotic behavior is

$$x_n \sim cn \lg n$$

for some as yet unknown constant c. The proposed equality above does not determine c because we have lost too much information in going from equation

(3.11) to equation (3.21) by lumping all except the leading term $\frac{1}{2}n^2 \lg n$ into a term $o(n^2 \log n)$.

Using equation (3.11) in its entirety to obtain a further term in the asymptotic expansion of $\sum k \lg k$ gives an additional constraint that will determine c:

$$cn \lg n \;\overset{?}{=}\; \frac{2c}{n}\left[\frac{n^2}{2}\lg n - \frac{n^2}{4 \ln 2} + O(n \log n)\right] + an + b,$$

and hence

$$cn \lg n \;\overset{?}{=}\; cn \lg n + \left(a - \frac{c}{2 \ln 2}\right)n + O(\log n).$$

To match the linear terms on each side, we must choose c such that

$$a - \frac{c}{2 \ln 2} = 0,$$

or

$$c = a \ln 4 \approx 1.386a.$$

We now know that the solution to recurrence relation (3.20) has the form

$$x_n = a \ln 4\, n \lg n + o(n \log n),$$

which, for most practical purposes, is enough information. Additional terms of the asymptotic expansion of x_n can be obtained by increasing the algebraic manipulation, but they are rarely of interest.

We have studied recurrence relation (3.20) merely to demonstrate techniques for solving such equations. In Chapters 6 and 7 we will encounter it in different contexts, where its solution will be used in the analysis of some searching and sorting techniques.

The solution procedure that we have presented is not rigorous. Much more needs to be established in order to complete the proof that $x_n \sim a \ln 4\, n \lg n$ as $n \to \infty$. Assumptions were left unstated when we used the relative sizes of the left and right sides of a proposed equation to guide our search for solutions that grew faster or slower than the current guess. Primarily, we have omitted the assumption that we were looking for a monotonically increasing solution. The justification for this lack of mathematical rigor is that we are only trying to discover a reasonable candidate for a solution to a complicated equation. Anything goes for this purpose, and the only relevant criterion is success—that is, whether we come up with a correct guess. This is only the first part of the solution process: when one is sufficiently confident that the guess is correct, one tries to prove that the guess is indeed a solution. This second phase is often a matter of mathematical technique—laborious but straightforward. For recurrence relation (3.20), it is pursued in Exercise 11 (see also Exercise 12).

3.3   GENERATING FUNCTIONS

As the preceding sections showed, "counting" often means computing or determining properties of some sequence of numbers $x_1, x_2, \ldots$, where x_k is associated with problem P_k in some sequence of problems $P_1, P_2, \ldots$. The structure of these problems induces a relationship among the elements of the corresponding counting sequence, and the manipulation of such sequences is an important aspect of many problems that deal with counting.

In this section we introduce a useful tool for operating with sequences. The idea is to associate with each sequence of numbers a function of a real or complex variable in such a way that common operations on sequences correspond to simple operations on the corresponding functions. Analytical techniques for operating on functions of a real or complex variable are frequently simpler and more powerful than combinatorial techniques for operating on sequences directly; consequently, this approach often leads to elegant derivations where a direct approach would involve cumbersome calculations.

There are many possible ways to map sequences onto functions, but the most common one in combinatorics is to associate with a sequence $x_0, x_1, x_2, \ldots$ the function

$$X(s) = \sum_{k=0}^{\infty} x_k s^k$$

of a real variable s. X is called the *generating function* of the sequence $x_0, x_1, x_2, \ldots$. Table 3.1 contains examples that the reader can check by expanding the generating function as a Taylor series around $s = 0$ (Exercise 14).

Since a sequence can be reconstructed from its generating function, any knowledge about the function gives some information about the sequence. However, the main usefulness of generating functions is that many common operations on sequences correspond to simple algebraic or analytic operations on the corresponding functions. In discussing the most important operations in the remainder of this section, we will denote sequences by $x_k, y_k, z_k, k = 0, 1, 2, \ldots$, and their respective generating functions by $X(s)$, $Y(s)$, $Z(s)$.

Linear Operations

If a, b are constants, then the sequence $z_k = ax_k + by_k$ has generating function $Z(s) = aX(s) + bY(s)$.

Shift of Origin

The sequence defined by

$$y_k = 0 \qquad\qquad \text{for } k = 0, \ldots, i - 1$$

$$y_k = x_{k-i} \qquad\qquad \text{for } k = i, i + 1, \ldots$$

Table 3.1 Some common generating functions.

Sequence $x_k, k = 0, 1, 2, \ldots$	Generating function $X(s)$
(1) $x_k = 1$	$\dfrac{1}{1-s}$
(2) $x_k = k + 1$	$\dfrac{1}{(1-s)^2}$
(3) $x_0 = 0$ $\quad$ $x_k = \dfrac{1}{k}, \quad k \geq 1$	$\ln\left(\dfrac{1}{1-s}\right)$
(4) $x_0 = 0$ $\quad$ $x_k = \dfrac{1}{k}$ $\;$ if k is odd $\quad$ $x_k = -\dfrac{1}{k}$ $\;$ if $k > 0$ is even	$\ln(1 + s)$
(5) $x_k = \dbinom{r}{k}$, r arbitrary	$(1 + s)^r$
(6) $x_k = \dfrac{1}{k!}$	e^s
(7) $x_k = \dfrac{r^k}{k!}$, r arbitrary	e^{rs}

has generating function $Y(s) = X(s)s^i$. Similarly, the sequence defined by

$$y_k = x_{k+i} \qquad \text{for } k = 0, 1, 2, \ldots$$

has generating function

$$Y(s) = \left[X(s) - \sum_{k=0}^{i-1} x_k s^k \right] s^{-i}.$$

As an example of the use of the linear operations and the shift of origin, consider the evaluation of

$$\sum_{i=1}^{t} \binom{n-i}{j},$$

which arises in the analysis of an algorithm in Section 5.2.2. Let

$$x_{ij} = \binom{n-i}{j},$$

and for each fixed value of i, consider x_{ij} as a sequence indexed by j, with generating function

$$X_i(s) = (1 + s)^{n-i}.$$

We are interested in the sequence y_j defined by

$$y_j = \sum_{i=1}^{t} x_{ij}$$

whose generating function is obtained by using linear operations

$$Y(s) = \sum_{i=1}^{t} X_i(s) = \sum_{i=1}^{t} (1 + s)^{n-i} = (1 + s)^n \sum_{i=1}^{t} \left(\frac{1}{1+s}\right)^i.$$

This last term is easily evaluated as the sum of a geometric series, and so we find

$$Y(s) = \frac{(1 + s)^n}{s} - \frac{(1 + s)^{n-t}}{s}.$$

Both $(1 + s)^n/s$ and $(1 + s)^{n-t}/s$ are simple shifts of origin of $(1 + s)^n$ and $(1 + s)^{n-t}$, respectively, and we conclude that

$$y_j = \sum_{i=1}^{t} \binom{n-i}{j} = \binom{n}{j+1} - \binom{n-t}{j+1}. \tag{3.22}$$

Partial Sums

If $y_k = \sum_{i=0}^{k} x_i, k = 0, 1, 2, \ldots$, then

$$Y(s) = \frac{X(s)}{1 - s}.$$

This follows from the rules for linear operations and shifts of origin (Exercise 15).

Complementary Partial Sums

If $X(1) = \sum_{k=0}^{\infty} x_k$ exists, and if $y_k = \sum_{i=k}^{\infty} x_i, k = 0, 1, 2, \ldots$, then

$$Y(s) = \frac{X(1) - sX(s)}{1 - s}.$$

Scaling

The sequence $y_k = kx_k, k = 0, 1, 2, \ldots$, has generating function $Y(s) = sX'(s)$, and the sequence $y_k = x_k/(k + 1)$ has generating function

$$Y(s) = \frac{1}{s} \int_0^s X(t)\, dt.$$

These last two rules, combined with those of linear operations and shifts of origin, allow us to compute $Y(s)$ from $X(s)$ for any transformation of sequences

$$y_k = R(k)x_k, \quad k = 0, 1, 2, \ldots,$$

where $R(k)$ is any rational function of k.

Scaling is also useful in evaluating sums. Consider the sum

$$\sum_{k=2}^{n-1} (n-k)(n-k)\binom{n-1}{n-k},$$

which can be rewritten as

$$\sum_{k=1}^{n-2} k^2\binom{n-1}{k}$$

by reversing the order of the terms. Letting

$$x_k = k^2\binom{n-1}{k}, \quad y_k = k\binom{n-1}{k}, \quad \text{and} \quad z_k = \binom{n-1}{k},$$

we have $y_k = kz_k$, and so

$$Y(s) = sZ'(s).$$

Similarly, since $x_k = ky_k$, we have

$$X(s) = sY'(s) = sZ'(s) + s^2Z''(s).$$

But we know from Table 3.1 that $Z(s) = (1+s)^{n-1}$, so that

$$X(s) = s(n-1)(1+s)^{n-2} + s^2(n-1)(n-2)(1+s)^{n-3}.$$

Obviously,

$$X(1) = \sum_{k=1}^{n-1} k^2\binom{n-1}{k},$$

which implies

$$\sum_{k=2}^{n-1} (n-k)(n-k)\binom{n-1}{n-k} = n(n-1)2^{n-3} - (n-1)^2.$$

This and similar summations occur in the analysis of an algorithm in Section 4.1.7.

Convolution

The sequence $z_k = \sum_{i=0}^{k} x_i y_{k-i}$, $k = 0, 1, 2, \ldots$, has generating function $Z(s) = X(s)Y(s)$.

This rule is probably the single most important reason for using generating functions, since convolution is a common occurrence in combinatorial problems and, when treated directly, results in a complicated sum. By contrast, the corresponding operation on generating functions is the simple operation of multiplication.

To illustrate the use of generating functions, we will now discuss the challenging problem of counting the number of binary trees with n nodes. The convolution formula plays a prominent role in the solution to this problem.

Recall from Chapter 2 that a binary tree T_n of n nodes is empty if $n = 0$. If $n > 0$, it is a triple (T_l, r, T_{n-l-1}), where r is a distinguished node called the root of T_n, T_l is a binary tree of l nodes for some $l = 0, 1, \ldots, n - 1$, called the left subtree of T_n, and T_{n-l-1} is a binary tree of $n - l - 1$ nodes, called the right subtree of T_n.

What is the number x_n of distinct binary trees T_n of n nodes? From the definition it follows that $x_0 = 1$, and obviously $x_1 = 1, x_2 = 2, x_3 = 5$. From the definition of T_n we deduce the recurrence relation

$$x_n = \sum_{l=0}^{n-1} x_l x_{n-l-1}. \tag{3.23}$$

This recurrence relation is rather difficult to solve by the techniques of Section 3.2. We would easily discover that x_n grows faster than a polynomial in n but that an exponential function e^{cn} grows too fast. It would then be a matter of luck to find a functional form that satisfies the recurrence relation.

We notice, however, that the right-hand side of (3.23) is a convolution and hence consider the generating function

$$X(s) = \sum_{n=0}^{\infty} x_n s^n.$$

Knowing that convolution of sequences corresponds to multiplication of generating functions, we write

$$X^2(s) = \left(\sum_{n=0}^{\infty} x_n s^n \right)^2 = \sum_{k=0}^{\infty} \left(\sum_{l=0}^{k} x_l x_{k-l} \right) s^k.$$

Since replacing n by $k + 1$ in (3.23) yields

$$x_{k+1} = \sum_{l=0}^{k} x_l x_{k-l},$$

we can simplify the right-hand side to obtain the equation

$$X^2(s) = \frac{1}{s} \sum_{k=0}^{\infty} x_{k+1} s^{k+1} = \frac{1}{s}[X(s) - x_0]$$

for the generating function. Using $x_0 = 1$, we obtain the solution

$$X(s) = \frac{1}{2s}(1 \pm \sqrt{1 - 4s}). \tag{3.24}$$

The binomial expansion of $\sqrt{1 - 4s}$ gives us

$$(1 - 4s)^{1/2} = \sum_{k=0}^{\infty} \binom{\frac{1}{2}}{k} (-4s)^k$$

$$= 1 - \frac{1}{2} 4s + \frac{\frac{1}{2}(-\frac{1}{2})}{2!} (4s)^2 - \frac{\frac{1}{2}(-\frac{1}{2})(-\frac{3}{2})}{3!} (4s)^3 + \cdots.$$

The coefficient of s^k, $k \geq 1$, in this series can be written as

$$-\frac{(\frac{1}{2})^k 1 \cdot 3 \cdot 5 \cdot \cdots \cdot (2k - 3) 4^k}{k!} = -\frac{2}{k} \cdot \frac{1 \cdot 3 \cdot 5 \cdot \cdots \cdot (2k - 3) 2^{k-1} (k - 1)!}{(k - 1)!(k - 1)!}$$

$$= -\frac{2}{k} \binom{2k - 2}{k - 1},$$

so that

$$(1 - 4s)^{1/2} = 1 - \sum_{k=1}^{\infty} \frac{2}{k} \binom{2k - 2}{k - 1} s^k. \tag{3.25}$$

Since the result of substituting (3.25) into (3.24) must be a series in nonnegative powers of s, the $+$ solution $X(s) = (1/2s)(1 + \sqrt{1 - 4s})$ in (3.24) is extraneous. Substituting (3.25) into the solution

$$X(s) = \frac{1}{2s}(1 - \sqrt{1 - 4s})$$

yields

$$X(s) = \sum_{n=0}^{\infty} \frac{1}{n + 1} \binom{2n}{n} s^n.$$

Therefore the number of binary trees of n nodes is

$$x_n = \frac{1}{n + 1} \binom{2n}{n} \qquad \text{for } n \geq 0. \tag{3.26}$$

It is unusual to find a simple closed-form expression for a nontrivial combinatorial sequence. Closed-form expressions are elegant and occasionally useful, but they do not always provide the answer in the most convenient form. For example, to determine the growth rate of x_n as a function of n, we rewrite (3.26) as

$$x_n = \frac{1}{n + 1} \frac{(2n)!}{n! n!}$$

and use Stirling's formula, equation (3.12), to obtain

$$x_n \sim \frac{4^n}{n\sqrt{\pi n}} \text{ as } n \to \infty. \tag{3.27}$$

It would have been difficult to obtain this formula by trial-and-error substitutions into the recurrence relation (3.23).

★3.4   COUNTING EQUIVALENCE CLASSES: POLYA'S THEOREM

The first difficulty in counting the elements in a set is often to define clearly just what the elements are. For example, consider the problem of Section 3.3 concerning the number of binary trees of n nodes. If we choose not to distinguish between left and right, then the following four pictures, which represented distinct trees in our earlier formulation, now each represent the same tree.

More generally, if we ask how many graphs of n nodes and e edges there are, the usual underlying assumption implies that the two pictures

represent the same graph, a conclusion that is not visually obvious. These examples suffice to point out that counting objects must not be confused with counting their representations because one object can often have many representations.

Whenever conceptual difficulties occur in precisely defining the objects to be counted, the standard approach is to construct an easily counted set of representations for these objects—that is, a set of representations in which it is easy to determine how many distinct representations there are. The problem is then reduced to determining an equivalence relation on this set of representations, such that all representations in the same equivalence class correspond to the same object.

This section introduces an elegant technique, known as *Polya's theory of counting*, for counting equivalence classes of a set. We will state the most important special case of this theorem and outline its proof. The general statement of Polya's theorem, although not difficult, requires much special terminology, and we leave it to be pursued by the interested reader (see Section 3.5). The concepts involved in this theorem are explained by means of an example in the following section.

3.4.1   An Example: Coloring the Nodes of a Binary Tree

Consider a complete binary tree of seven nodes and all the different ways that its nodes can be colored white or black, under the assumption that we do not distinguish left from right. The following examples are different representations of the same tree-coloring:

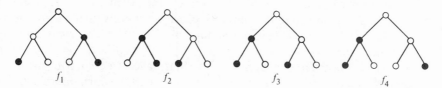

By contrast, each of the following pictures represents a different tree-coloring.

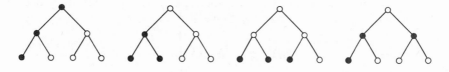

To clarify the set of objects to be counted, let us define the *set S of representations* that we want to consider. We associate a picture of the type above with a function from a domain D of seven nodes into a range R of two colors:

$$D = \{1, 2, 3, 4, 5, 6, 7\},$$

$$R = \{\circ, \bullet\}$$

$$S = \{f | f : D \to R\}.$$

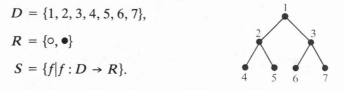

Notice that S has $2^7 = 128$ elements, called representations. For example, the representations f_3 and f_4 shown above correspond to the functions

$$f_3(1) = f_3(3) = f_3(5) = f_3(7) = \circ, \qquad f_3(2) = f_3(4) = f_3(6) = \bullet$$

and

$$f_4(1) = f_4(3) = f_4(5) = f_4(6) = \circ, \qquad f_4(2) = f_4(4) = f_4(7) = \bullet.$$

We define an equivalence relation γ on the set S of all functions from D to R, and we identify each equivalence class with one of the objects (tree-colorings) to be counted.

We know intuitively that f_3 and f_4 must be in the same equivalence class of γ. Why? There is a permutation π on the domain D that interchanges nodes 6 and 7, the two sons of node 3, and the two functions f_3 and f_4 differ only "modulo π"; that is, $f_4 = f_3\pi$, or, equivalently, the function f_4 is obtained by first applying the permutation π to D and then mapping D to R by means of f_3. Since our problem stated that left and right should not be distinguished, and π only interchanges the left and right sons of node 3, two functions like f_3 and f_4 must be equivalent.

Our first task, then, in defining the equivalence relation γ is to write down the group of all permutations of D that leave our original problem invariant. It is important to understand that it is not a matter of proving mathematically whether the permutations we write down are the correct ones; it is simply a matter of intuition and common sense, since we are merely giving a precise meaning to the somewhat vague original statement that "left and right must not be distinguished."

The group G of permutations that leave a tree-coloring invariant is generated by the three permutations shown in Figure 3.2, which interchange the subtrees of the nodes 1, 2, and 3, respectively. The cycle representation is a compact way of representing a permutation. A factor of the form $(e_1 e_2 \ldots e_k)$ in this representation is called a *cycle* and means that the permutation maps e_1 to e_2, e_2 to $e_3, \ldots, e_{k-1}$ to e_k, and e_k to e_1.

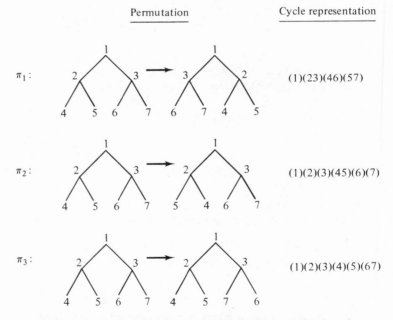

Figure 3.2 The permutations that leave a tree-coloring invariant.

When the permutations π_1, π_2, and π_3 are combined in all possible ways, they generate a group G consisting of the eight permutations shown in Table 3.2. In this table the column labeled *cycle index* associates with each permutation a term that describes the cycle structure of that permutation. These terms involve

Table 3.2 The group G of permutations generated by π_1, π_2, and π_3.

Permutation	Cycle Representation	Cycle Index
identity	(1)(2)(3)(4)(5)(6)(7)	x_1^7
π_1	(1)(23)(46)(57)	$x_1 x_2^3$
π_2	(1)(2)(3)(45)(6)(7)	$x_1^5 x_2$
π_3	(1)(2)(3)(4)(5)(67)	$x_1^5 x_2$
$\pi_2\pi_3 = \pi_3\pi_2$	(1)(2)(3)(45)(67)	$x_1^3 x_2^2$
$\pi_1\pi_2 = \pi_3\pi_1$	(1)(23)(4657)	$x_1 x_2 x_4$
$\pi_1\pi_3 = \pi_2\pi_1$	(1)(23)(4756)	$x_1 x_2 x_4$
$\pi_1\pi_2\pi_3$	(1)(23)(47)(56)	$x_1 x_2^3$

indeterminates $x_1, x_2, x_3, \ldots$, and the exponent of x_k is the number of cycles of length k in the permutation. Thus the cycle index $x_1 x_2^3$ of π_1 indicates that π_1 consists of one cycle of length 1 and three cycles of length 2. The cycle index of π does not preserve any information about which elements of D are in which cycle of π.

We are now ready to define the central concept involved in Polya's theorem. Given a permutation group G, the *cycle index P_G of G* is a polynomial in the indeterminates $x_1, x_2, \ldots$ that is the average of the cycle indices of the permutations π in G. In our example,

$$P_G = \frac{1}{8}(x_1^7 + 2x_1 x_2^3 + 2x_1^5 x_2 + 2x_1 x_2 x_4 + x_1^3 x_2^2).$$

The cycle index of a permutation group is a function that describes some of the properties of the group, much as a generating function describes properties of a sequence. However, whereas a generating function describes its associated sequence uniquely, the cycle index does not define a unique permutation group. Different, and nonisomorphic, permutation groups can have the same cycle index.

However, as Polya's theorem will show, the cycle index P_G does contain enough information to count the number of equivalence classes of the equivalence relation γ induced by the group G. The number of equivalence classes on S, the set of functions from D to R, under the equivalence relation γ (induced by a permutation group G on D), is obtained by assigning the value of $|R|$, the cardinality of R, to each indeterminate x_i in the cycle index P_G of G. In our example, $|R| = 2$, and hence we find

$$P_G(2, 2, 2, 2) = \frac{1}{8}(2^7 + 2^5 + 2^7 + 2^4 + 2^5) = 42.$$

The reader is invited to check by systematic construction that the number of different tree-colorings for our example is indeed 42.

3.4.2   Polya's Theorem and Burnside's Lemma

Here we state more formally the special case of Polya's theorem illustrated by means of the example in Section 3.4.1, and we outline its proof. The proof involves a result known as *Burnside's lemma*, which is an interesting counting technique in its own right. Although Polya's theorem is easier to use mechanically, Burnside's lemma more clearly reveals the mathematical idea on which the counting of equivalence classes is based.

Given a finite set D called the domain, a finite set R called the range, and a group G of permutations on D, we define

1. $S = \{f | f : D \to R\}$.

2. The equivalence relation γ on S induced by G: for f, g in S, let $f \gamma g$ if and only if there exists a permutation π in G such that $f = g\pi$.

3. The cycle index $P_G(x_1, x_2, \ldots, x_{|D|})$ of G (i.e., the polynomial in indeterminates $x_1, x_2, \ldots, x_{|D|}$):

$$P_G(x_1, x_2, \ldots) = \frac{1}{|G|} \sum_{\pi \in G} x_1^{l_1(\pi)} x_2^{l_2(\pi)} \ldots x_{|D|}^{l_{|D|}(\pi)},$$

where $l_k(\pi)$ is the number of cycles of length k in π.

Theorem (Polya)

The number of equivalence classes of γ is equal to $P_G(|R|, |R|, |R|, \ldots, |R|)$, where $|R|$ is the cardinality of the set R.

Given a finite set S and group G' of permutations on S, we define the equivalence relation γ on S induced by G' by $f \gamma g$ if and only if there is a π in G' such that $f = \pi(g)$. We define $I(\pi)$ as the number of elements of S that are invariant under π [f in S is invariant under π if and only if $\pi(f)$].

Lemma (Burnside)

The number of equivalence classes of γ is equal to

$$\frac{1}{|G'|} \sum_{\pi \in G'} I(\pi).$$

We have chosen the notation so that the objects to be identified in Polya's theorem and in Burnside's lemma have the same name. Notice that the permutation groups G and G' act on sets that do not correspond to each other: G acts on D and G' acts on S.

As an example, we can use Burnside's lemma to count the number of tree-colorings in the problem of Section 3.4.1. Recall that $D = \{1, 2, 3, 4, 5, 6, 7\}$ is the set of nodes, $R = \{\circ, \bullet\}$ is the set of colors, and S is the set of functions. The permutation group G on D defines a permutation group G' on S that is isomorphic to G, and so we use the same symbol to denote a permutation from G on D and the corresponding permutation from G' on S. We obtain Table 3.3.

Table 3.3   An example of Burnside's lemma.

Permutation π	$I(\pi) =$ number of tree-colorings invariant under π
identity	$2^7 = 128$
π_1	$2^4 = 16$
π_2	$2^6 = 64$
π_3	$2^6 = 64$
$\pi_2\pi_3 = \pi_3\pi_2$	$2^5 = 32$
$\pi_1\pi_2 = \pi_3\pi_1$	$2^3 = 8$
$\pi_1\pi_3 = \pi_2\pi_1$	$2^3 = 8$
$\pi_1\pi_2\pi_3$	$2^4 = 16$
	$\sum\limits_{\pi \in G'} I(\pi) = 336$

The numbers in the right-hand column of Table 3.3 are computed quite simply. For example, π_1, considered as a permutation on D, interchanges nodes 2 and 3, 4 and 6, and 5 and 7:

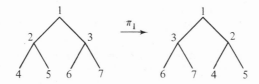

π_1, considered as a permutation on the set S of tree-colorings, simply maps one coloring into another, where the colors of nodes 2 and 3 are interchanged, as well as those of 4 and 6 and 5 and 7. Thus a coloring is invariant under π_1 if and only if the colors of 2 and 3 are identical, those of 4 and 6 are identical, and those of 5 and 7 are identical. As a result of these three constraints, 2^4 out of all 2^7 colorings are invariant under π_1. Similar considerations determine the other numbers.

Thus Burnside's lemma gives the number of distinct tree-colorings as

$$\frac{1}{|G'|} \sum_{\pi \in G'} I(\pi) = \frac{336}{8} = 42.$$

Outline of Proof of Burnside's Lemma: Burnside's lemma is easy to prove by using some elementary facts of group theory. We only give the major steps of the proof, leaving the details as Exercise 23.

(a) In the cartesian product $G' \times S$, consider all "invariances", that is, all pairs (π, f) with $\pi f = f$, and express this number in two ways:

$$\sum_{\pi \in G'} I(\pi) = \sum_{f \in S} J(f),$$

where $J(f)$ is the number of permutations that leave f invariant and $I(\pi)$, as before, is the number of elements of S left invariant under π. This equation is simply a special case of the identity $\sum_i \sum_j a_{ij} = \sum_j \sum_i a_{ij}$, as can be seen by recording the invariances by check marks in the matrix of Figure 3.3.

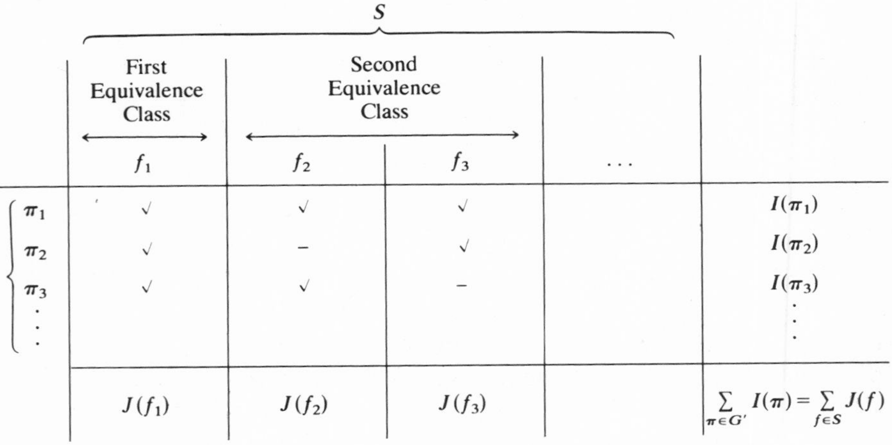

	S				
	First Equivalence Class	Second Equivalence Class			
	f_1	f_2	f_3	$\ldots$	
π_1	$\checkmark$	$\checkmark$	$\checkmark$		$I(\pi_1)$
π_2	$\checkmark$	$-$	$\checkmark$		$I(\pi_2)$
π_3	$\checkmark$	$\checkmark$	$-$		$I(\pi_3)$
	$J(f_1)$	$J(f_2)$	$J(f_3)$		$\sum_{\pi \in G'} I(\pi) = \sum_{f \in S} J(f)$

Figure 3.3   Step (a) in the proof of Burnside's lemma.

(b) If f and g are equivalent, then $J(f) = J(g)$; that is, each column within an equivalence class in the matrix above has the same number of invariances.

(c) If f and g are equivalent, then the number of permutations that map f into g is equal to $J(f)$, the number of permutations that leave f invariant.

(d) Within each equivalence class in the matrix above, there are exactly $|G'|$ invariances.

(e) Burnside's lemma follows immediately from step (d):

$$\sum_{\pi \in G'} I(\pi) = \sum_{f \in S} J(f) = |G'| \cdot (\text{number of equivalence classes}).$$

The special case of Polya's theorem stated at the beginning of this section can readily be reduced to Burnside's lemma, since the sums that express the number of equivalence classes in these two theorems can be equated term by term. When each of the indeterminates x_i has been assigned the value $|R|$, Polya's theorem becomes

$$\text{Number of equivalence classes} = \frac{1}{|G|} \sum_{\pi \in G} |R|^{\, l_1(\pi)+l_2(\pi)+\cdots+l_{|D|}(\pi)}$$

$$= \frac{1}{|G|} \sum_{\pi \in G} |R|^{(\text{number of cycles of } \pi)}$$

and Burnside's lemma becomes

$$\text{Number of equivalence classes} = \frac{1}{|G'|} \sum_{\pi \in G'} I(\pi).$$

Now observe that

1. if $|R| \geqq 1$, then G and G' are isomorphic, and

2. a function $f : D \rightarrow R$ is invariant under π (π considered as a permutation on S) if and only if f is constant on every cycle of π (π considered as a permutation on D).

It follows that

$$I(\pi) = |R|^{(\text{number of cycles of } \pi)},$$

thereby establishing the equivalence of the two theorems.

This brief introduction conveys the most important ideas in Polya's theory of counting. The theory is applicable primarily to problems that contain much symmetry, expressed by a rich permutation group. The more general form of Polya's theorem concerns two topics that we have omitted: counting when different elements may have different weights and explicit listing of all elements. Several references in Section 3.5 discuss these two subjects.

3.5 REMARKS AND REFERENCES

Short expositions of the techniques of asymptotic expansion of functions can be found in many textbooks on applied mathematics. Three comprehensive texts are

DE BRUIJN, N. G., *Asymptotic Methods in Analysis*, North-Holland Publishing Co., Amsterdam, 1961,

DINGLE, R. B., *Asymptotic Expansions: Their Derivation and Interpretation*, Academic Press, New York, 1973,

and

COPSON, E. T., *Asymptotic Expansions*, Cambridge University Press, New York, 1971.

Recurrence relations occur naturally in many areas of mathematics besides combinatorics. Any book on the numerical solution of differential equations deals extensively with the special type of recurrence relations known as difference equations. One such book is

HILDEBRAND, F. B., *Finite-Difference Equations and Simulations*, Prentice-Hall, Englewood Cliffs, N.J., 1968.

Recurrence relations are also prominent in the generation of random numbers. See Chapter 3 of

KNUTH, D. E., *The Art of Computer Programming*, Vol. 2 (*Seminumerical Algorithms*), Addison-Wesley, Reading, Mass., 1969.

Most books on combinatorics contain a section on recurrence relations and generating functions. A fairly extensive treatment of these two topics can be found in

RIORDAN, J., *Combinatorial Identities*, Wiley, New York, 1968.

Generating functions are also a standard technique in probability and statistics. See, for example, Chapter 11 of

FELLER, W., *An Introduction to Probability Theory and Its Applications*, Vol. 1 (3rd edition), Wiley, New York, 1968.

Polya's theorem was published in a paper on "Combinatorial Determination of the Number of Groups, Graphs, and Chemical Compounds":

POLYA, G., "Kombinatorische Anzahlbestimmungen fur Gruppen, Graphen und chemische Verbindungen," *Acta Math.*, **68** (1937), 145–254.

Some aspects of this theory of counting had been anticipated in

REDFIELD, J. H., "The Theory of Group-Reduced Distributions," *Am. J. Math.*, **49** (1927), 433–455.

There are expository surveys of Polya's counting theory in many textbooks. Two good ones are

DE BRUIJN, N. G., "Polya's Theory of Counting," Chapter 5 in E. F. Beckenbach (Ed.), *Applied Combinatorial Mathematics*, Wiley, New York, 1964

and Chapter 5 in

LIU, C. L., *Introduction to Combinatorial Mathematics*, McGraw-Hill, New York, 1968.

3.6 EXERCISES

1. Prove the asymptotic equations (3.1) through (3.6).

2. Verify equation (3.7) for $k = 1$ and then prove it in general by induction on k.

3. Push the asymptotic expansion of the sum of powers one term further than equation (3.7). That is, determine $f(k)$ in the equation.

$$\sum_{i=1}^{n} i^k = \frac{1}{k+1} n^{k+1} + \frac{1}{2} n^k + f(k) n^{k-1} + O(n^{k-2}) \text{ as } n \to \infty.$$

4. By repeated integration by parts of $\int_1^x (e^t/t)\, dt$, show that

$$f(x) = e^{-x} \int_1^x \frac{e^t}{t} dt \sim \sum_{k=1}^{\infty} \frac{(k-1)!}{x^k} \text{ as } x \to \infty$$

and that the asymptotic series on the right converges for no value of x. Investigate the error that is incurred when using the truncated series $\sum_{k=1}^{n} (k-1)!/x^k$ to approximate $f(x)$ as a function of x and n. What value of n (as a function of x) gives the best approximation?

5. Show that

$$\int_1^{n-1} x \ln x\, dx \leq \sum_{k=1}^{n-1} k \ln k \leq \int_1^{n} x \ln x\, dx$$

and then, using asymptotic expansion of the integrals, prove that

$$\sum_{k=1}^{n-1} k \ln k = \frac{n^2}{2} \ln n - \frac{n^2}{4} + O(n \log n).$$

6. Push the asymptotic expansion of equation (3.8) further to obtain

$$\sum_{k=1}^{n} \ln k = \left(n + \frac{1}{2} \right) \ln n - n + O(1) \text{ as } n \to \infty.$$

7. Use the result of Exercise 6 to obtain Stirling's formula

$$n! \sim \sigma n^{n+1/2} e^{-n} \text{ as } n \to \infty$$

for some constant σ. It can be shown that $\sigma = \sqrt{2\pi}$.

8. Find a closed-form expression for the sequence of numbers defined by

$$x_0 = 1, \quad x_1 = 0.5,$$

$$x_{n+1} = 2.5 x_n - x_{n-1} \quad \text{for } n \geq 1.$$

In order to see the problems that can occur when recurrence relations are used for numerical calculation, compute the first 10 to 15 terms of the sequence x_n according to the recurrence relation. Use 2- and 3-digit decimal arithmetic; that is, round every x_n to 2, respectively 3, significant decimal digits before calculating the next term in the sequence. Compare the computed values to the exact solution. How do you explain the difference?

9. Find the most general solution to the recurrence relation

$$x_n = x_{n-1} - \tfrac{1}{4} x_{n-2}.$$

10. Find the most general solution to the recurrence relation

$$x_n = x_{n-1} - \tfrac{1}{4}x_{n-2} + 2^{-n}.$$

11. In this exercise we pursue the solution of the special case $a = 1$, $b = -1$ of recurrence relation (3.20). The general case can be handled in the same way.

    (a) Show that $g(t) = 2t \ln t - t + 1$ is a solution to the integral equation

    $$g(t) = \frac{2}{t} \int_1^t g(s)\, ds + t - 1.$$

    (b) Show that $f(t) = 2(t + 2) \ln (t + 2) - (\tfrac{5}{3} + 2 \ln 3)(t + 2) + 5$ is a solution to the integral equation

    $$\left(1 + \frac{2}{t}\right) f(t) = \frac{2}{t} \int_1^t f(s)\, ds + t - 1.$$

    (c) Show that the sequence x_n, determined by the recurrence relation

    $$x_0 = 0,$$
    $$x_n = \frac{2}{n} \sum_{k=0}^{n-1} x_k + n - 1$$

    satisfies the inequalities

    $$f(n) \le x_n \le g(n) \qquad \text{for } n \ge 1.$$

    (d) Prove that

    $$x_n = (\ln 4)n \lg n + o(n \log n).$$

12. Completely solve the recurrence equation (3.20). [*Answer:* $2a(n + 1)H_n + (x_0 - 3a + b)n + x_0$.] [*Hint:* Subtract $(n - 1)x_{n-1}$ from nx_n and divide by $n(n + 1)$ to get an equation of the form

    $$\frac{1}{n + 1}x_n - \frac{1}{n}x_{n-1} = \frac{1}{n + 1}\alpha + \frac{1}{n}\beta.$$

    Sum terms of this form to get a telescoping series on the left-hand side.]

13. Estimate the difference between

    $$\lg n!, \quad \sum_{k=1}^{n} \lceil \lg k \rceil, \quad \text{and} \quad \sum_{k=1}^{n} \lfloor \lg k \rfloor.$$

14. Derive the generating functions shown in Table 3.1.

15. Prove that the composition rules for generating functions corresponding to the operations on sequences given in Section 3.3 (linear operations, shift of origin, partial sums, scaling, and convolution) are correct.

16. Let **x** be a random variable that assumes the values $0, 1, 2, \ldots$ with probabilities $x_0, x_1, x_2, \ldots$, respectively $(x_k \geq 0, \sum_{k=0}^{\infty} x_k = 1)$, and let $X(s) = \sum_{k=0}^{\infty} x_k s^k$ be its generating function. Let **y** be a random variable obtained by adding n independent samples of the random variable **x** and let $y_0, y_1, y_2, \ldots$ denote the probabilities with which **y** assumes the values $0, 1, 2, \ldots$, respectively. Show that the generating function $Y(s) = \sum_{k=0}^{\infty} y_k s^k$ is $X(s)^n$. [*Hint* (if the terminology is unfamiliar to you): Compute the probabilities of throwing $0, 1, 2, \ldots, 12$ with a toss of two dice.]

17. Show that $a_{nk} \sim b_{nk}$, as $n \to \infty$, *does not imply* that

$$\sum_k a_{nk} \sim \sum_k b_{nk} \text{ as } n \to \infty.$$

★18. The object of this exercise is to determine the average internal path length (see Section 2.3.3) of an extended binary tree with n internal nodes, assuming that each of the $\binom{2n}{n} \Big/ (n + 1)$ binary trees is equally probable. Let x_{nm} be the number of extended binary trees with n internal nodes and internal path length m. Prove that

$$x_{nm} = \sum_{\substack{i+j=n-1 \\ k+l+n-1=m}} x_{ik} x_{jl}$$

and use this to obtain a formula for

$$X(s, t) = \sum_{\substack{n \geq 0 \\ m \geq 0}} x_{nm} s^n t^m,$$

the generating function for x_{nm}. Use that formula to evaluate the generating function

$$Y(s) = \left. \frac{\partial X(s, t)}{\partial t} \right|_{t=1}$$

and relate this generating function to binary trees. Combine it with equation (3.26) to determine the average internal path length of a binary tree with n nodes and then use Stirling's formula, equation (3.12), to show that the average is asymptotically $n\sqrt{\pi n} - 3n + O(\sqrt{n})$.

★19. In how many different ways can you color with two colors the six faces of a cube that may be freely rotated in space. [*Hint:* The following pictures will help you determine the group of permutations of the domain (the set of faces) that describes the symmetry of this problem.]

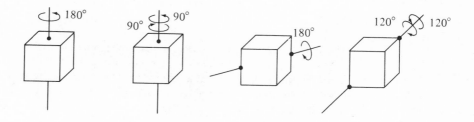

★20. Compute the number of the essentially different ways in which eight indistinguishable markers can be placed on an 8 by 8 chessboard, so that no square contains more than one marker. Two ways of placing markers are essentially different if they cannot be transformed into each other by a rotation of the board or by reflection on any of the following axes:

(a) Carefully define the permutation group involved.
(b) Solve the problem using Burnside's lemma alone.
(c) Solve the problem using Polya's counting formula.

★21. An integrated circuit manufacturer builds chips with 16 elements arranged in a 4 × 4 array:

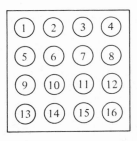

To realize different circuits, he needs different patterns for interconnecting the elements. Direct interconnections are made only between horizontally or vertically adjacent elements—for example,

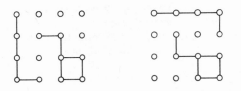

(Closed loops do not usually occur, but we will ignore this for simplicity.) To deposit interconnections on the chip, one needs a photo-mask of the interconnection pattern. Notice that the same photo-mask will do for the two interconnection patterns shown above (flip about a diagonal). How many photo-masks are needed in order to lay out all possible interconnection patterns on these chips?

★22. The following two figures are called the Star of David (D) and Mohammed's Scimitars (M), respectively:

Imagine they are manufactured out of pieces of wire, which are soldered together at the points of crossover. How many different stars and scimitars are there from the point of view of the pattern of crossovers? Call two figures equivalent if one can be moved in space so that the two are indistinguishable in their pattern of crossovers. For example,

Use Polya's theorem or Burnside's lemma. List carefully the group of transformations involved for each figure.

★23. Fill in the details in the proof of Burnside's lemma in Section 3.4.2.

chapter *4*

Exhaustive Search

Using a computer to answer such questions as "How many ways are there to ...," "List all possible ...," or "Is there a way to ...," usually requires an exhaustive search of the set of all potential solutions. For example, if we want to find all primes less than 10^4, no method is known that does not require, in some way, an examination of each of the integers between 1 and 10^4; or if we want to find all ways to thread a maze, we must examine all paths starting at the entrance.

Two general techniques for organizing such searches are presented in this chapter. The first, backtrack, works by continually trying to extend a partial solution. At each stage of the search, if an extension of the current partial solution is not possible, we "backtrack" to a shorter partial solution and try again. Backtracking is used in a wide range of search problems, including parsing, game playing, and scheduling. The second technique, sieves, is the logical complement of backtrack in that we try to eliminate the nonsolutions instead of trying to find solutions. Sieves are useful primarily in number-theoretic computations.

It must be kept in mind, however, that backtrack and sieves are only *general techniques*. Their straightforward application will typically result in algorithms whose time requirements are prohibitive; at current computer speeds it is not practical to search exhaustively more than about 10^8 elements. So in order for these techniques to be useful, they must be regarded only as a framework within which to approach the problem. The framework must be custom-tailored, often with great ingenuity, to fit the particular problem so that the resulting algorithm will be practical to use.

4.1 BACKTRACK

The idea behind backtrack can be most easily understood in the context of threading a maze: we start in some specified square with our goal being to arrive at some other specified square by a sequence of moves from one square to the next. The difficulty is that we are constrained by the existence of barriers that prohibit certain moves, as illustrated in Figure 4.1. One way to thread the maze is to travel from the starting square according to two rules:

1. From the current square, take any path not previously explored.

2. If the current square has no unexplored paths, go backward one square on the last path that you followed to get to this square.

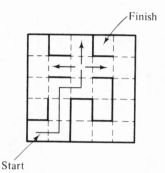

Figure 4.1 A maze—maze threading reveals the essence of backtrack. The arrows indicate possible extensions to the current path.

The first rule tells us how to extend the current path, if possible, and the second rule tells us how to back up when we get stuck. This is the essence of backtrack: keep extending the current solution as long as possible; and when the solution can no longer be extended, backtrack and try an alternative at the most recent stage for which there is an untried alternative.

4.1.1 The Generalized Algorithm

In the most general case, we assume that the solution to a problem consists of a vector $(a_1, a_2, \ldots)$ of finite but undetermined length, satisfying certain constraints. Each a_i is a member of a finite, linearly ordered set A_i. Thus the exhaustive search must consider the elements of $A_1 \times A_2 \times \cdots \times A_i$, for $i = 0, 1, 2, \ldots$, as potential solutions. Initially, we start with the null vector $(\;)$ as our partial solution, and the constraints tell us which of the members of A_1 are candidates for a_1; call this subset S_1. We choose the least element of S_1 as a_1, and now we have the partial solution (a_1). In general, the various constraints that describe the solutions tell us which subset S_k of A_k constitutes candidates for the extension of the partial solution $(a_1, a_2, \ldots, a_{k-1})$ to $(a_1, a_2, \ldots, a_{k-1}, a_k)$. If the partial solution $(a_1, a_2, \ldots, a_{k-1})$ admits no possibilities for a_k, then $S_k = \varnothing$, and so we backtrack and make a new choice for a_{k-1}. If there are no new choices for a_{k-1}, we backtrack still farther and make a new choice for a_{k-2}, and so on.

It is helpful to picture this process in terms of the depth-first tree traversal of Section 2.3.2. The subset of $A_1 \times A_2 \times \cdots \times A_i$ for $i = 0, 1, 2, \ldots$ that is searched is represented as a *search tree* as follows. The root of the tree (the 0th level) is the null vector. Its sons are the choices for a_1, and, in general, the nodes as the kth level are the choices for a_k, given the choices made for $a_1, a_2, \ldots, a_{k-1}$ as indicated by the ancestors of those nodes. In the tree shown in Figure 4.2, backtrack traverses the nodes of the tree as indicated by the dashed lines. In asking whether a problem has a solution $(a_1, a_2, \ldots)$, we are asking whether any nodes in the tree are solutions. In asking for all solutions, we want all such nodes.

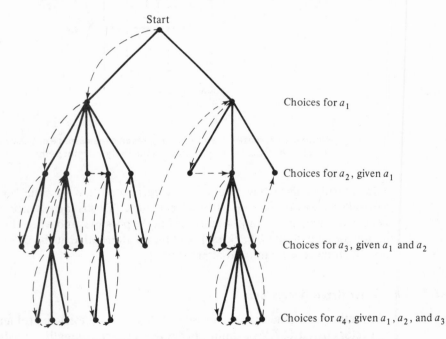

Figure 4.2   The search tree of partial solutions. The depth-first order in which backtrack explores it is shown by the dashed arrows.

The backtrack algorithm to find all solutions is described formally by Algorithm 4.1. If only one solution is to be found, then the program should quit after recording a solution; in this case, termination at the end of the **while** loop means that no solution exists. When the process halts, the value of *count* is the number of nodes in the tree that have been examined.

$$S_1 \leftarrow A_1$$

$$count \leftarrow 0$$

$$k \leftarrow 1$$

$$\textbf{while } k > 0 \textbf{ do} \begin{cases} \textbf{while } S_k \neq \emptyset \textbf{ do} \begin{cases} [\![\text{advance}]\!] \\ a_k \leftarrow \text{an element in } S_k \\ S_k \leftarrow S_k - \{a_k\} \\ count \leftarrow count + 1 \\ \textbf{if } (a_1, a_2, \ldots, a_k) \text{ is a solution } \textbf{then} \text{ record it} \\ k \leftarrow k + 1 \\ \text{compute } S_k \end{cases} \\ [\![backtrack]\!] \\ k \leftarrow k - 1 \end{cases}$$

$[\![$all solutions have been found; the value of *count* indicates how many nodes in the tree have been examined$]\!]$

Algorithm 4.1   Generalized backtrack search.

To see how this general algorithm can be applied to specific problems, consider the placement of as many queens as possible on an $n \times n$ chessboard so that no queen is attacking any other queen. Since a queen attacks all squares on its row, column, and diagonals, clearly at most n queens can be placed on the chessboard without having two attacking each other. Thus the problem reduces to determining whether n nonattacking queens can be placed on a chessboard and, if so, in how many ways they can be placed.

Exactly one queen must be in every column, which means that the solution can be represented as a vector $(a_1, a_2, \ldots, a_n)$ in which a_i represents the row of the queen in the ith column. Moreover, exactly one queen must be in every row, and so $a_i \neq a_j$ if $i \neq j$. Finally, since queens may not attack each other in the diagonals, we must have $|a_i - a_j| \neq |i - j|$ if $i \neq j$. Thus in order to determine whether a_k can be added to $(a_1, a_2, \ldots, a_{k-1})$, we simply test a_k with each a_i, $i < k$:

$$flag \leftarrow \textbf{true}$$

$$i \leftarrow 1$$

$$\textbf{while } i < k \textbf{ and } flag \textbf{ do} \begin{cases} \textbf{if } a_i = a_k \textbf{ or } |a_i - a_k| = |i - k| \textbf{ then } flag \leftarrow \textbf{false} \\ \\ i \leftarrow i + 1 \end{cases}$$

Using this procedure, the backtrack program to find all solutions to the $n \times n$ nonattacking queens problem is

$s_1 \leftarrow 1$

$k \leftarrow 1$

while $k > 0$ **do**
$\quad$ **while** $s_k \leq n$ **do**
$\qquad a_k \leftarrow s_k$
$\qquad s_k \leftarrow s_k + 1$
$\qquad$ **while** $s_k \leq n$ **and** a queen cannot be
$\qquad\qquad$ put in row s_k of column k **do** $s_k \leftarrow s_k + 1$
$\qquad$ **if** $k = n$ **then** record $(a_1, a_2, \ldots, a_n)$
$\qquad\qquad\qquad$ as a solution
$\qquad k \leftarrow k + 1$
$\qquad s_k \leftarrow 1$
$\qquad$ **while** $s_k \leq n$ **and** a queen cannot be
$\qquad\qquad$ put in row s_k of column k **do** $s_k \leftarrow s_k + 1$
$\quad k \leftarrow k - 1$

〚all solutions have been found〛

Notice that in customizing the generalized backtrack for the queens problem, we do not explicitly compute and store the sets S_k. It is easier in this case to store only the smallest value in S_k—that is, s_k—and compute the next value when needed. The test for $S_k \neq \emptyset$ corresponds to $s_k \leq n$; since the value computed for s_{n+1} is at least $n + 1$, S_{n+1} is always empty.

4.1.2 *Refinements*

In the backtrack program for finding all solutions to the $n \times n$ nonattacking queens problem, we did not attempt to examine all $\binom{n^2}{n}$ (about 4.4×10^9 for $n = 8$) possible ways of placing n queens on an $n \times n$ board. Instead we observed that each column could contain at most one queen, which leaves only n^n (about 1.7×10^7 for $n = 8$) possible placements. Furthermore, we observed that no two queens could be placed on the same row; and so in order for $(a_1, a_2, \ldots, a_n)$ to be a solution, it must be a permutation of $(1, 2, \ldots, n)$, leaving only $n!$ (about 4.0×10^4 for $n = 8$) possibilities. The observation that no two queens can be on the same diagonal cuts the number of possibilities still further until, for $n = 8$, only 2056 nodes remain in the forest. By our series of observations, we have eliminated from consideration a sizable number of the possible placements of n queens on an $n \times n$ board. Using this type of analysis to shorten the search process is called *preclusion* (or *branch pruning*, for the effect is to remove subtrees from the tree). For example, the placement of a queen on the $(1, 2)$ square (first column, second row) *precludes* the placement of a queen on, say, the $(2, 3)$ square.

Preclusion in the queens problem can be carried further. Two solutions can be considered equivalent if we can transform one into the other by a series of rotations and/or reflections. Clearly, if we find all inequivalent solutions, we can easily produce the set of all solutions. Notice that a queen in any corner attacks the other corners, and thus there are no solutions with queens in more than one corner. Consequently, any solution with a queen in the $(1, 1)$ square can be transformed by rotations and/or reflections into an equivalent solution in which the $(1, 1)$ square is empty. Thus we can start with $a_1 = 2$ and still obtain all inequivalent solutions. We have pruned a large subtree from the tree.

Another refinement is *fusion* or *branch merging*. The idea is to avoid doing the same work twice: if two or more subtrees of the tree are isomorphic, we want to search only one of them. In the queens problem, we can use fusion by observing that if $a_1 > \lceil n/2 \rceil$, the solution can be reflected to obtain an equivalent solution for which $a_1 \leq \lceil n/2 \rceil$. Therefore the trees in the forest corresponding, for example, to $a_1 = 2$ and $a_1 = n - 1$ are isomorphic. As a result, without bothering to examine the trees corresponding to $a_1 = \lceil n/2 \rceil + 1, a_1 = \lceil n/2 \rceil + 2, \ldots,$ $a_1 = n$, we still find all inequivalent solutions. Moreover, since use of this fusion does not interfere with the preclusion of the $(1, 1)$ square (why?), we need only search for solutions with $2 \leq a_1 \leq \lceil n/2 \rceil$. When n is odd and $a_1 = \lceil n/2 \rceil$, we may restrict $a_2 \leq \lceil n/2 \rceil - 2$ by the same principle. (In fact, when n is odd, we do better by restricting $2 \leq a_1 \leq \lfloor n/2 \rfloor$. Why?) The savings that result from these techniques are not trivial; for the 8×8 case, the forest is reduced to only 801 nodes, and the resulting computation can be done by hand in a few hours.

In addition to preclusion and fusion, which clearly can shorten the search if the entire tree is to be searched for all solutions, there is a heuristic technique that *might* be useful when the existence of a solution is in doubt or when only one solution is to be found instead of all solutions. This technique, called *tree arrangement*, can be used in several ways. First, if evidence indicates that all solutions will have a particular form, then it is prudent to structure the search so that potential solutions of that form are inspected before other solutions. Second, if possible, the tree should be rearranged so that nodes of low degree (i.e., those with relatively few sons) are near the top of the tree; for example, the tree in Figure 4.3(a) is preferable to the one in Figure 4.3(b). Why might this be helpful? Generally, in order to discover that a path cannot lead to a solution, several constraints must be accumulated, and normally this happens at a fixed depth from the top of the trees. The result is that more nodes will be precluded if a large part of the fanout occurs near the leaves rather than near the roots. Of course, it must be understood that this type of tree rearrangement may be of no help if the entire tree must be searched. Moreover, even if the entire tree need not be searched, rearranging the search may move the solutions not closer but farther away from the beginning of the search.

One final refinement is the technique of *problem decomposition* as described in Section 1.4. Decompose the problem into k subproblems, solve the subproblems, and then compose the solutions to the subproblems into a solution of the

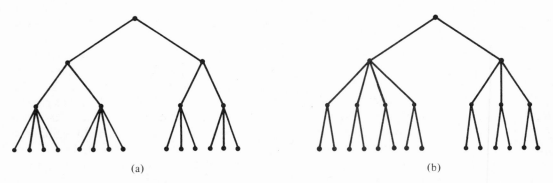

(a)   (b)

Figure 4.3 Two possible trees to be examined with backtrack.

original problem. The storage demand can be quite high because all the subsolutions must be saved, but the increase in speed can be considerable. For example, if it takes time $c \cdot 2^n$ to solve the problem of size n, then the time is reduced to $k \cdot c \cdot 2^{n/k} + T$, where T is the time required to compose the subsolutions. If the number of subsolutions is small and it is not too difficult to compose them, then T will be relatively small and the technique will result in huge savings.

It is important to realize that these refinements should be attempted not only in the preprogramming analysis of a problem but also on-line as the program is being run. In the queens problem, for example, the analysis allowed us to preclude quite a bit just in setting up the program; but the preclusion based on queens attacking along the diagonals can be applied only as the configurations are generated because no known analysis exists. Similarly, fusion and search rearrangement can be applied at run time in certain situations. Applying fusion at run time means being able to recognize duplicate subtrees from the partial solution at the root of the subtree. Applying search rearrangement at run time means either making an educated guess that the current subtree is not likely to be profitable, and hence its search should be postponed, or, when at the partial solution $(a_1, a_2, \ldots, a_k)$ searching for the remaining components in some dynamically determined order *instead* of the usual order $a_{k+1}, a_{k+2}, \ldots .$

4.1.3. *Estimation of Performance*

Typically, backtrack gives rise to algorithms that are exponential in their parameters. This is a simple consequence of the fact that the tree has roughly $\prod_{i=1}^{n} |A_i|$ nodes to be examined, assuming all solutions have length at most n, where $|A_i|$ is the number of elements in A_i. Even with extensive preclusion and fusion, the best we could expect would be for $|A_i|$ to be a constant, which produces trees with about C^n nodes for some $C > 1$. Since the size of the tree grows so rapidly, it is reasonable to try to determine the feasibility of the search by estimating the number of nodes in the tree.

Analytical estimation is seldom possible because it is difficult to predict how the various constraints interact as they build up going deeper in the tree. There is, however, an interesting Monte Carlo method for estimating the size of a tree. The idea is to run a number of experiments, each experiment consisting of performing the backtrack search with randomly chosen values of the a_i. Suppose that we have the partial solution $(a_1, a_2, \ldots, a_{k-1})$ and that the number of choices for a_k, based on whatever preclusion or fusion is done, is $x_k = |S_k|$. If $x_k \neq 0$, then a_k is chosen at random from S_k, each element having probability $1/x_k$ of being chosen; if $x_k = 0$, the experiment ends. Thus if $x_1 = |S_1|$, $a_1 \in S_1$ is chosen at random with probability $1/x_1$; if $x_2 = |S_2|$, given that a_1 was chosen from S_1, $a_2 \in S_2$ is chosen at random with probability $1/x_2$, and so on. The expected value of

$$x_1 + x_1x_2 + x_1x_2x_3 + x_1x_2x_3x_4 + \cdots$$

is precisely the number of nodes in the tree other than the root; that is, it is the number of cases that will be examined by the backtrack algorithm.

To prove this, let T be a fixed tree whose size is to be estimated. Define two functions on the nodes $x \in T$ as follows.

$$p(x) = \begin{cases} 1 & \text{if } x \text{ is the root of } T, \\ k \cdot p \,(\text{FATHER}\,(x)) & \text{if } x \text{ is not the root of } T \\ & \text{and FATHER}\,(x) \text{ has } k \text{ sons.} \end{cases}$$

$$I(x) = \begin{cases} 1 & \text{if } x \text{ is one of the nodes of } T \text{ traversed} \\ & \text{in the experiment,} \\ 0 & \text{otherwise.} \end{cases}$$

Clearly, then, the outcome of the experiment is a random variable

$$\mathbf{X} = x_1 + x_1x_2 + x_1x_2x_3 + \cdots$$

$$= \sum_{\substack{x \in T \\ x \neq \text{root}}} p(x)I(x),$$

and the expected value is

$$E(\mathbf{X}) = E\Big[\sum_{\substack{x \in T \\ x \neq \text{root}}} p(x)I(x) \Big].$$

But $p(x)$ is fixed with respect to $x \in T$ and hence is independent of the experiment, so that

$$E(\mathbf{X}) = \sum_{\substack{x \in T \\ x \neq \text{root}}} p(x)E(I(x)).$$

By construction, the probability that a node $x \in T$ is traversed in the experiment is $1/p(x)$; that is, $E(I(x)) = 1/p(x)$. Thus

$$E(\mathbf{X}) = \sum_{\substack{x \in T \\ x \neq \text{root}}} 1 = \text{number of nodes in } T \text{ other than the root}$$

as desired.

The generalized backtrack algorithm is easily modified to perform such experiments, for instead of backtracking when $S_k = \varnothing$, we simply end the experiment. Algorithm 4.2 performs N experiments to approximate the number of nodes in the tree.

$$
\begin{aligned}
&average \leftarrow 0 \\
&\textbf{for } i = 1 \textbf{ to } N \textbf{ do}
\left\{
\begin{aligned}
&sum \leftarrow 0 \\
&product \leftarrow 1 \\
&S_1 \leftarrow A_1 \\
&k \leftarrow 1 \\
&\textbf{while } S_k \neq \varnothing \textbf{ do}
\left\{
\begin{aligned}
&product \leftarrow product * |S_k| \\
&sum \leftarrow sum + product \\
&a_k \leftarrow \text{random element of } S_k \\
&k \leftarrow k + 1 \\
&\text{compute } S_k
\end{aligned}
\right. \\
&average \leftarrow average + sum
\end{aligned}
\right. \\
&average \leftarrow average/N
\end{aligned}
$$

Algorithm 4.2   Monte Carlo estimation of the size of the tree in a backtrack search.

This Monte Carlo calculation can be used to estimate the performance of the backtrack algorithm by comparing it to a benchmark obtained from smaller cases. For example, a run of the queens problem with preclusion and fusion as described earlier had to search a tree of 14566 nodes in order to find all solutions in the 10×10 case. With 1000 trials, a Monte Carlo estimate of the number of nodes in the 11×11 case was 70806, and so we can expect the computation to take about 4.9 times as long as the 10×10 case. In fact, when the 11×11 case was actually run, the tree had 70214 nodes, and the computation took about 5.3 times as long as the 10×10 case.

4.1.4   Two Programming Techniques

So far we have only expressed the backtrack algorithm as two nested loops—the inner loop for extending the solution and the outer loop for backing up. The same algorithm can be programmed in two other substantially different ways, one using a high-level programming language with recursion and the other using an assembly language with macro expansion facilities. In the latter case, we assume that only leaves of the tree can be solutions.

The recursive approach is useful primarily for graph algorithms (see Section 8.2.2 on depth-first search). The general backtrack algorithm can be expressed recursively as in Algorithm 4.3. The symbol $\parallel$ is used to denote vector concatenation:

$$(a_1, \ldots, a_n) \parallel (b_1, \ldots, b_m) = (a_1, \ldots, a_n, b_1, \ldots, b_m), (\,) \parallel (a) = (a),$$

where ( ) is the null vector. Notice that all the backtracking is hidden in the mechanism used to implement the recursion.

> $BACKTRACK$ $((\,), 1)$
>
> **procedure** $BACKTRACK$ $(vector, i)$
>
>   **if** $vector$ is a solution **then** record it
>
>   compute S_i
>
>   **for** $a \in S_i$ **do** $BACKTRACK$ $(vector \parallel (a), i + 1)$
>
>   **return**

Algorithm 4.3 Recursive implementation of backtrack.

The assembly language approach to backtrack is based on the assumption that the most important aspect of the program is that it be fast. Many backtrack applications require relatively little storage, and so it is not unreasonable to increase greatly the storage requirements in order to decrease the time requirements by a substantial amount. One idea is to use the macro expansion facilities of an assembler to produce a highly customized program in which some or all of the loops are unwound, thereby eliminating certain logical tests and removing the overhead of the instructions to control the loops. If all solutions have length n, this can be done, for example, by writing a macro called, say, CODE$_i$, whose body consists of

> compute S_i
>
> $L_i : $ **if** $S_i = \varnothing$ **then goto** L_{i-1}
>
> $a_i \leftarrow$ an element in S_i
>
> $S_i \leftarrow S_i - \{a_i\}$

This macro CODE$_i$ is expanded for $i = 1, 2, \ldots, n$ to produce a program like

> CODE$_1$
>
> CODE$_2$
>
> $\vdots$
>
> CODE$_n$
>
> record $(a_1, \ldots, a_n)$ as a solution
>
> **goto** L_n
>
> $L_0 : [\![$ all solutions have been found $]\!]$

Notice that this approach *requires* that all solutions have the same fixed length. This is frequently the case, and the advantage of this approach is obvious: the macro $CODE_i$ can be arranged so that the steps are customized for S_i. For example, in the nonattacking queens problem, we need the restrictions $2 \leq a_1 \leq \lceil n/2 \rceil$, and when n is odd and $a_1 = \lceil n/2 \rceil$, then $1 \leq a_2 \leq \lceil n/2 \rceil - 2$. Including these tests in the general backtrack algorithm written with **while** loops is expensive, since some tests must be made each time through the inner loop, even though they will rarely be successful. Using the macro approach, the instructions for the tests could be included in the expansion only where needed. Of course, this approach can be used without macro facilities, but it is inconvenient.

Essentially, separate programs are used for each level in the tree. In the queens problem this allows us—for example, on an IBM System/360—to do almost all the work in the registers instead of in the (relatively) slower memory. Thus registers 1 through 15 can be used, respectively, to hold the positions of the queen in columns 1 through 15 in the 15×15 case. Using such a technique, the 15×15 case was run in only 25 minutes on a 360/75 (as compared to 170 minutes for the obvious program), and the 16×16 case was run in 168 minutes.

There are other ways to use macros to speed up backtrack programs. In tiling problems [such as Exercises 1(h), 1(k), and 1(l)] a macro can be written that produces a separate section of code for each position to be tiled, or for each tile. The macro approach described above for the queen's problem can be viewed in this way: $CODE_i$ places the ith queen on the board. An example of the use of macros in backtrack that does *not* require that all solutions have the same length is described at the end of the next section.

4.1.5 An Example: Optimal Difference-Preserving Codes

In Section 1.2 we discussed difference-preserving codes and gave a table of the maximal ranges that can be obtained for various code lengths and thresholds. Here we will discuss how the entries in that table can be calculated by using the various techniques discussed in the previous sections of this chapter. In particular, we will concentrate on the ranges for optimal difference-preserving codes of n bits with threshold 1; the same techniques apply equally well for other threshold values.

The most obvious backtrack method for this problem would be to examine all possible 2^{2^n} subsets of codewords, analogous to examining all $\binom{n^2}{n}$ possible placements of n queens on an $n \times n$ board in the nonattacking queens problem. Even for $n = 5$ this would be impractical. Instead we make use of some symmetries inherent in such codes.

We define two codes to be *isomorphic* if and only if one can be obtained from the other by the complementation and/or interchange of columns. It should be clear that the difference properties of DP-t codes are not affected by these operations. This notion of isomorphism leads to two useful symmetries.

Symmetry 1

Any DP-t code with range $N \geq t + 3$ is isomorphic to a code whose first $t + 3$ codewords are

$$0 \ldots 00 \ldots 000$$
$$0 \ldots 00 \ldots 001$$
$$0 \ldots 00 \ldots 011$$
$$0 \ldots 00 \ldots 111$$
$$\begin{array}{ccc} \vdots & \vdots & \vdots \\ \vdots & \vdots & \vdots \end{array}$$
$$0 \ldots 001 \ldots 111$$
$$0 \ldots 011 \ldots 111$$
$$|\longleftrightarrow|$$
$$t + 2 \text{ bits}$$

Proof: We can force the first codeword to all zeros, since we can complement all columns in which the first codeword has a one. For $t = 1$, the second codeword must consist of all zeros with a single one, and we can interchange columns so that it is rightmost. Furthermore, we know that the third codeword must have two ones (by comparing it to the first and second codewords) and that one of these ones must be in the rightmost position (by comparing it to the second codeword). Thus we can move the other one into the next to the rightmost column by means of an interchange without affecting the first two codewords. A similar argument shows that the fourth codeword can be transformed to $00 \ldots 0111$ without affecting the first codewords and hence the symmetry holds for $t = 1$. The proof continues by induction. Suppose that the symmetry holds for $t - 1$; we want to show that it holds for t. Since any DP-t code is also a DP-$(t - 1)$ code, the first $t + 2$ codewords can be assumed to be as prescribed by the symmetry. Consider the $(t + 3)$rd codeword. The code is DP-t, and so we can compare the $(t + 3)$rd codeword with the first and $(t + 2)$nd codewords to conclude that it must have $t + 1$ ones on its right and another one somewhere else. Thus a single interchange transforms it to the desired form without disturbing the earlier codewords.

Symmetry 2

Any DP-t code is isomorphic to a code in which the first one to appear in column i (counting from the right) occurs before the first one to appear in column $i + 1$ for $1 \leq i < n$.

Proof: This symmetry is obvious, for we can always interchange the appropriate columns. For example, the DP-1 codes

0000000		0000000
0000001		0000001
0000011		0000011
0000111	and	0000111
1000111		0001111
1001111		0011111
1101111		0111111

are isomorphic, since we can permute the leftmost four columns of one to yield the other.

Applying these two symmetries to the problem of finding optimal DP-1 codes prunes the search tree significantly. The first symmetry says that we must start with, say, {000000, 000001, 000011, 000111} for $n = 6$. Figure 4.4 shows the effect of the second symmetry; the dashed lines show which subtrees are precluded. This preclusion occurs near the root, and therefore the number of nodes pruned from the tree is significant.

Using these two methods of preclusion, a program was written and run for the cases $n = 5$ and $n = 6$, examining 348 and 651138 nodes, respectively. For $n = 5$, the optimal code found had range 14; for $n = 6$, the optimal code found had range 27. In an attempt to reduce the size of the search so that $n = 7$ could be made feasible, the entire tree for $n = 5$ was examined for further symmetries. A frequently observed pattern in the tree is shown in Figure 4.5. After some trial and error, Symmetry 3 was discovered.

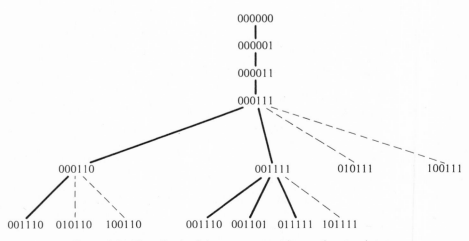

Figure 4.4 The effects of the two symmetries on the search tree.

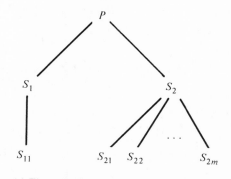

(a) The general pattern, $S_{11} = S_{2i}$ for some i.

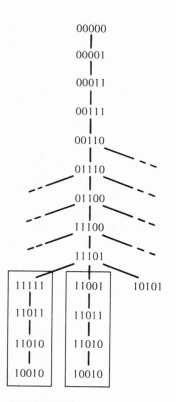

(b) An example of the pattern: $P = 11101$,
$S_1 = 11111, S_2 = 11001, S_{11} = S_{21} = 11011$.

Figure 4.5   A frequently observed pattern in the search tree for optimal DP-1 codes of length S.

Symmetry 3

Consider the search tree for an n-bit DP-1 code. Suppose that a node P has sons S_1 and S_2 such that S_1 has only one son, S_{11}, and S_{21}, one of the sons of S_2, is the same codeword as S_{11}. Then any code in the subtree rooted at S_{21} is a code in the subtree rooted at S_{11}.

Proof: Suppose that the sequence $V_1, V_2, \ldots, V_h$ can be added below the node S_{21} in Figure 4.5(a) to form a DP-1 code and assume that adding $V_1, V_2, \ldots, V_h$ below the node S_{11} does *not* form a DP-1 code. The codewords above S_1 and S_2 are the same; thus the only reason that adding $V_1, V_2, \ldots, V_h$ below the node S_{11} might not form a DP-1 code is that one of the V_i is too close to S_1; that is $H(S_1, V_i) \leqq 1$. Clearly, $H(S_1, V_i) \neq 0$; otherwise $S_1 = V_i$ and then $H(P, V_i) = 1$, which contradicts the supposition that $V_1, V_2, \ldots, V_h$ can be added below S_{21} to form a DP-1 code; $V_i = S_{11} = S_{21}$ is similarly impossible. So we must have $H(S_1, V_i) = 1$ and $V_i \neq S_{11} = S_{21}$; but in this case V_i would also have to be a son of S_1, thereby contradicting the fact that S_1 has only one son, S_{11}.

We conclude that any extension of the code from the node S_{21} is no longer than some extension of the code from the node S_{11}. Our interest is only in finding a single optimal code, which means that we can prune the subtree rooted at S_{21} in such cases.

It might be expected that Symmetry 3 is relatively weak, but, in fact, it reduced the number of nodes in the tree by about 60%; for example, when $n = 5$, only 145 nodes were left in the tree. Using the three symmetries, a Monte Carlo calculation was done to estimate the number of nodes in the tree for $n = 7$. In two runs with 3500 and 10000 trials, respectively, the number of nodes was estimated. at about 2×10^{12}. Since the program processes nodes at a rate of 6000 per second, the search for optimal DP-1 codes of length $n = 7$ is not feasible; it would require over ten years of computing.

In an attempt at a faster program, a macro was used to generate a short subroutine for each of the 2^n possible codewords. For example, if W is a codeword and $W_1, W_2, \ldots, W_n$ are the codewords adjacent to W, the macro would produce a subroutine like

> add W to the code
>
> **if** this code is longer than the previously
>   generated longest code **then** record it
>
> **if** W_1 can be added **then call** W_1
>
> **if** W_2 can be added **then call** W_2
>
>   ⋮
>
> **if** W_n can be added **then call** W_n
>
> remove W from the code
>
> **return**

This technique increased the rate of processing to about 20000 nodes per second, still much too slow to attempt the case $n = 7$.

4.1.6 *Branch and Bound*

The well-known variation on backtrack called *branch and bound* is actually only a specific type of preclusion. The preclusion is based on the assumption that each solution has a cost associated with it and that the optimal solution (the one of least cost) is to be found. In order for branch and bound to be applicable, the cost must be well defined for partial solutions; and for all partial solutions $(a_1, a_2, \ldots, a_{k-1})$ and all extensions $(a_1, a_2, \ldots, a_{k-1}, a_k)$, we must have

$$cost\,(a_1, a_2, \ldots, a_{k-1}) \leq cost\,(a_1, a_2, \ldots, a_{k-1}, a_k)$$

When the cost has these properties, we can discard a partial solution $(a_1, a_2, \ldots, a_k)$ if its cost is greater than or equal to the cost of a previously computed solution. This preclusion is easily incorporated into the general backtrack algorithm, as seen in Algorithm 4.4. In most cases, the cost function is nonnegative and satisfies the even stronger condition that

$$cost\,(a_1, a_2, \ldots, a_k) = cost\,(a_1, a_2, \ldots, a_{k-1}) + C(a_k),$$

where $C(a_k) \geq 0$ is a function defined for all a_k. This condition, of course, slightly simplifies Algorithm 4.4.

```
lowcost ← ∞
cost ← 0
S₁ ← A₁
k ← 1

while k > 0 do
    while Sₖ ≠ ∅ and
          cost < lowcost do
        aₖ ← an element in Sₖ
        Sₖ ← Sₖ − {aₖ}
        cost ← cost (a₁, a₂, ..., aₖ)
        if (a₁, a₂, ..., aₖ)
           is a solution and
           cost < lowcost then
               save (a₁, a₂, ..., aₖ)
               as the solution of
               lowest cost so far
               lowcost ← cost
        k ← k + 1
        compute Sₖ
    k ← k − 1
    cost ← cost (a₁, a₂, ..., aₖ)
〚the last solution saved is one of the lowest cost〛
```

Algorithm 4.4   Generalized branch and bound.

The *traveling salesman problem* is typical of the type of optimization problem that can be solved with branch-and-bound techniques. In this problem the salesman must visit n cities, returning to his starting point, and is required to minimize the total cost of his trip. In going from city i to city j, he incurs a cost C_{ij}.

For example, if the cost matrix is

$i \searrow^{j}$	1	2	3	4	5	6	7
1	∞	3	93	13	33	9	57
2	4	∞	77	42	21	16	34
3	45	17	∞	36	16	28	25
4	39	90	80	∞	56	7	91
5	28	46	88	33	∞	25	57
6	3	88	18	46	92	∞	7
7	44	26	33	27	84	39	∞

then the cost of the path 1–2–5–7–3–6–4–1 is 227, whereas the cost of an optimal path, say 1–4–6–7–3–5–2–1, is only 126.†

It is easy to specialize the general branch-and-bound algorithm to this problem, producing an algorithm that backtracks whenever the cost of the current partial solution equals or exceeds the cost of the best solution found so far. This test does preclude the search of parts of the tree but is actually rather weak, allowing deep penetration into the tree before branches are pruned. Moreover, the arbitrary fixed order of the cities causes much time to be wasted exploring paths that begin 1–2 if cities 1 and 2 are far apart. This illustrates how the technique of tree rearrangement applies to branch-and-bound methods; near-optimal solutions should be found early in the search.

A relatively successful technique for tree rearrangement in the traveling salesman problem is to split all remaining solutions into two groups at each stage—those that include a particular arc and those that exclude that arc. The arc used to split the solution space is chosen according to a heuristic (described below) that is intended to prune the tree as much as possible. Of course, every node in the tree, now obviously a binary tree, will have associated with it a lower bound on the cost of all solutions descended from it.

The method uses the fact that if a constant is subtracted from any row or any column of the cost matrix, the optimal solution does not change. The *cost* of the optimal solution changes, of course, but not the path itself. In fact, the cost of the optimal solution is diminished by exactly the amount subtracted from the row or column. Thus if such subtraction is done, so that each row and column contains a zero and yet the C_{ij} all remain nonnegative, then the total amount subtracted will be a lower bound on the cost of any solution. For example, the cost matrix given above can be reduced by subtracting 3, 4, 16, 7, 25, 3, and 26 from rows 1 through 7, respectively, and then subtracting 7, 1, and 4 from columns 3, 4, and 7,

†It may seem strange to have $C_{ii} = \infty$ (instead of $C_{ii} = 0$), but it is convenient for the discussion in this section. We will use the more natural $C_{ii} = 0$ in Section 4.3.

respectively, leaving the reduced matrix

$i \backslash j$	1	2	3	4	5	6	7
1	∞	0	83	9	30	6	50
2	0	∞	66	37	17	12	26
3	29	1	∞	19	0	12	5
4	32	83	66	∞	49	0	80
5	3	21	56	7	∞	0	28
6	0	85	8	42	89	∞	0
7	18	0	0	0	58	13	∞

Since a total of 96 was subtracted, 96 is a lower bound on the cost of any solution. Initially, then, we have the root of the binary tree and its associated lower bound:

$$\left(\text{All Solutions}\right) \quad \text{lower bound} = 96.$$

How does this node split into subtrees? Suppose that we choose the arc 4–6 on which to split the root. The right subtree will contain all solutions that exclude the arc 4–6; knowing that 4–6 is excluded, we can change the cost matrix by setting $C_{4,6} = \infty$. The resulting matrix can then have 32 subtracted from the fourth row, and thus the right subtree has a lower bound of $96 + 32 = 128$. The left subtree will contain all solutions that include the arc 4–6, and so the fourth row and the sixth column must be deleted from the cost matrix because now we can never go from 4 to anywhere else nor arrive at 6 from anywhere else. The result is a cost matrix of one less in dimension. Furthermore, since all solutions in this subtree use the arc 4–6, the arc 6–4 is no longer usable and we must set $C_{6,4} = \infty$. Finally, we can now subtract 3 from the fourth row of the resulting matrix ($i = 5$), giving the left subtree a lower bound of $96 + 3 = 99$. The binary tree looks as shown in Figure 4.6.

The arc 4–6 was used to split the root because, of all arcs, it caused the greatest increase in the lower bound of the right subtree. This rule is used to split a node because we would prefer to find the solution by following left edges rather than right edges; the left edges reduce the dimension of the problem, whereas the right edges only add another ∞ and perhaps a few new zeros without changing the dimension. It is not difficult to determine which arc gives the largest increase in the lower bound of the right subtree: choose the zero that, when changed to infinity, allows the most to be subtracted from its row and column.

In applying this branch-and-bound method to the sample cost matrix given above, we notice a complication. The first edge included is 4–6, the next is 3–5, and the third is 2–1 (see the leftmost path from the root of the tree in Figure 4.7). At this node we have a lower bound of 112 and a reduced cost matrix

$i \backslash j$	2	3	4	7
1	∞	74	0	41
5	14	∞	0	21
6	85	8	∞	0
7	0	0	0	∞

The zero that when changed to infinity allows the most to be subtracted from its row and column is in position $i = 1, j = 4$. Thus we split this node using the edge 1–4. In the left subtree we get a lower bound of 126 and reduced cost matrix

$i \backslash j$	2	3	7
5	14	∞	21
6	85	8	0
7	0	0	∞

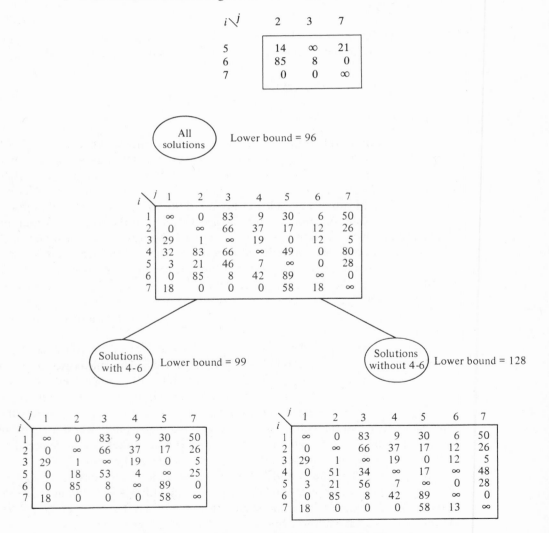

Figure 4.6 The splitting of the root of the binary tree.

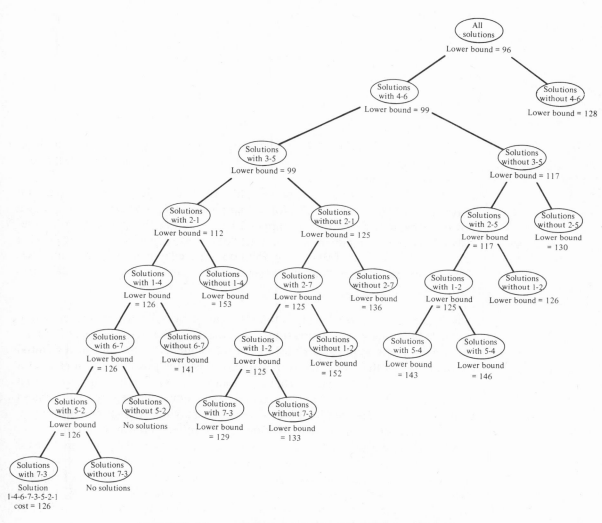

Figure 4.7   The complete search tree for the sample traveling salesman problem.

Now, according to our description of the process, we want to prevent the use of the edge 4–1, but the cost matrix does not have an element corresponding to $i = 4$, $j = 1$. What do we do?

The answer is found in the purpose of preventing the use of the edge 4–1. That purpose is to prevent the tour from returning to an already visited city before the very last edge which returns us to the starting city. In this case, the partial tour constructed so far consists of the path 2–1–4–6 and the edge 3–5; clearly it is the edge 6–2 whose use must be prevented. Thus we would change the reduced cost matrix to

$i\diagdown^j$	2	3	7
5	14	∞	21
6	∞	8	0
7	0	0	∞

In general then, if the edge added to the partial tour is from i_u to j_1 and the partial tour contains paths $i_1-i_2-\cdots-i_u$ and $j_1-j_2-\cdots-j_v$, then the edge whose use is to be prevented is j_v-i_1.

Applying this method to the sample cost matrix given earlier, yields the binary tree shown in Figure 4.7. Notice that the first solution found is optimal; we obviously cannot expect that to happen in general. For this particular example, only 27 nodes are examined; if we use the "obvious" specialization of the general branch-and-bound algorithm, then 559 nodes are examined. Experimental evidence suggests that, for random $n \times n$ distance matrices, the number of nodes examined is $O(1.26^n)$.

α-β Pruning. A considerably different type of branch-and-bound technique occurs in the evaluation of game trees. A *game tree* is the tree that arises from the backtrack examination of all possible move sequences: the root is the initial configuration of the game; the sons of the root are the possible positions after a move by the first player; the sons of those nodes are the possible positions after an answering move by the second player, and so on. For example, Figure 4.8 shows sections of the game tree for tic-tac-toe with x making the first move; the tree shows only inequivalent (under rotation and/or reflection) sons of a node.

Each leaf of a game tree represents a possible ending of the game; in the case of tic-tac-toe, it could be a win for x, a win for o, or a draw. We are interested in evaluating the game tree in terms of the first player: each leaf is assigned a value in terms of its payoff for the first player—$+1$ for a win, -1 for a loss, 0 for a draw. The value of other nodes is defined by the values of its sons. For a node N with sons $N_1, N_2, \ldots, N_k$, the value, $V(N)$, is given by

$$V(N) = \begin{cases} \max\,(V(N_1), \ldots, V(N_k)) & \text{if the level of } N \text{ is even,} \\ \min\,(V(N_1), \ldots, V(N_k)) & \text{if the level of } N \text{ is odd.} \end{cases}$$

This means that the first player is assumed to be trying to maximize his gain, whereas the second player is assumed to be trying to minimize his loss. Under these conditions it is well known that the value of the root of the tree shown in Figure 4.8 is zero, meaning that under the minimax assumption the game always ends in a draw.

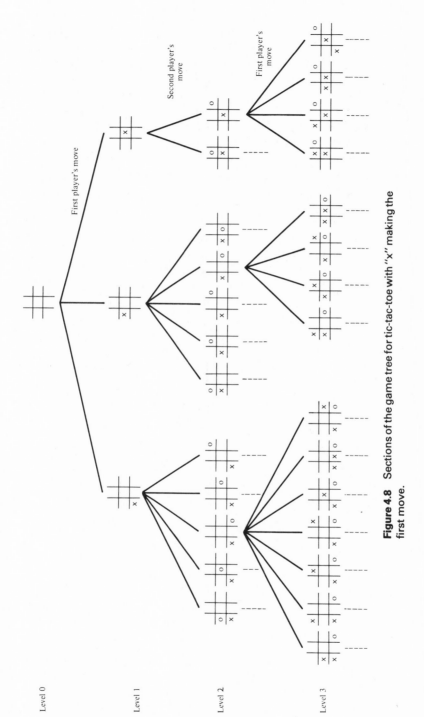

Figure 4.8 Sections of the game tree for tic-tac-toe with "x" making the first move.

Of course, we could simply use the obvious backtrack technique to evaluate all the nodes in the tree and thus determine the value of the root. However, a better method, called α-β *pruning*, uses the branch-and-bound philosophy to prune subtrees from the game tree. Note that as soon as a node is evaluated, something is known about the value that will be assigned to its father. In particular, if its father is on a max level in the tree, then the value assigned to the node is a lower bound on the value of its father. Such a lower bound at a max level is called an α-value. In the symmetrical situation where the father lies on a min level, the resultant upper bound on its value is called a β-value.

In the example in Figure 4.9, as soon as node d is assigned its value of 3, we immediately have a β-value of 3 for node m, and so we know that the value at m will not be larger than 3. With this information we reason as follows. If the value of any other descendant of node m can be shown to be greater than 3, we may safely ignore the rest of the tree below that descendant. Thus we can then act as if these branches and nodes had been pruned from the tree. Such a situation is termed a β-*cutoff*, and it occurs whenever a node two levels below a node with a

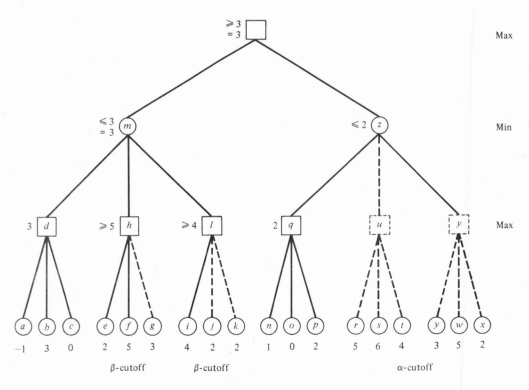

Figure 4.9 α–β pruning.

β-value is found to have a value larger than that β. The situation at a max level is similar. In the example again, having found the value 3 for node m, we have an α-value of 3 for the root of the tree, so that if node z can be shown to have a value less than 3, we can ignore anything below it. Thus after establishing the value of 2 at node q, we cut off the rest of the tree below node z, for its evaluation will no longer influence the ultimate value of the root. This is called an α-*cutoff*.

The advantage of this procedure over the exhaustive backtrack algorithm is obvious; with it we need check only a fraction of the nodes of the tree instead of all of them. We immediately notice, however, that the amount of work saved is greatly influenced by the ordering of the nodes in the tree, for this ordering determines the time at which α- and β-cutoffs occur. Despite this obvious dependence on the order of the search, we can say something about the relative efficiency of α-β pruning over the straightforward backtrack algorithm. Consider the tree pictured in Figure 4.10. Here the value A is a lower bound for the value of the root C. Assuming, as above, that evaluation of the nodes B_i takes place from left to right, from B_1 towards B_n, we may define d, the "distance to α-cutoff," as the least $i \leq n$ such that $B_i \leq A$. We want to see how the quantity d behaves as n varies; in order to do so, we make an assumption about the values for the nodes. We know that among the values B_i some fraction p of them will be greater than A and $(1 - p)$ of them less than or equal to A. We assume more generally that there is a p, independent of n, such that each B_i has probability p of being greater than A, independently of any of the other values B_j. Then we define the "average distance to α-cutoff":

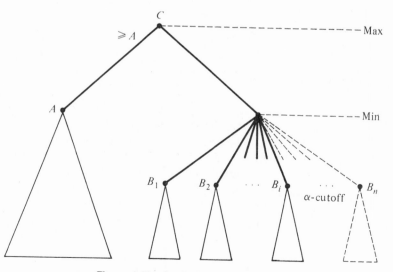

Figure 4.10 Analysis of α–β pruning.

$$D(n, p) = \sum_{k=1}^{n} k \Pr (\text{cutoff at } k)\dagger$$

$$= \sum_{k=1}^{n} k \Pr (B_1 > A, B_2 > A, \ldots, B_{k-1} > A, \text{ and } B_k \leqq A)$$

$$= \sum_{k=1}^{n} kp^{k-1}(1 - p)$$

$$= (1 - p) \sum_{k=1}^{n} kp^{k-1}$$

As $n \to \infty$, we have partial sums of a convergent series, so that $D(n, p)$ is always bounded above by its limit,

$$D(\infty, p) = (1 - p) \sum_{k=1}^{\infty} kp^{k-1}$$

$$= \frac{1}{1 - p}.$$

In other words, on a tree satisfying the preceding assumptions, α-β pruning has the effect of limiting the branching of the tree to some value that is independent of the actual branching of the tree. For instance, suppose that each B_i has an equal probability of being larger or smaller than A; that is, $p = 1 - p = \frac{1}{2}$. This assumption may be reasonable to make in the absence of any specific knowledge about the distribution of values in the tree. Then, according to our analysis, we may expect an average distance to cutoff to be no greater than $D(\infty, \frac{1}{2}) = 2$ regardless of how many B_i there are.

4.1.7 Dynamic Programming

The traveling salesman problem of the previous section is an example of a *multistage decision process*: a process in which a sequence of decisions is made, the choices available being dependent on the current state of the system—that is, on the previous decisions. For the traveling salesman problem, the decision at each stage is which city to visit next. In such processes, the problem is to determine the optimal sequence of decisions—that is, those that minimize (or perhaps maximize) some objective function. The objective function to be minimized in the traveling salesman problem, is the cost of the tour.

---

†$\Pr(x)$ denotes the probability of event x.

In the solution of such problems by *dynamic programming*, we rely on the *principle of optimality*:

> An optimal sequence of decisions has the property that whatever the initial state and initial decision are, the remaining decisions must be an optimal sequence of decisions with regard to the state resulting from the first decision.

Applying this principle to the solution of combinatorial problems essentially means using the decomposition principle: the solutions to subproblems are found and then used to find the solutions to larger subproblems and, finally, to the problem itself.

For the traveling salesman problem, define

$$T(i\ ; j_1, j_2, \ldots, j_k) = \begin{cases} \text{cost of the optimal tour from } i \\ \text{to 1 that goes through each of the} \\ \text{cities } j_1, j_2, \ldots, j_k \text{ exactly once, in} \\ \text{any order, and through no others.} \end{cases}$$

The principle of optimality tells us that

$$T(i; j_1, j_2, \ldots, j_k) = \min_{1 \le m \le k} \{C_{ij_m} + T(j_m; j_1, j_2, \ldots, j_{m-1}, j_{m+1}, \ldots, j_k)\},$$

$$(4.1)$$

where, as in the previous section, C_{ij} is the cost of going directly from city i to city j. Furthermore, by definition we have

$$T(i\ ; j) = C_{ij} + C_{j1}. \tag{4.2}$$

We are interested in computing $T(1\ ; 2, 3, \ldots, n)$ and a tour of that length. Equations (4.1) and (4.2) define a recursive computation for $T(1\ ; 2, 3, \ldots, n)$; the sequence of values of j_m arising from the recursive application of equation (4.1) gives the optimal tour. The sequence of recursive calls for $n = 5$ is shown as a tree in Figure 4.11.

Notice that there is considerable duplication at the leaves; for larger values of n, trees of greater height occur, with duplications at higher levels in the tree. The higher in the tree the duplications occur, the more costly they are. In the recursive, depth-first computation of the tree (see Section 4.1.4), it is difficult to recognize identical subtrees and prevent duplicate computations. Using the principles of branch merging and search rearrangement, we organize the computation bottom-up, level by level. Since the values of T at one level depend only on the values of T at the level below, we can begin at the leaves (level $n - 2$), whose values we know from equation (4.2), and work up to the root (level 0), level by

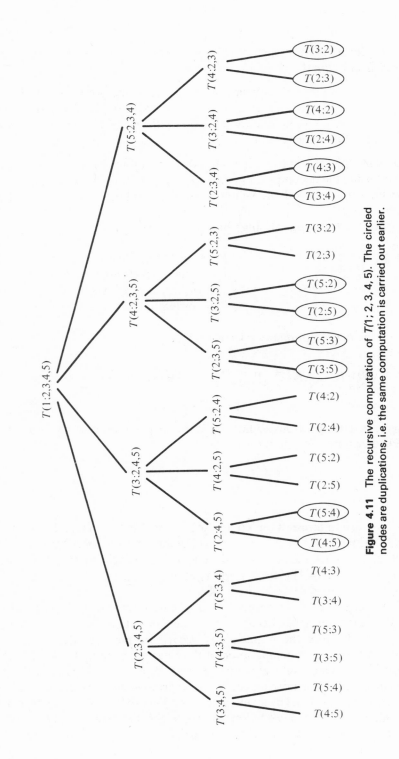

Figure 4.11 The recursive computation of $T(1; 2, 3, 4, 5)$. The circled nodes are duplications, i.e. the same computation is carried out earlier.

level. Of course, this approach involves more storage than the depth-first approach. If we use one word of memory for each distinct node in the tree, a total of†

$$\sum_{i=0}^{n-2} (\text{number of nodes at level } i) = 1 + \sum_{k=2}^{n-1} (n - k)\binom{n - 1}{n - k}$$

$$= (n - 1)2^{n-2} - n + 2$$

words is needed. This relatively large memory requirement is typical of the level-by-level approach to dynamic programming.

The number of additions required for the evaluation of equations (4.1) and (4.2) is 1 for each leaf, 2 for each node whose sons are leaves, 3 for each node whose grandsons are leaves, and so on, that is, $n - i - 1$ for a node at level i, for a total of

$$\sum_{i=0}^{n-2} (\text{number of nodes at level } i)(n - i - 1)$$

$$= (n - 1) + \sum_{k=2}^{n-1} (n - k)(n - k)\binom{n - 1}{n - k}$$

$$= n(n - 1)2^{n-3} - (n - 1)(n - 2).$$

The number of comparisons is almost the same—namely, $n - i - 2$ for a node at level i—giving a total of

$$(n - 2) + \sum_{k=2}^{n-1} (n - k - 1)(n - k)\binom{n - 1}{n - k} = (n - 1)(n - 2)2^{n-3} - (n - 2)^2.$$

Notice that the amount of work required in the bottom-up, level-by-level approach is a function *only* of the number of cities; it is independent of the entries in the cost matrix, in contrast to the (depth-first) branch-and-bound methods, which depend heavily on the order of the rows and columns as well as on the entries of the matrix.

Unfortunately, applying the principle of optimality and dynamic programming to the traveling salesman problem did not significantly reduce the computation time; it only reorganized it. However, for other problems, the principle of optimality can drastically reduce the size of the space to be searched, and so dynamic programming can greatly decrease the computation time (see Exercises 5, 6, and 7). We will see such an example in Section 6.3.2, in which the computation time is reduced from being proportional to $4^n/n^{3/2}$ for the obvious exhaustive algorithm to $O(n^3)$ for the dynamic programming approach, a significant improvement indeed.

Our failure in this chapter to find an efficient method for finding an optimal solution to the traveling salesman problem will be examined in detail in Chapter 9.

†The summations in this section can all be evaluated by the techniques of Section 3.3.

4.2 SIEVES

A *sieve*, as the name suggests, is a combinatorial programming technique that takes a finite set and eliminates all the members of that set that are *not* of interest. It is the logical complement of the backtrack process described in the previous section, which enumerates all the members of a set that *are* of interest.

Sieves are primarily useful in number-theoretic computations. For example, the sieve of Eratosthenes is one of the best-known methods of finding prime numbers; this sieve enumerates the composite (nonprime) numbers between N and N^2 for some N, as illustrated in Figure 4.12 for $N = 6$. The sieve begins by writing down the integers from N to N^2 and then removing the composite numbers in stages. First all multiples of 2 are removed; then all multiples of 3 are removed and so on. The process stops after sifting with largest prime less than N.

Stage 0 (initially)
6 7 8 9 10 11 12 13 14 15 16 17 18 19 20 21 22 23 24 25 26 27 28 29 30 31 32 33 34 35 36

Stage 1 (multiples of 2 eliminated)
6̸ 7 8̸ 9 1̸0̸ 11 1̸2̸ 13 1̸4̸ 15 16 17 1̸8̸ 19 2̸0̸ 21 2̸2̸ 23 2̸4̸ 25 2̸6̸ 27 2̸8̸ 29 3̸0̸ 31 3̸2̸ 33 3̸4̸ 35 3̸6̸

Stage 2 (multiples of 3 eliminated)
6̸ 7 8̸ 9̸ 1̸0̸ 11 1̸2̸ 13 1̸4̸ 1̸5̸ 1̸6̸ 17 1̸8̸ 19 2̸0̸ 2̸1̸ 2̸2̸ 23 2̸4̸ 25 2̸6̸ 2̸7̸ 2̸8̸ 29 3̸0̸ 31 3̸2̸ 3̸3̸ 3̸4̸ 35 3̸6̸

Stage 3 (multiples of 5 eliminated)
6̸ 7 8̸ 9̸ 1̸0̸ 11 1̸2̸ 13 1̸4̸ 1̸5̸ 1̸6̸ 17 1̸8̸ 19 2̸0̸ 2̸1̸ 2̸2̸ 23 2̸4̸ 2̸5̸ 2̸6̸ 2̸7̸ 2̸8̸ 29 3̸0̸ 31 3̸2̸ 3̸3̸ 3̸4̸ 3̸5̸ 3̸6̸

Figure 4.12 The sieve of Eratosthenes used to find all primes between 6 and 36: 7, 11, 13, 17, 19, 23, 29, 31.

It is easy to understand why sieves can be a useful technique. If the elements in the set of potential solutions can be conveniently indexed with the natural numbers, then only a characteristic vector (see Section 2.2.1) need be kept; the ith bit is zero if the ith set element is known not to be a solution, and it is one otherwise. In this manner, sets with literally millions of elements can be searched without explicitly generating and examining each element of the set. Moreover, on most computers, boolean operations can be used to operate on many bits in parallel, thus providing a significant savings in time.

4.2.1 Nonrecursive Modular Sieves

We can interpret the sieve of Eratosthenes as searching for all numbers between N and N^2 that are simultaneously members of one arithmetic progression in each of the following sets:

$$\{2k + 1\},$$

$$\{3k + 1, 3k + 2\},$$

$$\{5k + 1, 5k + 2, 5k + 3, 5k + 4\},$$

$$\{7k + 1, \ldots, 7k + 6\},$$

$$\{11k + 1, \ldots, 11k + 10\},$$

$$\vdots$$

$$\{pk + 1, \ldots, pk + p - 1\},$$

where p is the largest prime less than or equal to N. The fact that a number is in the progression $2k + 1$ means that it is not even; the fact that it is in one of the progressions $3k + 1$ or $3k + 2$ means that it is not a multiple of 3, and so on. Thus the only numbers between N and N^2 satisfying all these conditions are the primes.

Similarly, consider the medieval puzzle of the woman and the eggs. On her way to the market to sell her eggs, a horseman knocks her down by accident, causing all the eggs to be broken. He offers to pay the damages and wants to know how many eggs she had. She says she does not remember the exact number, but when she took them 2 at a time, 1 was left over. One was also left over when they were taken 3, 4, 5, and 6 at a time; but when they were taken 7 at a time, it came out even. Clearly, she could have had N eggs if and only if N was simultaneously a member of each of the arithmetic progressions

$$2k + 1, 3k + 1, 4k + 1, 5k + 1, 6k + 1, \text{ and } 7k.$$

How should the old woman determine all values less than 1000 that are possibilities? A sieve can be used to solve this problem by listing the integers $1, 2, 3, 4, \ldots, 1000$ and then removing all elements not in the various progressions: first all the even numbers, then numbers of the form $3k$ or $3k + 2$, and so on. The numbers that remain are the possible solutions.

Both the sieve of Eratosthenes and the sieve solution to the medieval puzzle are special cases of the *generalized modular sieve*. Let $m_1, m_2, m_3, \ldots, m_t$ be a set of t positive integers (called the *moduli*). For each m_i, we consider n_i arithmetical progressions

$$m_i k + a_{ij}, \qquad j = 1, 2, \ldots, n_i.$$

The problem is to find all integers, between given limits A and B, that simultaneously satisfy, for each m_i, one of the n_i progressions. In the sieve of Eratosthenes, we have $m_1 = 2, m_2 = 3, m_3 = 5, \ldots, m_t = p$ (the largest prime less than or equal to N), $n_i = m_i - 1$, and $a_{ij} = j$. In the problem of the old woman and the eggs, we have $m_i = i + 1, n_i = 1, 1 \leq i \leq 6$, and $a_{i1} = 1, 1 \leq i \leq 5, a_{6,1} = 0$.

If the moduli are relatively prime in pairs, we can combine all the progressions to obtain *one* set of Πn_i progressions with modulus Πm_i. This is done by repeated application of the following technique to combine $m_1 k + a_1$ and $m_2 k + a_2$ into a single progression $m_1 m_2 k + a$. Since, by assumption, the greatest common divisor of m_1 and m_2 is 1, the Euclidean algorithm† guarantees the existence of integers u and v such that $m_1 u + m_2 v = 1$. Choosing $a = m_1 u a_2 + m_2 v a_1$, we find that $x \equiv a \pmod{m_1 m_2}$ if and only if $x \equiv a_1 \pmod{m_1}$ and $x \equiv a_2 \pmod{m_2}$. Thus to determine which integers in the range between A and B satisfy one of the n_i progressions for each m_i, we can take each integer in that range, divide it by Πm_i and see if the remainder is one of the Πn_i acceptable ones. Of course, the size of this new set of progressions grows rapidly, so rapidly, in fact, that having only 10 sets of three progressions each yields a single set of $n = 3^{10} = 59049$ progressions. Therefore we would have to compute these 59049 values and then search them for each remainder that we calculate. This is quite a bit of work, and it is usually more practical to use a sieve.

The usefulness of the generalized modular sieve can be appreciated by considering the problem of testing the first one million Fibonacci numbers to see which are squares. Since

$$F_n \approx \frac{1}{\sqrt{5}}\left(\frac{1 + \sqrt{5}}{2}\right)^n,$$

we know that

$$F_{1000000} > 10^{200000};$$

that is, $F_{1000000}$ has over 200000 decimal digits. It is obvious that the brute-force method of generating the first one million Fibonacci numbers and testing each one to see if it is a square is unfeasible; but the computation can easily be done in a few minutes with a sieve technique.

The calculation begins by setting aside one million bits in memory for the characteristic vector. The ith bit will represent the integer F_i; if that bit is 1, then F_i might be a square; if it is 0, then F_i cannot be a square. Initially, all the bits will be 1, and during the sifting process certain of these bits are set to 0 at each stage. When the sifting is completed, if any bit position is 1, the corresponding Fibonacci number must be examined to see if it is a square. The one million bits of memory required for the characteristic vector is large but not excessive; for example, on a 32-bit word machine, it is only 31250 words of storage.

---

†The Euclidean algorithm finds the greatest common divisor of m and n as follows. Let $r_0 = m, r_1 = n$, and, for $i \geq 2$, define r_i, q_i so that $r_{i-2} = r_{i-1} q_i + r_i$, where $0 \leq r_i < r_{i-1}$. If r_t is the first remainder to be zero, then r_{t-1} is the greatest common divisor of m and n. Back substitution yields integers u and v such that $mu + nv = r_{t-1}$. For more details, see any introductory number theory textbook.

Consider the Fibonacci sequence modulo p for a prime p. Let this sequence be $P_1, P_2, P_3, \ldots$, defined by

$$P_1 = P_2 = 1$$

$$P_{i+1} = P_{i-1} + P_i \pmod{p}.$$

This sequence is periodic, the period starting with $P_1 = 1$. To prove it, consider the $p^2 + 1$ ordered pairs of integers

$$(P_1, P_2), (P_2, P_3), \ldots, (P_{p^2}, P_{p^2+1}), (P_{p^2+1}, P_{p^2+2}).$$

Since there are only p^2 different ordered pairs of integers modulo p, two of these pairs must be equal. Let (P_r, P_{r+1}) and (P_s, P_{s+1}) be the leftmost equal pairs

$$(P_r, P_{r+1}) = (P_s, P_{s+1}), \qquad r < s.$$

If $r > 1$, then we have

$$(P_{r-1}, P_r) = (P_{s-1}, P_s),$$

contradicting the assumption that (P_r, P_{r+1}) and (P_s, P_{s+1}) are the leftmost equal pair. Thus the Fibonacci numbers modulo p are periodic starting at P_1, so that $P_i = P_{i+nT_p}$ for some period T_p and all $n \geqq 0$.

Modulo any number n there are certain squares (usually called *quadratic residues*† modulo n) and certain nonsquares. For example, 0, 1, 2, and 4 are quadratic residues modulo 7, whereas 3, 5, and 6 are *quadratic nonresidues*. Thus any number $m \equiv 3, 5$ or 6 (mod 7) cannot be a square because it is a quadratic nonresidue modulo 7. Since, for every prime p, the Fibonacci numbers modulo p are periodic with period T_p, we sift by using the arithmetical progressions $T_p k + i$, for each i, such that P_i is a quadratic nonresidue modulo p. To illustrate the sieve, consider $p = 7$. The Fibonacci series modulo 7 is the sequence of 16 elements

$$1, 1, 2, 3, 5, 1, 6, 0, 6, 6, 5, 4, 2, 6, 1, 0$$

repeated over and over again, so that $T_7 = 16$. Since 3, 5, and 6 are quadratic nonresidues, we know that $F_{4+16n}, F_{5+16n}, F_{7+16n}, F_{9+16n}, F_{10+16n}, F_{11+16n}$, and F_{14+16n} are not squares for $n \geqq 0$, and so the bits corresponding to those Fibonacci numbers can be changed from 1 to 0.

After sifting with the quadratic nonresidues of the first 32 prime numbers, we find that all the bits are changed to zeros except the first, second, and twelfth. These bits correspond to the three numbers $F_1 = F_2 = 1$ and $F_{12} = 144$; therefore we can conclude that they are the only squares among the first one million Fibonacci numbers. In fact, it has been proven that they are the only square Fibonacci numbers.

---

†Actually, $a \neq 0$ is a quadratic residue modulo n if and only if a is a perfect square modulo n [i.e., $x^2 \equiv a \pmod{n}$ has a solution] *and* a and n are relatively prime. We can ignore the latter condition because, for our purposes, n will always be a prime number.

4.2.2 Recursive Sieves

There are many sieves in which the moduli $m_1, m_2, \ldots$ are not predetermined; the value of m_i will depend on the numbers *not* eliminated after sifting with m_{i-1}. Many sieves are constructed in this recursive manner; in fact, the sieve of Eratosthenes is usually so constructed: after writing down the integers $2, 3, \ldots, N$, we cancel all multiples of 2 except 2. Then since the smallest remaining number whose multiples have not been removed is 3, all multiples of 3, except 3, are removed, and so on. Notice that at each stage the first number removed is the square of the sifting number, and thus the first number eliminated by 2 is 4, by 3, 9, and so on. When the sifting number becomes larger than $\sqrt{N}$, no other numbers can be removed and the process ends.

Typically, we double the range of the sieve of Eratosthenes by *presifting* with 2; in other words, we begin with only the odd integers, sifting out multiples of 3, 5, 7, 11, and so forth. Letting X be the bit string, the recursive version of the sieve of Eratosthenes, with presifting by 2, for odd primes up to $2N + 1$ is shown in Algorithm 4.5.

$X \leftarrow (1, 1, \ldots, 1)$

for $k = 3$ **to** $\sqrt{2N + 1}$ **by** 2 **do**

    if $X_{(k-1)/2} = 1$ **then for** $i = k^2$ **to** $2N + 1$ **by** $2k$ **do** $X_{(i-1)/2} = 0$

for $k = 1$ **to** N **do if** $X_k = $ **then** output $2k + 1$

Algorithm 4.5 Sieve of Eratosthenes for odd primes.

Another well-known recursive sieve produces the *lucky numbers*. From the list of numbers $1, 2, 3, 4, 5, \ldots$ remove every second number, leaving the list $1, 3, 5, 7, 9, \ldots$. Since 3 is the first number (excepting 1) that has not been used as a sifting number, we remove every third number *from those remaining numbers*, yielding $1, 3, 7, 9, 13, 15, 19, 21, \ldots$ Now every seventh number is removed, leaving $1, 3, 7, 9, 13, 15, 21, \ldots$. Numbers that are never removed from the list are considered to be "lucky."

In spite of the great similarity between the sieve of Eratosthenes and the lucky number sieve, the latter is harder to implement. The difficulty occurs because the numbers sifted out at the kth stage depend *positionally* on the elements not yet eliminated, not on the elements of the initial set. Assuming that a bit representation is used, the lucky number sieve will have to count, say, every seventh 1-bit instead of every seventh bit. This task is expedited considerably by the use of *tags*. Instead of using an entire computer word to represent the potential solutions through which we must sift, we use a portion of the word to store a count of the number of 1-bits in the remainder of the word. For example, on a computer

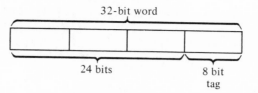

Figure 4.13   Using the rightmost byte of a word as the *tag*, i.e. a count of the number of ones in the leftmost three bytes.

whose words consist of four bytes with eight bits each, we can use a word, as shown in Figure 4.13. The rightmost byte contains a count of the number of 1-bits in the leftmost three bytes. With tagging, finding the tth uncanceled element (the tth 1-bit) is easily done by summing the tags until the sum S is greater than or equal to t; the appropriate element is the $(S - t + 1)$st 1-bit from the right in the leftmost bytes of the last word whose tag was summed.

Obviously, tagging saves no time if one or more bits are eliminated from each word. For this reason, it is usually worthwhile to run the sieve for several stages without tags, beginning the use of tags when the elements to be sifted out become infrequent enough. The exact number of stages run without tags depends on the type of sieve and on the word size. For example, in the lucky number sieve with 32-bit words and 8-bit tags, it is better to run the first three or four stages without tags.

Tagging can be extended to higher-order tags by choosing an integer $m \geq 2$ and computing the sum of the tags in blocks of m words. These tags would normally be stored outside the original area, since the values can get large. Higher-and-higher-order tags can be used, imposing a kind of tree structure on the set, but after a while the process of diminishing returns sets in and little is gained. As with first-order tags, higher-order tags become worthwhile only if less than one bit is eliminated per block.

A recursive sieve of somewhat different character can be used to compute the following sequence U. Initially, $U = (1, 2)$. If membership has been decided for all integers less than n, then $n \in U$ if and only if it is the sum of a unique pair of distinct elements in U. Thus $U = (1, 2, 3, 4, 6, 8, 11, 13, 16, 18, 26, \ldots)$. This sequence can be computed by running two sieves in parallel; an integer must be sifted in (must be a sum in at least one way) and must not be sifted out (must be a sum in at most one way). Two bit vectors are used. For $i > 2$,

$X_i = 1$ if and only if i is representable as a sum in at least one way,

$Y_i = 1$ if and only if i is representable as a sum in at most one way.

The "double sieve" to compute the elements of U up to N is given in Algorithm 4.6. The counter k is incremented until it reaches the first integer beyond the previous member of U that is representable in exactly one way. This integer is in U, and so all integers $k + i$ for $i < k$ must have their bits updated.

$$X \leftarrow (1, 1, 0, 0, \ldots, 0)$$
$$Y \leftarrow (1, 1, 1, 1, \ldots, 1)$$
$$k \leftarrow 1$$

$$\textbf{while } k < N \textbf{ do} \begin{cases} k \leftarrow k + 1 \\ \textbf{while } X_k \wedge Y_k = 0 \textbf{ and } k < N \textbf{ do } k \leftarrow k + 1 \\ \textbf{for } i = 1 \textbf{ to } \min(k - 1, N - k) \textbf{ do} \begin{cases} Y_{k+i} \leftarrow Y_{k+i} \wedge \overline{(X_{k+i} \wedge X_i \wedge Y_i)} \\ X_{k+i} \leftarrow X_{k+i} \vee (X_i \wedge Y_i) \end{cases} \end{cases}$$

$$\textbf{for } i = 1 \textbf{ to } N \textbf{ do if } X_i \wedge Y_i = 1 \textbf{ then } \text{output } i$$

Algorithm 4.6   Double sieve for the sequence U.

4.2.3 Isomorph-Rejection Sieves

It is sometimes necessary to extract from a given set of objects a maximal subset of the objects that are pairwise nonisomorphic. We have, say, objects

$$o_1 \quad o_2 \quad o_3 \quad o_4 \quad o_5 \quad \cdots \quad o_n.$$

We find all $o_i \cong o_1, i > 1$, and cancel them. Then if o_{a_1} is the first uncanceled object beyond o_1, we find all $o_i \cong o_{a_1}, i > a_1$, and cancel them. Continuing in this fashion, we are left with a maximal set of inequivalent objects. Thus isomorph-rejection sieves are a special case of recursive sieves.

Consider, for example, the $n \times n$ nonattacking queens problem. If we have found N solutions and we want to sift through them to find all solutions that are inequivalent under rotations and/or reflections, we can do so with an isomorph-rejection sieve. In this case, however, the sieve need not be explicitly performed, since the solutions (permutations) are found in increasing lexicographic order.† As a permutation is found, we generate all those that are equivalent. If any of those with $a_1 \neq 1$ is lexicographically less than the permutation, then it has been found previously and we can ignore it. In the 7×7 case, for instance, 16

---

†If $A = (a_1, a_2, \ldots, a_n)$ and $B = (b_1, b_2, \ldots, b_n)$ are permutations of $(1, 2, \ldots, n)$, then A is *lexicographically less than* B if and only if, for some $k \geq 1$, $a_i = b_i$ when $i < k$ and $a_k < b_k$. For example, $(3, 4, 1, 2, 5)$ is lexicographically less than $(3, 4, 2, 5, 1)$. See Section 5.1.1.

permutations are found to be solutions. They are found in the order

$$2, 4, 1, 7, 5, 3, 6$$

$$2, 4, 6, 1, 3, 5, 7$$

$$2, 5, 1, 4, 7, 3, 6$$

$$2, 5, 3, 1, 7, 4, 6$$

$$2, 5, 7, 4, 1, 3, 6$$

$$\vdots$$

$$4, 1, 3, 6, 2, 7, 5$$

$$4, 1, 5, 2, 6, 3, 7$$

$$4, 2, 7, 5, 3, 1, 6$$

Finding $(2, 4, 1, 7, 5, 3, 6)$, we record it; since it is the first one found, no equivalent solution with $a_1 \neq 1$ is lexicographically less than it. Continuing, we find $(2, 4, 6, 1, 3, 5, 7)$. Upon computing all equivalent solutions, we find only $(1, 3, 5, 7, 2, 4, 6)$ less than it. Since this has $a_1 = 1$, we record $(2, 4, 6, 1, 3, 5, 7)$. Next, $(2, 5, 1, 4, 7, 3, 6)$ is found. When all equivalent solutions are computed, we find that none of them is less than it, so it is recorded. Next, $(2, 5, 3, 1, 7, 4, 6)$ is found, and since it is equivalent to $(2, 4, 1, 7, 5, 3, 6)$, it is discarded. This process continues until all 16 permutations have been examined and the six inequivalent solutions have been found.

4.3 APPROXIMATION TO EXHAUSTIVE SEARCH

The traveling salesman problem introduced in Section 4.1.6 is one of a number of combinatorial optimization problems for which no efficient algorithms are known. The branch-and-bound method illustrated in Figure 4.7 appears, on the basis of empirical evidence, to require searching trees with proportional to 1.26^n nodes for an average (i.e., random) n-city problem. The dynamic programming approach of Section 4.1.7 always requires searching trees of about $n2^n$ nodes. All known algorithms require searching a space whose size is exponential in the number of cities.

When dealing with such computationally difficult problems, one approach is to relax the requirement that an algorithm produce an optimal result and insist only that the result be reasonably close to optimal. The relaxation of the optimality constraint frequently allows algorithms to be more efficient, since the exhaustive search is only "approximated." It is hoped that there will be only a moderate sacrifice in the cost of the (suboptimal) solutions found.

It is usually easy to devise heuristic algorithms that find solutions quickly; in fact, we may find that most, but not all, of the solutions produced by the heuristic are good. For example, in Section 6.2.3 we'will discuss heuristic algorithms for constructing binary search trees; the quality of the *average* tree constructed is quite good, but sometimes very poor trees are constructed. It is exceptional to be able to prove that a heuristic algorithm *always* finds solutions that are close to being optimal.

In this section we examine heuristic algorithms for the traveling salesman problem. The algorithms require only time $O(n^2)$ for an n-city problem and can be shown to produce tours whose cost is guaranteed not to be too far from the cost of an optimal tour if the cost matrix is symmetric and satisfies the triangle inequality. Such approximate algorithms are of great practical importance for problems in which the time needed to find an optimal result grows rapidly as a function of the problem's size.

A typical heuristic technique in optimization problems is to search for solutions that are, in some sense, *locally* optimal instead of being *globally* optimal. In particular, if we view the construction of a tour of n cities as a sequence of decisions, we can make each decision according to a local optimality rule. For example, consider the obvious *nearest-neighbor* method of constructing a tour of n cities. Arbitrarily pick one of the cities as the starting point. Among all the cities not yet visited, choose as the next city in the tour the one that is closest to the current city (i.e., the last city added to the tour); if all the cities have been visited, return to the starting city. In this case, the local optimality condition is that the city chosen at each step is the unvisited one closest to the last chosen city. This condition is not necessarily a global optimality condition because going slightly out of the way early in the tour may result in considerable savings later on.

The nearest-neighbor algorithm can easily be implemented in time proportional to n^2 (Exercise 9). If N_n is the tour of length $|N_n|$ produced by this algorithm and O_n is an optimal tour of length $|O_n|$, then we have

Theorem 4.1

If C is an $n \times n$ cost matrix, $n \geq 2$, that is symmetric ($C_{ij} = C_{ji}$ for any i and j) and satisfies the triangle inequality ($C_{ij} \leq C_{ik} + C_{kj}$ for any i, j, and k), then

$$\frac{|N_n|}{|O_n|} \leq \frac{1}{2}(\lceil \lg n \rceil + 1).$$

Proof: Let the lengths of the edges of the nearest-neighbor tour N_n be $l_1 \geq l_2 \geq \cdots \geq l_n$, so that $\sum_{i=1}^{n} l_i = |N_n|$. To prove the theorem, we will prove three inequalities:

$$|O_n| \geq 2l_1. \tag{4.3}$$

$$|O_n| \geq 2 \sum_{i=k+1}^{2k} l_i, \qquad 1 \leq k \leq \left\lfloor \frac{n}{2} \right\rfloor. \tag{4.4}$$

$$|O_n| \geq 2 \sum_{i=\lceil n/2 \rceil+1}^{n} l_i. \tag{4.5}$$

When n is even, inequality (4.5) is a special case of inequality (4.4). The theorem follows easily from these inequalities: Taking $k = 1, 2, 2^2, \ldots, 2^{\lceil \lg n \rceil - 2}$ in (4.4) and adding these $\lceil \lg n \rceil - 1$ inequalities to inequalities (4.3) and (4.5) yield

$$(\lceil \lg n \rceil + 1)|O_n| \geqq 2 \sum_{i=1}^{n} l_i = 2|N_n|,$$

from which the theorem follows.

Inequality (4.3) is a trivial consequence of the triangle inequality, the symmetry of the cost matrix C, and the fact that $n \geqq 2$. To prove inequality (4.4), let c_i be the city at which the edge l_i was added in the construction of the nearest-neighbor tour. Given $k, 1 \leqq k \leqq \lfloor n/2 \rfloor$, let A_k be the set of cities $\{c_1, c_2, \ldots, c_{2k}\}$ and T_k be the tour of the cities in A_k taken in the same circular order in which the optimal tour visits them. By the triangle inequality, we have $|O_n| \geqq |T_k|$.

We need only show that the right-hand side of inequality (4.4) is a lower bound on $|T_k|$. Let c_i and c_j be cities in A_k and $c_i\text{-}c_j$ be an edge in T_k. According to the nearest-neighbor construction, if c_i was added to N_n before c_j, then $C_{c_ic_j} \geqq l_i$. On the other hand, if c_j was added to N_n before c_i, then $C_{c_jc_i} = C_{c_ic_j} \geqq l_j$. Since one of c_i and c_j must be added before the other, we have

$$C_{c_ic_j} \geqq \min(l_i, l_j).$$

Summing this inequality over all the edges in T_k, we find

$$|T_k| = \sum_{\substack{c_i\text{-}c_j \\ \text{in } T_k}} C_{c_ic_j} \geqq \sum_{\substack{c_i\text{-}c_j \\ \text{in } T_k}} \min(l_i, l_j). \tag{4.6}$$

The smallest edge in T_k has length at least l_{2k}, the next smallest edge has length at least l_{2k-1}, and so on. Thus the smallest value of $\min(l_i, l_j)$ in equality (4.6) is l_{2k}, the next smallest is l_{2k-1}, and so on. Since an edge $l_i, 1 \leqq i \leqq 2k$, can appear in the sum at most twice, and there are $2k$ edges in T_k, the sum in inequality (4.6) is at least twice the sum of the smallest k edges:

$$|T_k| \geqq \sum_{\substack{c_i\text{-}c_j \\ \text{in } T_k}} \min(l_i, l_j) \geqq 2(l_{2k} + l_{2k-1} + \cdots + l_{k+1}) = 2 \sum_{i=k+1}^{2k} l_i,$$

and therefore the right-hand side of inequality (4.4) is a lower bound on $|T_k|$ as desired.
     The proof of inequality (4.5) is similar and is left as Exercise 11.

This theorem only gives us an *upper bound* on $|N_n|/|O_n|$ and does not tell us how bad the ratio can actually be. It is possible to construct a symmetric $n \times n$ distance matrix that satisfies the triangle inequality for which the nearest-neighbor algorithm produces a tour whose cost is more than $\frac{1}{3} \lg n$ times the cost of an optimal tour.

We can obtain better results by using a more elaborate locally optimal strategy. The *closest insertion* algorithm maintains a tour of a subset of the n cities; at each stage a new city is added to the tour, and the process continues until all n cities are included in the tour. The algorithm begins with a tour consisting of a single, arbitrarily chosen city. The new city added at each stage is chosen, among the cities not yet in the tour, to be the city that is closest to a city already on the tour; the new city is then inserted into the tour next to the city on the tour to which it is closest.

Although more complicated than the nearest-neighbor algorithm, the closest insertion algorithm can also be implemented in time proportional to n^2 by maintaining, for each city not yet on the tour, a record of the city on the tour to which it is closest. Determining the city to be added to the tour, deciding where it is to be inserted, and updating the records of the closest cities can all be done in time proportional to n. Since n cities are inserted, the total number of operations is proportional to n^2.

In the closest insertion algorithm, the local optimality condition is more sophisticated than in the nearest-neighbor algorithm, and so we expect it to produce a tour whose cost is guaranteed to be closer to the cost of the optimal tour than the upper bound of Theorem 4.1 is. This is indeed the case. As before, let O_n be an optimal tour of length $|O_n|$. Let I_n be the tour produced by the closest insertion algorithm and $|I_n|$ be its length. We have

Theorem 4.2

If C is an $n \times n$ cost matrix, $n \geqq 2$, that is symmetric and satisfies the triangle inequality, then

$$\frac{|I_n|}{|O_n|} < 2.$$

Proof: Let $c_1, c_2, \ldots, c_n$ be the cities in the order in which they are added to the tour by the closest insertion algorithm. We will prove the theorem by constructing a one-to-one correspondence between cities $c_2, \ldots, c_n$ and all but the largest edge of O_n in such a way that the cost of inserting city c_k between cities c_i and c_j, that is, $C_{c_i c_k} + C_{c_k c_j} - C_{c_i c_j}$,† is at most twice the cost of the edge in O_n that corresponds to c_k. Thus we will actually prove that $|I_n| \leqq 2(|O_n| - |l_{\max}|)$, where $l_{\max}$ is the largest edge in O_n and $|l_{\max}|$ is its length.

To establish the correspondence, imagine the closest insertion algorithm keeping track not only of the current tour of $c_1, c_2, \ldots, c_t, 1 \leqq t < n$, but of a spiderlike configuration that also includes a subset of the edges of $O_n - \{l_{\max}\}$ connecting the cities $c_{t+1}, c_{t+2}, \ldots, c_n$ to the current tour. Thus the algorithm maintains configurations as shown in Figure 4.14. The initial configuration appears in Figure 4.14(a) and a typical configuration of later stages in Figure 4.14(b); in each case, the current closest insertion tour is shown in boldface.

As each city c_i is inserted into the current tour, the algorithm will delete from the configuration one of the edges of $O_n - \{l_{\max}\}$. The edge deleted is chosen so that at each stage of the algorithm the spiderlike configuration has a "body" (the current closest insertion tour, shown in boldface in Figure 4.14) and "legs" (the subset of the edges of $O_n - \{l_{\max}\}$, shown in lightface in Figure 4.14) such that only one end of each leg is connected to the body. Furthermore, we will show that the cost incurred by inserting c_i is at most twice the cost of the length of the edge deleted. In this way, the correspondence between the cities $c_2, \ldots, c_n$ and the edges of $O_n - \{l_{\max}\}$ will be established and the theorem proved.

---

†In order for it to be correct initially, we must define $C_{ii} = 0$, in contrast to the definition of $C_{ii} = \infty$ used in Section 4.1.6.

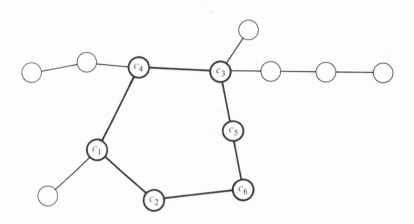

(a) Initially c_1 is the only city in the closest insertion tour; the edges shown are those of $O_n - \{l_{max}\}$.

(b) Later in the algorithm the configuration consists of the closest insertion tour (in boldface) and certain edges from $O_n - \{l_{max}\}$ (in lightface).

Figure 4.14 Examples of the spiderlike configuration in the closest insertion algorithm.

Suppose that city c_k is to be inserted in the closest insertion tour, c_k being closest to city c_m on the tour, as illustrated in Figure 4.15. The cost of inserting c_k between c_m and an adjacent city, say c_l, is $C_{c_l c_k} + C_{c_k c_m} - C_{c_l c_m}$. Let c_x be the point on the tour to which the leg containing c_k is attached and let c_x–c_y be the first edge of that leg (it is possible that $c_y = c_k$). Since the algorithm chooses the city closest to the tour to insert, we have

$$C_{c_k c_m} \leqq C_{c_x c_y}, \tag{4.7}$$

and by the triangle inequality

$$C_{c_l c_k} \leqq C_{c_l c_m} + C_{c_m c_k}.$$

Using symmetry, we can combine these to obtain

$$C_{c_l c_k} \leqq C_{c_l c_m} + C_{c_x c_y}. \tag{4.8}$$

Adding (4.7) and (4.8) gives

$$C_{c_l c_k} + C_{c_k c_m} \leqq C_{c_l c_m} + 2C_{c_x c_y},$$

which is equivalent to

$$C_{c_l c_k} + C_{c_k c_m} - C_{c_l c_m} \leqq 2C_{c_x c_y}.$$

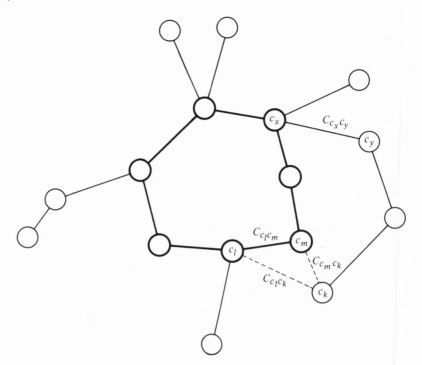

Figure 4.15 The spiderlike configuration as c_k is about to be added to the closest insertion tour (shown in boldface). The cost of inserting c_k between c_l and c_m is $C_{c_l c_k} + C_{c_k c_m} - C_{c_l c_m}$ which is at most $2C_{c_x c_y}$. After c_k is inserted, the edge c_x–c_y is deleted.

Consequently, inserting c_k between c_l and c_m costs at most $2C_{c_x c_y}$. After c_k is inserted, the edge c_x–c_y is deleted from the configuration (which thus remains spiderlike) and the algorithm continues.

Thus each city $c_k, k = 2, \ldots, n$, corresponds to a unique edge c_x–c_y of $O_n - \{l_{\max}\}$ in such a way that the insertion of c_k costs at most twice the length of the corresponding edge in $O_n - \{l_{\max}\}$. Since the cost of I_n is the sum of the costs of the insertions, we have

$$|I_n| \leq 2(|O_n| - |l_{\max}|)$$

and the theorem follows.

So we know that the closest insertion algorithm always produces tours that cost at most twice the cost of an optimal tour. Moreover, we can show that there is an n-city problem, $n \geq 6$, for which the closest insertion algorithm produces a tour whose length is almost twice that of an optimal tour. Consider the cost matrix C defined by

$$C_{ij} = C_{ji} = \min (j - i, n - j + i), \qquad i \leq j.$$

This matrix corresponds to having $C_{ij} = C_{ji} = 1$ for $j = i + 1 \pmod n$ and making $C_{ij} = C_{ji}$ the length of the shortest path from i to j following arcs of the

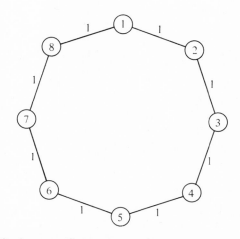

Figure 4.16 A traveling salesman problem for which the closest insertion algorithm produces a tour almost twice the length of an optimal tour. The distance from one city to another is the length of the shortest path on the circle between the two cities. Thus the distance from city 3 to city 8 is $C_{3,8} = C_{8,3} = 3$.

type k–$(k \pm 1)(\bmod n)$. This configuration is illustrated in Figure 4.16 for $n = 8$. The cities are arranged in a circle, each city at distance 1 from its successor and predecessor. The distance between two nonadjacent cities is the length of the shortest path on the circle between the two cities.

For this n-city problem, let the tour T_n consist of edges 1–2, $(n - 1)$–n, and i–$(i + 2)$ for $1 \leq i \leq n - 2$; T_8 is shown in Figure 4.17. T_n contains two edges of length 1 and $n - 2$ edges of length 2, for a total length of

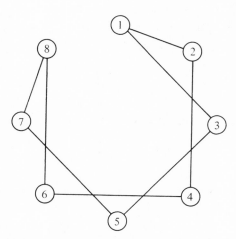

Figure 4.17 T_8, a possible outcome of the closest insertion algorithm when applied to the 8-city problem of Figure 4.16.

$|T_n| = 2n - 2$. Obviously, the optimal tour O_n consisting of edges $1-n$ and $i-(i + 1), 1 \leqq i \leqq n - 1$, has length n, and so

$$\frac{|T_n|}{|O_n|} = \frac{2n - 2}{n} = 2\left(1 - \frac{1}{n}\right).$$

Assuming that T_n can be the result of the closest insertion algorithm, this ratio is the largest permitted by Theorem 4.2, since $|l_{\max}| = 1$.

To prove that T_n can be the result of the closest insertion algorithm, observe that the algorithm can start with city 1, then add city 2, then add city 3, and so on, because each city added is distance 1 (the smallest possible) from a city on the tour. Let $T_k, 3 \leqq k \leqq n$, be a tour resulting from the closest insertion algorithm after the insertion of city k. We will be finished if we can prove by induction that T_k can consist of edges $1-2, (k - 1)-k$, and $i-(i + 2)$ for $1 \leqq i \leqq k - 2$. This is obviously true for $k = 3$, since T_3 is the unique path $1-2-3-1$. Assuming that it is true for $3 \leqq i \leqq k$, observe that T_{k+1} is obtained from T_k by inserting city $k + 1$ between cities $k - 1$ and k.

4.4 REMARKS AND REFERENCES

The name "backtrack" was coined by D. H. Lehmer in the 1950s, but backtrack techniques have been discovered and rediscovered innumerable times. One early description of such a method is given in

LUCAS, E., *Récréations Mathématiques*, Gauthier-Villars, Paris, 1891,

where it is used to solve mazes. The first description of the generalized algorithm is in

WALKER, R. J., "An Enumerative Technique for a Class of Combinatorial Problems," in *Combinatorial Analysis* (*Proceedings of Symposia in Applied Mathematics*, Vol. X), American Mathematical Society, Providence, R.I., 1960.

The recursive form of backtrack was used by Robert Tarjan in various graph algorithms. His paper

TARJAN, R. E., "Depth-First Search and Linear Graph Algorithms," *SIAM J. Comput.*, **1** (1972), 146–160

is discussed in detail in Section 8.2.2.

The Monte Carlo estimation technique given in Section 4.1.3 was proposed in

HALL, M., and D. E. KNUTH, "Combinatorial Analysis and Computers," *Amer. Math. Monthly*, **72**, Part II (February 1965), 21–28.

It was analyzed in depth in

KNUTH, D. E., "Estimating the Efficiency of Backtrack Programs," *Math. Comp.*, **29** (1975), 121–136.

The use of macros to customize backtrack programs was first applied to finding tilings with pentominoes [see Exercise 1(h)] in

FLETCHER, J. G., "A Program to Solve the Pentomino Problem by the Recursive Use of Macros," *Comm. ACM*, **8** (1965), 621–623.

Since then it has been used successfully to solve many combinatorial problems; see

BITNER, J. R., and E. M. REINGOLD, "Backtrack Programming Techniques," *Comm. ACM*, **18** (1975), 651–656.

Section 4.1.5, the example of backtrack applied to finding optimal distance-preserving codes, is based on the unpublished work of James R. Bitner. The question of the length of the optimal n-bit DP-t code is still open, even for $t = 1$. However, in

EVDOKIMOV, A. A., "Maximal Length of Circuit in a Unitary n-Dimensional Cube," *Mat. Zametki*, **6** (1969), 309–329 (Russian); English translation in *Math. Notes*, **6** (1969), 642–648

it is proven that the length of the optimal n-bit DP-1 code is $c2^n$ for some constant c.

Branch-and-bound techniques have been successfully used in optimization problems since the late 1950s. The name comes, presumably, from splitting the solution space into subsets (branching) and bounding the cost function over the subsets of the solution space. Many of the diverse applications are outlined in

LAWLER, E. L., and D. E. WOOD, "Branch-and-Bound Methods: A Survey," *Operations Research*, **14** (1966), 699–719

and in

LAWLER, E. L., *Combinatorial Optimization: Networks and Matroids*, Holt, Rinehart & Winston, New York, 1976.

The extremely clever approach to the traveling salesman problem (illustrated in Figure 4.7) is due to

LITTLE, J. D. C., K. G. MURTY, D. W. SWEENEY, and C. KAREL, "An Algorithm for the Traveling Salesman Problem," *Operations Research*, **11** (1963), 972–989.

That method and other approaches are nicely discussed in

GOMORY, R. E., "The Traveling Salesman Problem," in *Proceedings of the IBM Scientific Computing Symposium on Combinatorial Problems*, IBM Data Processing Division, White Plains, N.Y., 1966.

Additional material about game trees and their evaluation can be found in

NILSSON, N. J., *Problem-Solving Methods in Artificial Intelligence*, McGraw-Hill, New York, 1971.

A much deeper analysis of the α-β pruning algorithm and its extension is given in

KNUTH, D. E., and R. W. MOORE, "An Analysis of Alpha-Beta Pruning," *Artificial Intelligence*, **6** (1975), 293–326.

Dynamic programming is due to Richard Bellman, who has written extensively about it. A general discussion, together with a large number of exercises, can be found in

BELLMAN, R., *Dynamic Programming*, Princeton University Press, Princeton, N.J., 1957.

Detailed descriptions of its application to combinatorial problems are given in

BELLMAN, R., "Combinatorial Processes and Dynamic Programming," in *Combinatorial Analysis* (*Proceedings of Symposia in Applied Mathematics*, Vol. X), American Mathematical Society, Providence, R.I., 1960.

The application to the traveling salesman problem is due, independently, to

BELLMAN, R., "Dynamic Programming Treatment of the Travelling Salesman Problem," *J. ACM*, **9** (1962), 61–63.

and to

HELD, M., and R. M. KARP, "A Dynamic Programming Approach to Sequencing Problems," *J. Soc. Indust. Appl. Math.*, **10** (1962), 196–210.

Sieves, of course, go back to Eratosthenes of Cyrene, who lived around 200 B.C. More recently, the generalized modular sieve was presented in

LEHMER, D. H., "The Sieve Problem for All-Purpose Computers," *Math. Tables Aids Comput.*, **7** (1953), 6–14.

The technique used to combine pairs of progressions with relatively prime moduli is based on the famous "Chinese remainder theorem." For an interesting exposition of this and related topics, see Chapter 10 of

ORE, O., *Number Theory and Its History*, McGraw-Hill, New York, 1948.

The sieve to determine perfect-square Fibonacci numbers is presented in

WUNDERLICH, M. C., "On the Existence of Fibonacci Squares," *Math. Comp.*, **17** (1963), 455–457.

The proof that 1 and 144 are the only such squares can be found in

COHN, J. H. E., "Square Fibonacci Numbers," *Fibonacci Quart.*, **2** (1964), 109–113.

The lucky numbers and related sequences are described in

HAWKINS, D., "Mathematical Sieves," *Sci. Amer.*, **199**, No. 6 (December 1958), 105–112.

The use of tagging in the sieves to compute these sequences is discussed in the survey article

WUNDERLICH, M. C., "Sieving Procedures on a Digital Computer," *J. ACM*, **14** (1967), 10–19

along with a number of other applications of sieves.

The sequence *U*, discussed in Section 4.2.2, is due to S. Ulam, who suggested it during a number theory conference in 1963. The double-sieve computation of that and similar sequences is described in

WUNDERLICH, M. C., "The Use of Bit and Byte Manipulation in Computing Summation Sequences," *BIT*, **11** (1971), 217–224.

Some computational results on the asymptotic behavior of the sequence U are reported in

WUNDERLICH, M. C., "The Improbable Behavior of Ulam's Summation Sequence," in *Computers in Number Theory*, A. O. L. Atkin and B. J. Birch (Eds.), Academic Press, New York, 1971.

Isomorph-rejection sieves are discussed in

SWIFT, J. D., "Isomorph Rejection in Exhaustive Search Techniques," in *Combinatorial Analysis* (*Proceedings of Symposia in Applied Mathematics*, Vol. X), American Mathematical Society, Providence, R.I., 1960.

An important approach to difficult computational problems like the traveling salesman one is to look for good heuristic algorithms. The heuristic techniques and their analyses given in Section 4.3 are from

ROSENKRANTZ, D. J., R. E. STEARNS, and P. M. LEWIS, "Approximation Algorithms for the Traveling Salesperson Problem," *SIAM J. Comput.*, to appear,

which gives several other heuristic approaches and their analyses. The heuristic given in

CHRISTOFIDES, N., "Worst-case Analysis of a New Heuristic for the Travelling Salesman Problem," *Technical Report of the Graduate School of Industrial Administration*, Carnegie–Mellon University, Pittsburgh, Pa., 1976

produces tours whose cost is at most 50% above optimal; this heuristic requires time proportional to n^3 for an n-city problem.

Results on heuristic algorithms for some other combinatorial problems can be found in

GRAHAM, R. L., Bounds on the Performance of Scheduling Algorithms, Chapter 5 of *Computer and Job-Shop Scheduling Theory*, E. G. Coffman, Jr. (Ed.), Wiley, New York, 1976, 165–227.

4.5 EXERCISES

1. Use backtrack techniques to solve the following problems. After running as many cases of a problem as possible, use the Monte Carlo estimation techniques to see how much time would be needed for cases not solved.
    (a) *Mazes.* Find a path between two specified points. Find the shortest such path. Find all such paths.
    (b) *Postage stamp problem.* The post office in some distant country issues n denominations of stamps and prohibits the use of more than m stamps on one letter. For various values of n and m, compute the greatest consecutive range of postages from one upward and all possible sets of denominations that realize that range. For example, for four denominations of stamps and at most five stamps per letter, all postages $1, 2, 3, \ldots, 71$ can be formed with the set $\{1, 4, 12, 21\}$ or with the set $\{1, 5, 12, 28\}$.

(c) *Tiling with incomparable rectangles.*   Two rectangles are called incomparable if neither can be placed inside the other when they are lined up so that corresponding sides are parallel. Find tilings of various rectangular regions with sets of pairwise incomparable rectangles. For example, the smallest possible solution is a 13 × 22 rectangle tiled as follows.

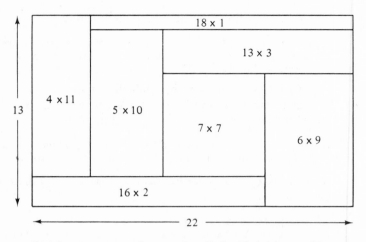

The smallest known square that can be tiled with incomparable rectangles is 27 × 27. Are there smaller ones?

(d) *Hi-Q.*   Thirty-two movable pieces are arranged on a board, as shown, with a blank space in the center. A piece moves on this board by jumping over one of its immediate neighbors (horizontally or vertically) into an empty space opposite. Whenever a piece is so "jumped" by another, it is removed from the board.

```
            x   x   x

            x   x   x

    x   x   x   x   x   x   x

    x   x   x   o   x   x   x

    x   x   x   x   x   x   x

            x   x   x

            x   x   x
```

Find a series of jumps so as to remove all pieces but one from the board and leave the remaining piece in the center space.

(e) *Symmetric n × n nonattacking queens problem.*   Find all solutions to n × n nonattacking queens problem that are invariant under rotations of 90°, 180°, and 270°. Each queen placed on the board at other than the center thus determines the placement of three other queens.

(f) *Minimal nonattacking domination of an* $n \times n$ *chessboard.* Determine the minimal number of nonattacking queens that are needed and all possible placements, so that every unoccupied square of the chessboard is under attack.

(g) *Knight's tour on an* $n \times n$ *chessboard.* Is there a way to move the knight around the chessboard so that it visits every square exactly once, returning to the starting square?

(h) *Pentominoes.* There are 12 pentominoes (rigid, connected, plane figures, each composed of five equal-sized squares joined along their edges) that can be put together as a jigsaw puzzle in many ways to form a 6×10 rectangle. For example, one solution is

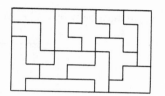

Find all such solutions for 6×10 and 5×12 rectangles.

(i) *Soma cube.* The following seven pieces can be put together in many ways to form a $3 \times 3 \times 3$ cube:

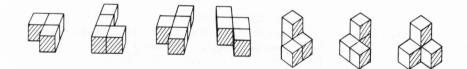

Find the number of inequivalent (under rotation and/or reflection) solutions.

(j) *Mikusiński cube.* The following pieces can be put together in only a few ways to form a $3 \times 3 \times 3$ cube:

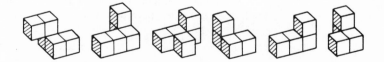

Find all inequivalent solutions.

(k) *Y-pentomino.* Find the smallest n such that a $12 \times 5n$ region can be tiled with copies of the Y-pentomino ⌐⊔⊔ . Find all values of n for which it is possible.

(l) The identity

$$1^2 + 2^2 + 3^2 + \cdots + 24^2 = 4900 = 70^2$$

suggests that it may be possible to tile a 70×70 square by using 24 tiles of sizes $1 \times 1, 2 \times 2, 3 \times 3, \ldots, 24 \times 24$, respectively. Is there a solution?

(m) Given an $n \times n \times n$ cube consisting of n^3 cells, fill the cells in the cube with white or black balls so as to minimize the number of instances of n balls of the same color in a line (diagonal lines included). Compute both the total number of solutions and the total number of solutions that are inequivalent under rotations and reflections.

(n) *Magic squares.*   A magic square of order n is an $n \times n$ array of the integers $1, 2, \ldots, n^2$ arranged so that the sum of every row, column, and the two main diagonals is the same. Since

$$\sum_{i=1}^{n^2} i = \tfrac{1}{2}n^2(n^2 + 1),$$

the sum must be $\tfrac{1}{2}n(n^2 + 1)$. Thus, for example,

1	15	24	8	17
23	7	16	5	14
20	4	13	22	6
12	21	10	19	3
9	18	2	11	25

is a magic square of order 5, each row, column, and main diagonal summing to $\tfrac{1}{2}5(5^2 + 1) = 65$. Determine the number of equivalent (under rotations and/or reflections) magic squares of order n.

(o) *Pandiagonal magic squares.*   The magic square shown in part (n) has the additional property that the "broken diagonals" also sum to 65. When a pandiagonal magic square of order n is used to tile the plane, any $n \times n$ square is a pandiagonal magic square. Determine the number of inequivalent pandiagonal magic squares of order n.

(p) *Latin squares.*   A Latin square of order n is an $n \times n$ array in which each position contains an integer between 1 and n and, furthermore, each row and column of the array is a permutation of $\{1, 2, \ldots, n\}$. Thus

1	2	3	4
3	4	1	2
4	3	2	1
2	1	4	3

is a Latin square of order 4. A latin square is said to be *reduced* if in the first row and column its elements occur in the natural order $1, 2, \ldots, n$. Determine the number of reduced Latin squares of order n.

(q) *Orthogonal Latin squares.* Two Latin squares of order n are said to be *orthogonal* if, when one is superimposed on the other, each of the n^2 possible pairs of elements occurs exactly once. For example, both

1	2	3	4
2	1	4	3
3	4	1	2
4	3	2	1

1	2	3	4
4	3	2	1
2	1	4	3
3	4	1	2

are orthogonal to the example in part (p), and, in fact, they are also orthogonal to each other. Find all such sets of mutually orthogonal Latin squares of order n.

2. Implement the more sophisticated branch-and-bound technique for solving the traveling salesman problem illustrated in Figures 4.6 and 4.7.

3. Use branch-and-bound and/or dynamic programming techniques to solve the problems below.

   (a) *Assignment problem.* There are n men to be assigned to n jobs. The cost of assigning the ith man to the jth job is C_{ij}. The problem is to find the assignment that assigns every job to a man and minimizes the total cost of the assignment.

   (b) *Circuit boards.* There are components to be placed in n locations on a circuit board. The number of connections between pairs of components is given by the matrix C where C_{ij} is number of connections between the ith component and the jth component. The distance between pairs of locations is given by the matrix D, where D_{kl} is the distance between location k and location l. Thus the cost, in terms of the total length of wire used, of putting component i at location k and component j at location l is $C_{ij}D_{kl}$. Each location has room for only one component and each component can be in only one location. Find the assignment of components to locations that minimizes the total length of wire used.

   (c) *Job scheduling.* Jobs $J_1, J_2, \ldots, J_n$ are to be executed successively by a single processor. The jobs require $\tau_1, \tau_2, \ldots, \tau_n$ units of time respectively, and with each J_k there is a monotone-increasing cost function $c_k(t)$ that gives the cost associated with completing J_k at time t. Find the schedule $J_{i_1}, J_{i_2}, \ldots, J_{i_n}$ that minimizes

$$\sum_{k=1}^{n} c_{i_k}\left(\sum_{j=1}^{k} \tau_{i_j}\right).$$

   (d) *Scheduling parallel processors.* There are n jobs to be done and k processors working in parallel to do them; the time required by the ith job is t_i. Furthermore, some jobs cannot be done until others have finished: there is a strictly upper triangular matrix $P = (p_{ij})$ such that $p_{ij} = -1$ if job i must be done before job j, $p_{ij} = +1$ if job j must be done before job i, and $p_{ij} = 0$ if the jobs are unrelated. Find the schedule of jobs to processors that minimizes the time from start to finish.

(e) *Packing problem.*   A box is to be packed with objects. The box can hold at most *W* pounds. There are at most m_i of the ith object, each of which has value v_i and weight w_i pounds. Find the collection of objects that maximizes the total value of the contents of the box.

4. Use α-β pruning to show that conventional tic-tac-toe always ends in a draw when played by minimax strategy. How about three-dimensional tic-tac-toe played on three 3×3 boards? How about on four 4×4 boards?

5. Consider the multiplication of *n* matrices

$$M = M_1 \times M_2 \times \cdots \times M_n,$$

where M_i has dimensions $r_{i-1} \times r_i$. Use a dynamic programming approach to obtain an $O(n^3)$ algorithm that determines the most efficient order in which to form subproducts when computing *M*.

6. Suppose we are given the coordinates of the *n* vertices of a convex *n*-sided polygon, and we are to triangulate the polygon by drawing $n-3$ nonintersecting diagonals. The cost of a triangulation is the sum of the lengths of the $n-3$ diagonals.
    (a) How many distinct triangulations are there? [Hint: See equation (3.26).]
    (b) Show how the dynamic programming approach to exhaustive search allows the triangulation of least cost to be found in $O(n^3)$ time.

7. Consider an $n \times n$ array of positive integers (a_{ij}), $0 \leq i, j < n$, rolled into a cylinder, as illustrated below.

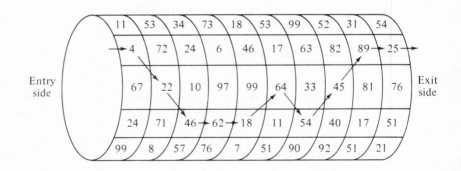

A path is to be threaded from the entry side of the cylinder to the exit side, subject to the restriction that from a given square it is possible to go only to one of the three positions in the next column adjacent to the current position. The path may begin at any position on the entry side and may end at any position on the exit side. The cost of such a path is the sum of the integers in the squares through which it passes. Thus the cost of the sample path shown above is 429.
    (a) How many distinct such paths are there?
    (b) Show now the dynamic programming approach to exhaustive search allows the path of least cost to be found in $O(n^2)$ time.

8. Use sieve techniques to compute the following sequences.

   (a) *Lucky numbers.* From the list of natural numbers 1, 2, 3, 4, 5, . . . remove every second number, leaving 1, 3, 5, 7, 9, ... Since 3 is the first number (excepting 1) that has not been used as the sifting number, remove every third number from those remaining, obtaining 1, 3, 7, 9, 13, 15, 19, 21, ... Now every seventh number is removed, and so on.

   (b) *Random sequences.* The primes and lucky numbers have the same asymptotic distribution among the integers, since they are special cases of the following: starting with the integers 1, 2, 3, 4, 5, . . . , we sift with 2 by removing or not removing each subsequent remaining number with probability 1/2. At each stage, if p is the first uncanceled number not yet used as a sifting number, each subsequent remaining number is canceled with probability $1/p$. Using a random number generator, compute some such sequences and study their asymptotic behavior.

   (c) *Summation sequences.* Modify the sieve for the sequence U (Section 4.2.2) so that after membership for integers less than n is decided, n is in the sequence if and only if it is representable as the sum of any distinct pair of elements in the sequence in (i) no way, (ii) at most two ways, (iii) exactly two ways, and so on.

   (d) *Pseudo-squares.* Let p be a prime and N_p be the least positive nonsquare integer of the form $8i + 1$ that is a nonzero quadratic residue for all primes less than or equal to p. The numbers N_p are called *pseudo-squares* because they behave like squares in some respects.

   (e) *Sprague numbers.* It has been proven that for every integer $n \geqq 2$ there exists a largest positive integer s_n that is *not* expressible as a sum of distinct nth powers of positive integers. Determine the values of s_i for small i ($s_2, s_3,$ and s_4 are known), by using the following result.

   Theorem. Let $m_1, m_2, \ldots$ be an infinite increasing sequence of positive integers such that for some positive integer k the inequality $m_{i+1} \leqq 2m_i$ holds for all $i > k$. If there exists a nonnegative integer a such that $a + 1, a + 2, \ldots,$ $a + m_{k+1}$ are all expressible as sums of distinct members of the set $\{m_1, m_2, \ldots, m_k\}$, then every integer greater than a is expressible as a sum of distinct members of the sequence $m_1, m_2, \ldots$.

   (f) *Sums of primes.* Let $P_0 = \{p_1, p_2, \ldots\} = \{2, 3, \ldots\}$ be the sequence of primes and let P_n be the subsequence of P_0 defined recursively by $P_n = \{p_i | i \in P_{n-1}\}$, $n \geqq 1$. Let t_n be the largest positive integer that is *not* representable as the sum of distinct elements of P_n. It is known that t_0 and t_1 exist, but not if t_i exists for $i \geqq 2$. Using the theorem stated in part (e), compute t_0 and t_1 and as many other t_i as possible. Note that, to use the theorem, you must show that P_n satisfies its hypothesis. Use the results that

   $$p_{i+1} < \tfrac{9}{8}p_i, \quad i > 15,$$

   and

   $$i \ln i < p_i < i (\ln i + \ln \ln i), \quad i > 6.$$

9. Describe how to implement the nearest-neighbor algorithm for the n-city traveling salesman problem in time proportional to n^2.

10. Find examples for which the nearest-neighbor algorithm produces optimal tours. Find examples for which it produces tours that are very far from optimal.

11. Prove inequality (4.5) for n odd.

12. One criticism of the nearest-neighbor algorithm is that the last edge added (from the last city back to the starting city) can be very long. How much effect can the last edge have on the ratio $|N_n|/|O_n|$?

13. Given an arbitrary symmetric cost matrix C, it is possible to force the triangle inequality to be true by adding the value max $\{C_{ij}\}$ to each element of C. Does this mean that Theorems 4.1 and 4.2 hold for symmetric cost matrices C for which the triangle inequality does *not* hold?

14. The *cheapest insertion* algorithm for the traveling salesman problem selects the city to be inserted as the one that can be inserted at the least cost; that is, city k (not on the current tour) is inserted between adjacent cities i and j (on the current tour) with i, j, and k chosen to minimize $C_{ik} + C_{kj} - C_{ij}$. Establish a result like that of Theorem 4.2 for the cheapest insertion algorithm. How many operations, as a function of n, are needed to implement the cheapest insertion algorithm for an n-city problem? How far from optimal can the cheapest insertion tour really be?

chapter **5**

Generating Elementary Combinatorial Objects

In combinatorial algorithms it is frequently necessary to generate and examine all members of a certain class of combinatorial objects. The most general techniques for such problems, based on backtrack, are discussed in Chapter 4, but in many instances the objects are simple enough that specialized methods are more practical. In this section we discuss the generation of some of the combinatorial objects that commonly occur in practice. In each case, there are two possible goals—the systematic generation of all possible configurations and the generation of uniformly distributed random configurations.

Problems requiring systematic generation arise frequently in the evaluation of combinatorial formulas. For example, it is often necessary to evaluate sums that look like

$$\sum_{\substack{\text{all sequences} \\ x_1, x_2, \ldots, x_k \text{ that} \\ \text{satisfy some constraints}}} f(x_1, x_2, \ldots, x_k).$$

Evaluating such a sum involves, in general, generating each of the possible configurations of the x_i's in some systematic order. Of course, the general backtrack technique of Chapter 4 can be used, but in many cases the configurations are sufficiently well understood that their generation can be considerably simplified. This is true of combinatorial objects discussed in this chapter: permutations, subsets, combinations, and compositions and partitions of integers.

There are three components to an algorithm for systematic generation: the initialization, the transformation from one object to the next, and the end condition telling when to stop. Throughout this chapter the general form of such an algorithm will be

initialize

while not end condition do $\begin{cases} \text{output object} \\ \text{transform to next object} \end{cases}$

Notice that within such a framework the initialization must take into account the fact that the algorithm attempts one more transformation *after* the last object is listed. Since this extraneous transformation must not cause a program error, some of the initializations may have to assign values to seemingly extraneous variables—for example, the 0th or $(n + 1)$st element of an array that would ordinarily have only elements 1 through n. A reference to, or a change in the value of, this extraneous variable can often be used as the end condition.

A procedure that produces successive objects on successive calls can be written in a straightforward manner from the given initialization, transformation, and end condition:

procedure *GENERATE* (object, *flag*)

if *flag* then $\begin{cases} \text{initialize} \\ flag \leftarrow \textbf{false} \end{cases}$

else transform to next object

if end condition then *flag* ← **true**
return

This procedure uses *flag* as a signal to restart the generation (a call with *flag* = **true**) and as a signal that the last object has been generated (a return with *flag* = **true**).

In an algorithm for generating all elements of a set, we are primarily interested in the total amount of time required for the generation of the entire set. In particular, in some of the algorithms, it is possible to generate the entire set in time proportional to its size. Such algorithms, which are called *linear*, are desirable because their efficiency is asymptotically the best possible. Also of interest is the amount of change that occurs between successive objects. It is sometimes desirable to have as little change as possible; for instance, in Section 5.1 we will examine an algorithm for generating permutations in which successive permutations differ only by a transposition of adjacent elements. Such algorithms are called *minimal change algorithms*. The precise definition of the minimal change depends on the combinatorial configuration being considered.

Problems requiring random generation of elements in a class of combinatorial objects occur in performing Monte Carlo calculations to estimate performance (see Section 4.1.3). For example, if we are unable to analyze the average behavior of an algorithm satisfactorily, we may be forced to try it on a large number of randomly chosen input configurations in order to observe its performance. Random generation is sometimes difficult in that we require each of the possible configurations to be equally probable.

One way to approach both systematic and random generation is to determine an explicit correspondence between the integers $0, 1, 2, \ldots, N - 1$ and the N objects. Systematic generation then amounts to counting from 0 to $N - 1$ and converting each number into an object, whereas random generation amounts to generating a random integer between 0 and $N - 1$ and converting it into an object. The conversion process, however, can be expensive and it is usually best to avoid it when possible.

5.1   PERMUTATIONS OF DISTINCT ELEMENTS

The $n!$ permutations of a set $\{a_1, a_2, \ldots, a_n\}$ of (distinct) elements are among the most commonly generated combinatorial objects. Without loss of generality, we assume the elements of the set to be the integers 1 to n; that is, we only consider permutations of the set $\{1, 2, \ldots, n\}$. This simplification will also be possible for other combinatorial objects, such as subsets and combinations.

5.1.1   *Lexicographic Order*

The sequence of permutations of $\{1, 2, \ldots, n\}$ in lexicographic order is the sequence of permutations in increasing numerical order. For example, the lexicographic sequence of permutations on three elements is 123, 132, 213, 231, 312, 321. In general, then, if $\sigma = (\sigma_1, \sigma_2, \ldots, \sigma_n)$ and $\tau = (\tau_1, \tau_2, \ldots, \tau_n)$ are permutations, we say that σ is *lexicographically less than* τ if and only if, for some $k \geq 1$, we have $\sigma_j = \tau_j$ for all $j < k$ and $\sigma_k < \tau_k$.

It is easy to specify explicitly the correspondence between the integers $0, 1, 2, \ldots, n! - 1$ and the $n!$ permutations of $\{1, 2, \ldots, n\}$ written in lexicographic order: the index of the permutation $\Pi = (\pi_1, \pi_2, \ldots, \pi_n)$ is defined by

$$\text{index}_1[(1)] = 0$$

$$\text{index}_n(\Pi) = (\pi_1 - 1)(n - 1)! + \text{index}_{n-1}(\Pi'),$$

where $\Pi' = (\pi_1', \pi_2', \ldots, \pi_{n-1}')$ is the permutation obtained from Π by deleting π_1 and reducing by one all elements greater than π_1.

When the recursive definition of index is iterated to produce $\Pi', \Pi'' = (\Pi')'$, and so on, the sequence of values $\pi_1 - 1, \pi_1' - 1, \pi_1'' - 1, \ldots$ consists of the digits of the factorial system representation of $\text{index}_n(\Pi)$. So given a value i, we can compute the ith permutation in the lexicographic list by converting i into the factorial representation (see Section 2.1)

$$i = d_1 1! + d_2 2! + \cdots + d_{n-2}(n-2)! + d_{n-1}(n-1)!$$

and proceeding recursively: form Π', the permutation of $\{1, \ldots, n-1\}$ whose index is $d_1 1! + \cdots + d_{n-2}(n-2)!$, and increment by one all elements greater than d_{n-1}; $d_{n-1} + 1$ is then added as the first element of the permutation. As above, the permutation of one element with index 0 is (1).

The obvious application of the backtrack algorithm (Section 4.1.1) to permutations generates them in lexicographic order, as observed in Section 4.2.3. The permutations may be generated one to the next as follows. Beginning with the permutation $(1, 2, \ldots, n)$, we go from $\Pi = (\pi_1, \pi_2, \ldots, \pi_n)$ to its successor by scanning Π from right to left, looking for the rightmost place where $\pi_i < \pi_{i+1}$. Having found it, we now look for π_j, the smallest element to the right of π_i and greater than it; π_j is interchanged with π_i, and the elements $\pi_{i+1}, \ldots, \pi_n$ (which are in decreasing order) are reversed. For example, for $n = 8$ and $\Pi = (2, 6, 5, 8, 7, 4, 3, 1)$, we have $\pi_i = 5$ and $\pi_j = 7$. Interchanging them and reversing $\pi_{i+1}, \ldots, \pi_n$ give $(2, 6, 7, 1, 3, 4, 5, 8)$, the permutation that follows Π in the lexicographic order. Algorithm 5.1 is an implementation of this procedure.

for $j = 0$ **to** n **do** $\pi_j \leftarrow j$

$i \leftarrow 1$

while $i \neq 0$ **do** $\left\{ \begin{array}{l} \text{output } \Pi = (\pi_1, \pi_2, \ldots, \pi_n) \\[4pt] [\![\text{find rightmost place where } \pi_i < \pi_{i+1}]\!] \\[4pt] i \leftarrow n - 1 \\[4pt] \textbf{while } \pi_i > \pi_{i+1} \textbf{ do } i \leftarrow i - 1 \\[4pt] [\![\text{find } \pi_j, \text{ the smallest element to the right} \\ \text{of } \pi_i \text{ and greater than it}]\!] \\[4pt] j \leftarrow n \\[4pt] \textbf{while } \pi_i > \pi_j \textbf{ do } j \leftarrow j - 1 \\[4pt] [\![\text{interchange and then reverse } \pi_{i+1}, \ldots, \pi_n]\!] \\[4pt] \pi_i \leftrightarrow \pi_j \\[4pt] r \leftarrow n \\[4pt] s \leftarrow i + 1 \\[4pt] \textbf{while } r > s \textbf{ do } \left\{ \begin{array}{l} \pi_r \leftrightarrow \pi_s \\ r \leftarrow r - 1 \\ s \leftarrow s + 1 \end{array} \right. \end{array} \right.$

Algorithm 5.1 Lexicographic generation of permutations.

Algorithm 5.1 obviously starts by listing $\Pi = (1, 2, \ldots, n)$, the first permutation in lexicographic order, and stops only when $i = 0$, which occurs if and only if $\pi_1 > \pi_2 > \cdots > \pi_n$—that is, after $\Pi = (n, n - 1, \ldots, 1)$, the last permutation in lexicographic order, has been listed. It thus remains to show that the transformation correctly produces, from $\Pi \neq (n, n - 1, \ldots, 1)$, the next permutation in lexicographic order. Consider what happens when the transformation is applied to $\Pi = (\pi_1, \pi_2, \ldots, \pi_n)$. At the end of the loop "**while** $\pi_i > \pi_{i+1}$" we have $i \geqq 1$, indicating the rightmost position π_i that has the property that at least one element to its right is greater than it. We leave it to the reader to show that, since π_i has this property, the remaining steps of the transformation correctly produce the next permutation in the lexicographic order.

Two operations determine the overall performance of this algorithm: the number of interchanges and the number of comparisons between elements of the permutation. Once we know how many times these operations are performed, it will be easy to determine the total performance (Exercise 2) of the algorithm and compare it with other algorithms that systematically generate permutations.

Notice that in the lexicographically ordered sequence of permutations there is a complete subsequence of the $k!$ permutations of the rightmost k elements before any other elements move. This observation allows us to get a recurrence relation for I_k and C_k, the number of interchanges "$\pi_r \leftrightarrow \pi_s$" or "$\pi_i \leftrightarrow \pi_j$" and comparisons "$\pi_i > \pi_{i+1}$" or "$\pi_i > \pi_j$," respectively, used by Algorithm 5.1 to generate the first $k!$ of the $n!$ permutations. In particular, notice that I_{n-1} interchanges are used for each of the n possible values of π_1—that is, for each subsequence of permutations corresponding to the $n - 1$ rightmost components. The transformation from the last of one of these subsequences to the first in the next subsequence requires $\lfloor (n + 1)/2 \rfloor$ interchanges, and, in total, there will be $n - 1$ such transformations. Thus except for the transformation performed when $i = 0$, there are $nI_{n-1} + (n - 1)\lfloor (n + 1)/2 \rfloor$ interchanges, and so

$$I_n = nI_{n-1} + (n - 1)\left\lfloor \frac{n + 1}{2} \right\rfloor$$

and

$$I_1 = 0.$$

This recurrence is difficult to solve directly, but a simple change in variables does the trick. Let

$$S_n = I_n + \left\lfloor \frac{n + 1}{2} \right\rfloor;$$

we have

$$S_1 = 1$$

$$S_n = n(S_{n-1} + \varepsilon_{n-1}), \qquad \varepsilon_n = \begin{cases} 0 & \text{if } n \text{ is odd} \\ 1 & \text{if } n \text{ is even} \end{cases}$$

whose solution can be easily obtained:

$$S_n = n!\left(1 + \frac{1}{2!} + \frac{1}{4!} + \frac{1}{6!} + \cdots + \frac{1}{(2\lfloor(n-1)/2\rfloor)!}\right).$$

Therefore

$$I_n = n!\left(\sum_{j=0}^{\lfloor(n-1)/2\rfloor} \frac{1}{(2j)!}\right) - \left\lfloor\frac{n+1}{2}\right\rfloor.$$

Adding the $\lfloor(n+2)/2\rfloor$ interchanges made when $i = 0$ in Algorithm 5.1 yields a total of

$$n!\left(\sum_{j=0}^{\lfloor(n-1)/2\rfloor} \frac{1}{(2j)!}\right) + \varepsilon_n$$

interchanges. Since

$$\sum_{j=0}^{\lfloor(n-1)/2\rfloor} \frac{1}{(2j)!} \approx \cosh 1 \approx 1.54308,$$

we find that the algorithm uses approximately $1.54308n!$ interchanges to generate the $n!$ permutations. Similarly, we find from Exercise 2 that $C_n \approx (\frac{3}{2}e - 1)n! \approx 3.07742n!$

A slightly more efficient algorithm for generating the permutations lexicographically is the object of Exercise 13. It should be emphasized, however, that the lexicographic order cannot be generated as efficiently as other orders; the transformation of one permutation to the next is an inherently involved process. The importance of the lexicographic order lies in its simplicity and naturalness.

5.1.2 Inversion Vectors

Let $X = (x_1, x_2, \ldots, x_n)$ be a sequence of numbers. A pair (x_i, x_j) is called an *inversion* of X if $i < j$ and $x_i > x_j$. The *inversion vector* of X is the sequence of integers

$$d_1 d_2 \ldots d_n$$

obtained by letting d_j be the number of x_i such that (x_i, x_j) is an inversion. In other words, d_j is the number of elements greater than x_j and to its left in the sequence, so $0 \leq d_j < j$. For example, the inversion vector of the permutation

$$(4, 3, 5, 2, 1, 7, 8, 6, 9)$$

is

k	1	2	3	4	5	6	7	8	9
d_k	0	1	0	3	4	0	0	2	0.

A simple proof by induction shows that an inversion vector $(d_1, d_2, \ldots, d_n)$ uniquely determines a permutation of $\{1, 2, \ldots, n\}$ (Exercise 7), as the following example makes clear. Suppose that we have the inversion vector

$$
\begin{array}{cccccc}
k & 1 & 2 & 3 & 4 & 5 \\
d_k & 0 & 1 & 2 & 0 & 1.
\end{array}
$$

We compute the associated permutation by noting that since $d_5 = 1$, we must have $\pi_5 = 4$. Then since $d_4 = 0$, $\pi_4 = 5$. Similarly, since $d_3 = 2$, $\pi_3 = 1$. Since $d_2 = 1$, $\pi_2 = 2$ and so $\pi_1 = 3$. Thus the permutation is $\Pi = (3, 2, 1, 5, 4)$.

Noticing that the restrictions $0 \le d_j < j, j = 1, 2, \ldots, n$ are almost those obeyed by the digits in the factorial number system, we get a different indexing of the permutations of $\{1, 2, \ldots, n\}$. Let

$$
\Pi = (\pi_1, \pi_2, \ldots, \pi_n)
$$

have inversion vector

$$
d_1 d_2 \ldots d_n.
$$

Then since $d_1 = 0$,

$$
\text{index}_n(\Pi) = d_2 1! + d_3 2! + \cdots + d_n (n - 1)!
$$

This indexing gives us another way of generating permutations, but it is a poor way because the construction of the permutation from its inversion vector is an n^2 process. Furthermore, the generation of the next permutation from the previous one in the order induced by inversion vectors is, as with the lexicographic order, a messy process but without the advantage of the latter's naturalness. Our interest in inversion vectors lies in the fact that the inversion vector (and particularly the sum of its entries) tells us about the amount of "disorder" in the permutation. This turns out to be useful in the analysis of some of the sorting algorithms in Chapter 7.

5.1.3 Nested Cycles

One of the simplest kinds of permutations is a single cycle, a permutation in which some of the elements have undergone a right cyclic shift or have been "rotated" or "end-around shifted" by some number of positions (see Section 3.4.1). In particular, a *cyclic permutation of order k and degree d* is one in which the leftmost k elements have been shifted (cyclically, to the right) by d positions and the remaining elements are fixed. For example,

$$
\begin{pmatrix}
1 & 2 & 3 & 4 & 5 & 6 & 7 \\
3 & 4 & 5 & 1 & 2 & 6 & 7
\end{pmatrix}
$$

is a cyclic permutation of order 5 and degree 3.

Such permutations are of interest because every permutation Π of $\{1, 2, \ldots, n\}$ can be written as a product

$$\Pi = P_n P_{n-1} \ldots P_2 P_1,$$

where P_i is a cyclic permutation of order i (clearly, P_1 is the identity permutation and can be ignored) and the multiplication is done from left to right. To understand this, consider how we would write the permutation $\Pi = (4, 3, 5, 2, 1)$ as such a product. We want

$$\Pi = P_5 P_4 P_3 P_2$$

or

$$\begin{pmatrix} 1 & 2 & 3 & 4 & 5 \\ 4 & 3 & 5 & 2 & 1 \end{pmatrix} = \begin{pmatrix} 1 & 2 & 3 & 4 & 5 \\ w_1 & w_2 & w_3 & w_4 & w_5 \end{pmatrix} \begin{pmatrix} 1 & 2 & 3 & 4 & 5 \\ x_1 & x_2 & x_3 & x_4 & 5 \end{pmatrix}$$
$$\begin{pmatrix} 1 & 2 & 3 & 4 & 5 \\ y_1 & y_2 & y_3 & 4 & 5 \end{pmatrix} \begin{pmatrix} 1 & 2 & 3 & 4 & 5 \\ z_1 & z_2 & 3 & 4 & 5 \end{pmatrix}.$$

Since P_4, P_3, and P_2 all leave 5 invariant, and $\Pi(3) = 5$, we must have $P_5(3) = 5$ and hence $P_5 = (3, 4, 5, 1, 2)$. We now have

$$\begin{pmatrix} 1 & 2 & 3 & 4 & 5 \\ 4 & 3 & 5 & 2 & 1 \end{pmatrix} = \begin{pmatrix} 1 & 2 & 3 & 4 & 5 \\ 3 & 4 & 5 & 1 & 2 \end{pmatrix} \begin{pmatrix} 1 & 2 & 3 & 4 & 5 \\ x_1 & x_2 & x_3 & x_4 & 5 \end{pmatrix}$$
$$\begin{pmatrix} 1 & 2 & 3 & 4 & 5 \\ y_1 & y_2 & y_3 & 4 & 5 \end{pmatrix} \begin{pmatrix} 1 & 2 & 3 & 4 & 5 \\ z_1 & z_2 & 3 & 4 & 5 \end{pmatrix},$$

and since P_3 and P_2 leave 4 invariant, and $\Pi(1) = 4$ and $P_5(1) = 3$, we must have $P_4(3) = 4$ and hence $P_4 = (2, 3, 4, 1, 5)$, or

$$\begin{pmatrix} 1 & 2 & 3 & 4 & 5 \\ 4 & 3 & 5 & 2 & 1 \end{pmatrix} = \begin{pmatrix} 1 & 2 & 3 & 4 & 5 \\ 3 & 4 & 5 & 1 & 2 \end{pmatrix} \begin{pmatrix} 1 & 2 & 3 & 4 & 5 \\ 2 & 3 & 4 & 1 & 5 \end{pmatrix}$$
$$\begin{pmatrix} 1 & 2 & 3 & 4 & 5 \\ y_1 & y_2 & y_3 & 4 & 5 \end{pmatrix} \begin{pmatrix} 1 & 2 & 3 & 4 & 5 \\ z_1 & z_2 & 3 & 4 & 5 \end{pmatrix}.$$

Similarly, since $\Pi(2) = 3$ while $P_4(P_5(2)) = 1$ and P_2 leaves 3 invariant, we must have $P_3(1) = 3$; that is, $P_3 = (3, 1, 2, 4, 5)$. Thus

$$\begin{pmatrix} 1 & 2 & 3 & 4 & 5 \\ 4 & 3 & 5 & 2 & 1 \end{pmatrix} = \begin{pmatrix} 1 & 2 & 3 & 4 & 5 \\ 3 & 4 & 5 & 1 & 2 \end{pmatrix} \begin{pmatrix} 1 & 2 & 3 & 4 & 5 \\ 2 & 3 & 4 & 1 & 5 \end{pmatrix}$$
$$\begin{pmatrix} 1 & 2 & 3 & 4 & 5 \\ 3 & 1 & 2 & 4 & 5 \end{pmatrix} \begin{pmatrix} 1 & 2 & 3 & 4 & 5 \\ z_1 & z_2 & 3 & 4 & 5 \end{pmatrix}.$$

Finally, since $\Pi(4) = 2$ while $P_3(P_4(P_5(4))) = 1$, we must have $P_2(1) = 2$, or $P_2 = (2, 1, 3, 4, 5)$, giving

$$\begin{pmatrix} 1 & 2 & 3 & 4 & 5 \\ 4 & 3 & 5 & 2 & 1 \end{pmatrix} = \begin{pmatrix} 1 & 2 & 3 & 4 & 5 \\ 3 & 4 & 5 & 1 & 2 \end{pmatrix} \begin{pmatrix} 1 & 2 & 3 & 4 & 5 \\ 2 & 3 & 4 & 1 & 5 \end{pmatrix}$$

$$\begin{pmatrix} 1 & 2 & 3 & 4 & 5 \\ 3 & 1 & 2 & 4 & 5 \end{pmatrix} \begin{pmatrix} 1 & 2 & 3 & 4 & 5 \\ 2 & 1 & 3 & 4 & 5 \end{pmatrix}.$$

We leave it to the reader to show by induction that, in general, every permutation can be written in a unique way as such a product.

Thus we can list all permutations of n elements by systematically running through all possibilities for $P_n, P_{n-1}, \ldots, P_2$. One simple way of doing this is to start with the permutation $(1, 2, \ldots, n)$ and successively apply one-place rotations of all n elements. When a rotation of the first n elements returns us to a previously generated permutation (i.e., P_n becomes the identity permutation), we rotate the first $n - 1$ elements by one position. If that step returns us to a previously generated permutation (i.e., P_{n-1} becomes the identity permutation), we rotate the first $n - 2$ elements by one position. After arriving at a new permutation, we again apply rotations of all n elements. The details are shown in Algorithm 5.2.

for $i = 1$ **to** n **do** $\pi_i \leftarrow i$
$k \leftarrow 0$

while $k \neq 1$ **do** $\left\{ \begin{array}{l} \text{output } \Pi = (\pi_1, \pi_2, \ldots, \pi_n) \\ k \leftarrow n \\ \text{rotate the first } k \text{ elements by one position} \\ \textbf{while } \pi_k = k \neq 1 \textbf{ do} \left\{ \begin{array}{l} k \leftarrow k - 1 \\ \text{rotate the first } k \text{ elements} \\ \text{by one position} \end{array} \right. \end{array} \right.$

Algorithm 5.2   Generation of permutations by rotations.

We can explicitly give the indexing defined by this algorithm by noticing that at each stage we are trying to increase the degree of P_n by 1. If doing so causes P_n to become the identity, we then try to increase the degree of P_{n-1} by 1. If it, too, becomes the identity, we try to increase the degree of P_{n-2} by 1, and so on. We are simply counting in the falling factorial system (see Section 2.1), and so if

$$\Pi = P_n P_{n-1} \dots P_2$$

and the degree of P_i is d_i, $0 \leq d_i < i$, then

$$\text{index}_n(\Pi) = d_n \frac{n!}{n!} + d_{n-1} \frac{n!}{(n-1)!} + \dots + d_2 \frac{n!}{2!}.$$

The order induced by these "nested cycles" cannot be efficiently generated. In the worst case, the transformation from one permutation to the next can require proportional to n^2 interchanges; it is always at least n interchanges. Thus it is grossly inefficient. Even if the rotation is considered an elementary operation, the worst-case transformation still requires proportional to n operations, which still does not yield a linear algorithm. It seems, then, that the method of nested cycles, although of some historic and theoretical interest, has little practical use.

5.1.4 *Transposition of Adjacent Elements*

A sequence of the $n!$ permutations of $\{1, 2, \dots, n\}$ in which successive permutations differ by as little as possible is the best that could be hoped for in terms of minimizing the amount of work needed to generate the sequence. In order to differ as little as possible, the sequence should have the property that a permutation differs from its predecessor by the transposition (interchange) of two adjacent elements. Such a sequence of permutations can be useful in some applications; for example, consider the evaluation of

$$\min_{\Pi} |(\Pi(x), a)|,$$

where x and a are fixed vectors of n elements, (x, a) is the inner product of the two vectors, and Π ranges over all the permutations of n elements. Successive inner products are most efficiently calculated if x changes only by one transposition at a time (why?). In cases in which the amount of work needed to interchange two elements is proportional to their distance, we would like to make interchanges of adjacent elements.

Such a sequence of permutations is easy to construct recursively. For $n = 1$, the single permutation (1) satisfies the requirements. Suppose that we have a sequence $\Pi_1, \Pi_2, \dots$ of the permutations of $\{1, 2, \dots, n-1\}$ in which successive permutations differ only by a transposition of adjacent elements. We will "expand" each of these $(n-1)!$ permutations by inserting the element n in each of the n possible locations. The trick however, is to arrange these n permutations in an order that satisfies the minimal change requirement. A moment's reflection reveals an obvious way to do this: the n is added to Π_i on successive positions from right to left if i is odd and from left to right if i is even. The order of the permutations thus generated is

$$
1 \begin{cases}
12 \begin{cases}
123 \begin{cases}
1234 \\
1243 \\
1423 \\
4123
\end{cases} \\
132 \begin{cases}
4132 \\
1432 \\
1342 \\
1324
\end{cases} \\
312 \begin{cases}
3124 \\
3142 \\
3412 \\
4312
\end{cases}
\end{cases} \\
21 \begin{cases}
321 \begin{cases}
4321 \\
3421 \\
3241 \\
3214
\end{cases} \\
231 \begin{cases}
2314 \\
2341 \\
2431 \\
4231
\end{cases} \\
213 \begin{cases}
4213 \\
2413 \\
2143 \\
2134
\end{cases}
\end{cases}
\end{cases}
$$

It is, of course, hardly practical to generate the entire list of permutations as a unit. However, the same sequence of permutations can be generated iteratively, each from its predecessor and a small amount of auxiliary information. This is done by maintaining three vectors: the current permutation $\Pi = (\pi_1, \pi_2, \ldots, \pi_n)$, its inverse $P = (p_1, p_2, \ldots, p_n)$, and a record of the direction d_i in which each element i is moving (-1 for left, $+1$ for right, 0 for no movement). An element keeps moving until it reaches an element greater than itself; then it stops. At this point its direction is reversed, and the next smallest movable element is moved. The inverse of Π is maintained so that it is easy to find the location in Π of the next smallest element. Algorithm 5.3 is an implementation of this method. Notice that $\pi_0 = \pi_{n+1} = n + 1$ to stop n from moving; $d_1 = 0$ to ensure that when m becomes 1 in the inner loop, the rest of the outer loop is well defined.

$$\textbf{for } i = 1 \textbf{ to } n \textbf{ do} \begin{cases} \pi_i \leftarrow p_i \leftarrow i \\ d_i \leftarrow -1 \end{cases}$$

$$d_1 \leftarrow 0$$

$$\pi_0 \leftarrow \pi_{n+1} \leftarrow m \leftarrow n + 1$$

$$\textbf{while } m \neq 1 \textbf{ do} \begin{cases} \text{output } \Pi = (\pi_1, \pi_2, \ldots, \pi_n) \\ m \leftarrow n \\ \textbf{while } \pi_{p_m + d_m} > m \textbf{ do} \begin{cases} d_m \leftarrow -d_m \\ m \leftarrow m - 1 \end{cases} \\ \pi_{p_m} \leftrightarrow \pi_{p_m + d_m} \\ [\![\text{at this point } \pi_{p_m + d_m} = m]\!] \\ p_{\pi_{p_m}} \leftrightarrow p_m \end{cases}$$

Algorithm 5.3   Generation of permutations in a minimal change order.

The correctness of Algorithm 5.3 follows by a simple induction on n once we have proven that when the permutation $\Pi = (\pi_1, \pi_2, \ldots, \pi_n)$ is listed, the values of d_i, $1 \leq i < n$, are exactly the same as the corresponding values when the algorithm lists $\Pi - n$ (the permutation Π with n deleted) during the generation of all permutations of $\{1, 2, \ldots, n - 1\}$. This property of the d_i can, in turn, be proven by induction on the number of permutations of $\{1, 2, \ldots, n\}$ that have been generated. It is obviously true after the first permutation $(1, 2, \ldots, n)$ is listed. Suppose that it is true when the kth permutation $\Pi = (\pi_1, \pi_2, \ldots, \pi_n)$ is listed. There are two cases to be considered. First, if $\pi_{p_n + d_n} \leq n$, then none of the d_i changes in value. On the other hand, if $\pi_{p_n + d_n} > n$, we set $d_n \leftarrow -d_n$ and continue the loop with $m = n - 1$. However, this means that the changes made will be precisely the ones made had we been listing $\Pi - n$, and hence the induction hypothesis for the successor to Π, that is, for the $(k + 1)$st permutation to be listed.

Algorithm 5.3 is one of the most efficient for generating permutations, although some improvement is possible. The inner loop can be eliminated at considerable expense to the simplicity of the algorithm, without speeding it up (see Exercise 11). The algorithm is linear, since the test in the inner loop is made a total of $\sum_{i=1}^{n} i! = n! + o(n!)$ times (Exercise 12). A more efficient algorithm is obtained by applying the macro expansion techniques described in Section 4.1.4, since on $n - 1$ out of n moves it is the element n that moves (Exercise 13).

The indexing of permutations generated by this algorithm is easily seen (from the recursive definition of the sequence) to be

$$\text{index}_n\,(\Pi) = n \cdot \text{index}_{n-1}\,(\Pi - n) + \begin{cases} n - k & \text{if index}_{n-1}\,(\Pi - n) \text{ is even,} \\ k - 1 & \text{if index}_{n-1}\,(\Pi - n) \text{ is odd,} \end{cases}$$

where k is defined by $\pi_k = n$.

5.1.5 Random Permutations

Any of the sequences discussed can be used to generate random permutations, for there is an explicit correspondence between the integers and the permutations. By choosing a random integer between 0 and $n! - 1$, we are implicitly choosing a permutation. However, in addition to the problem of choosing directly, say, a random integer between 0 and $52! - 1$, we have the problem of converting that number into a permutation. That conversion requires proportional to n^2 operations.

An efficient method for generating random permutations performs a series of $n - 1$ transpositions. Starting with any permutation $\Pi = (\pi_1, \pi_2, \ldots, \pi_n)$, the element π_n is interchanged with one of $\pi_1, \pi_2, \ldots, \pi_n$, chosen randomly. Then π_{n-1} is interchanged with one of $\pi_1, \pi_2, \ldots, \pi_{n-1}$, chosen randomly, and so on.

for $i = n$ **to** 2 **by** -1 **do** $\pi_i \leftrightarrow \pi_{\text{rand}(1,i)}$[†]

Algorithm 5.4   Generation of random permutations.

To show that each of the $n!$ permutations is equally probable, we use induction. For $n = 1$ the result is obvious. Suppose that the algorithm produces random permutations for $n = k - 1$ and let $\Sigma = (\sigma_1, \sigma_2, \ldots, \sigma_k)$ be any permutation of $\{1, 2, \ldots, k\}$. Then

$$\Pr(\Pi = \Sigma) =$$

$$\Pr(\pi_k = \sigma_k)\,\Pr[(\pi_1, \ldots, \pi_{k-1}) = (\sigma_1, \ldots, \sigma_{k-1}), \text{given that } \pi_k = \sigma_k].$$

By induction, the second factor is $1/(k-1)!$ and by construction the first is $1/k$. Thus

$$\Pr(\Pi = \Sigma) = \frac{1}{k!},$$

as desired.

---

†The function rand(k, l) produces uniformly distributed random integers in the range k through l.

5.2 SUBSETS OF SETS

Generating subsets of a set $\{a_1, a_2, \ldots, a_n\}$ is equivalent to generating n-bit strings: a_i is in the subset if and only if the ith bit is one. Thus the problem of generating all subsets of a set is reduced to the problem of generating all possible n-bit strings, and the problem of generating a random subset is reduced to the problem of generating a random n-bit string (i.e., a random integer i, $0 \le i < 2^n$).

Clearly, the most straightforward way of generating all n-bit strings is to count in base 2 as shown in Algorithm 5.5. Translating this algorithm into terms of subsets of $\{a_1, a_2, \ldots, a_n\}$ is simple. We add a dummy element a_{n+1} and arrive at Algorithm 5.6.

$$\textbf{for } i = 0 \textbf{ to } n \textbf{ do } b_i \leftarrow 0$$

$$\textbf{while } b_n \neq 1 \textbf{ do} \begin{cases} \text{output } (b_{n-1}b_{n-2} \ldots b_0) \\ i \leftarrow 0 \\ \textbf{while } b_i = 1 \textbf{ do} \begin{cases} b_i \leftarrow 0 \\ i \leftarrow i + 1 \end{cases} \\ b_i \leftarrow 1 \end{cases}$$

Algorithm 5.5   Counting in base 2 to generate all n-bit strings.

$$S \leftarrow \varnothing$$

$$\textbf{while } a_{n+1} \notin S \textbf{ do} \begin{cases} \text{output } S \\ i \leftarrow 0 \\ \textbf{while } a_i \in S \textbf{ do} \begin{cases} S \leftarrow S - \{a_i\} \\ i \leftarrow i + 1 \end{cases} \\ S \leftarrow S \cup \{a_i\} \end{cases}$$

Algorithm 5.6   Subset generation by counting in base 2.

Algorithm 5.6 is linear, since the test "$a_i \in S$" is made once for every bit string, once more for every other bit string, once more still for every fourth bit string, ..., or $2^n + 2^{n-1} + \cdots + 2 + 1 = 2^{n+1} - 1$ times. However, the algorithm is not a minimal change algorithm, and the order in which the subsets are generated has no special properties to recommend it (the natural lexicographic order is obtained by applying the backtrack algorithm to subset generation). The next section develops a slightly more efficient minimal change algorithm. The section following discusses the generation of all subsets of size k, that is, combinations of n things taken k at a time.

5.2.1   Gray Codes

The smallest change possible in going from one subset of a set to another is the addition or deletion of a single element of the set. In terms of bit strings, this means that successive strings must differ by a single bit. Such sequences of bit strings are known as *Gray codes*; more precisely, an n-bit Gray code is an ordered, cyclic sequence of the 2^n n-bit strings (*codewords*) such that successive codewords differ by the complementation of a single bit (compare these codes to the DP-t codes of Chapter 1). A Gray code can be conveniently represented by its *transition sequence*—that is, the ordered list of the bit positions (numbered from right to left) that change as we go from one codeword to the next. Normally the starting codeword is taken to be $(00 \ldots 0)$ although, since the code is cyclic, it does not matter.

There are many n-bit Gray codes, but we will be concerned only with one special class that has some useful properties. Suppose that

$$G(n) = \begin{pmatrix} G_0 \\ G_1 \\ G_2 \\ \vdots \\ G_{2^n-1} \end{pmatrix}$$

is an n-bit Gray code, written in the form of a $2^n \times n$ binary matrix so that the ith row of the matrix is the ith codeword. Then

$$G(n + 1) = \begin{pmatrix} 0G_0 \\ 0G_1 \\ 0G_2 \\ \vdots \\ 0G_{2^n-1} \\ 1G_{2^n-1} \\ 1G_{2^n-2} \\ \vdots \\ 1G_1 \\ 1G_0 \end{pmatrix}$$

is obviously an $(n + 1)$-bit Gray code. Throughout this section we will consider only the so-called *binary-reflected Gray codes* obtained when this recursive definition is applied, starting with the trivial 1-bit Gray code

$$G(1) = \begin{pmatrix} 0 \\ 1 \end{pmatrix}.$$

By numbering the columns of $G(n)$ from *right to left*, the transition sequence T_n of $G(n)$ can be defined recursively. If

$$T_n = t_1, t_2, \ldots, t_{2^n-1}$$

is the transition sequence for the n-bit code, obviously

$$T_1 = 1$$
$$T_{n+1} = T_n, n + 1, T_n^{\text{reversed}}$$
$$= t_1, t_2, \ldots, t_{2^n-1}, n + 1, t_{2^n-1}, t_{2^n-2}, \ldots, t_1.$$

Notice that $T_1^{\text{reversed}} = T_1$, and so, by induction, $T_n^{\text{reversed}} = T_n$ for all n. Thus the recursive definition simplifies to

$$T_1 = 1$$
$$T_{n+1} = T_n, n + 1, T_n.$$

For example, the 4-bit code

$$G(4) = \begin{pmatrix} 0000 \\ 0001 \\ 0011 \\ 0010 \\ 0110 \\ 0111 \\ 0101 \\ 0100 \\ 1100 \\ 1101 \\ 1111 \\ 1110 \\ 1010 \\ 1011 \\ 1001 \\ 1000 \end{pmatrix}$$

has

$$1, 2, 1, 3, 1, 2, 1, 4, 1, 2, 1, 3, 1, 2, 1$$

as its transition sequence.

The same transition sequences are obtained by a different recursive definition—namely,

$$T_1 = 1$$

$$T_{n+1} = 1, t_1 + 1, 1, t_2 + 1, 1, t_3 + 1, \ldots, 1, t_{2^n-1} + 1, 1, \qquad (5.1)$$

where $T_n = t_1, t_2, \ldots, t_{2^n-1}$. This yields another recursive definition for the same codes. If

$$G(n) = \begin{pmatrix} G_0 \\ G_1 \\ G_2 \\ \vdots \end{pmatrix}$$

is the n-bit code, then the $(n + 1)$-bit code can alternatively be defined by

$$G(n + 1) = \begin{pmatrix} G_0 0 \\ G_0 1 \\ G_1 1 \\ G_1 0 \\ G_2 0 \\ G_2 1 \\ \vdots \end{pmatrix}.$$

This recursive definition is very close to a recursive definition of the binary representation of integers. If we have

$$\begin{pmatrix} 0 \\ 1 \\ 2 \\ \vdots \\ 2^n - 1 \end{pmatrix} = \begin{pmatrix} B_0 \\ B_1 \\ B_2 \\ \vdots \\ B_{2^n-1} \end{pmatrix},$$

then

$$
\begin{pmatrix}
0 \\
1 \\
2 \\
3 \\
\vdots \\
2^{n+1} - 2 \\
2^{n+1} - 1
\end{pmatrix}
=
\begin{pmatrix}
B_0 0 \\
B_0 1 \\
B_1 0 \\
B_1 1 \\
\vdots \\
B_{2^n-1} 0 \\
B_{2^n-1} 1
\end{pmatrix}
$$

Let $K_n = k_1, k_2, \ldots, k_{2^n-1}$ be the sequence such that $k_i - 1$ is the bit at which the carry stops propagating when 1 is added to B_{i-1}. From the recursive definition of the $(n + 1)$-bit binary numbers, it is evident that

$$ K_{n+1} = 1, k_1 + 1, 1, k_2 + 1, 1, k_3 + 1, \ldots, 1, k_{2^n-1} + 1, 1; $$

and since $K_1 = 1$, we have $K_n = T_n$. Thus the bit that changes in going from G_i to G_{i+1} in $G(n)$ is one to the left of the bit at which the carry stops propagating when 1 is added to $i = (b_n b_{n-1} \ldots b_0)_2$. This observation offers a way of explicitly obtaining G_i from the binary representation of i. If

$$ i = (b_n b_{n-1} \ldots b_0)_2, \qquad 0 \leq i < 2^n, $$

then

$$ G_i = (g_n g_{n-1} \ldots g_1), $$

where

$$ g_j = b_j + b_{j-1} \,(\text{mod } 2), \qquad 1 \leq j \leq n. \tag{5.2} $$

Equation (5.2) allows us to compute the index of a codeword

$$ G_i = (g_n g_{n-1} \ldots g_1) $$

directly from G_i. Let the index i be represented in binary as

$$ i = (b_n b_{n-1} \ldots b_0)_2. $$

Since $0 \leq i < 2^n$, we have $b_n = 0$. Together with equation (5.2), this implies by induction that

$$ b_j = \sum_{m=j+1}^{n} g_m \,(\text{mod } 2), \qquad 0 \leq j < n. $$

The addition of two bits modulo 2 is the same as the exclusive or of the two bits, and so equation (5.2) allows us to convert $i = (b_n, b_{n-1}, \ldots, b_0)_2$,

$0 \leqq i < 2^n$, to G_i very rapidly in machine language on a binary computer. We simply shift the bits of i one position to the left and do an exclusive or:

$$b_n \quad b_{n-1} \ldots b_2 \, b_1 \, b_0$$

$$\text{exclusive or} \quad \frac{b_n \, b_{n-1} \, b_{n-2} \ldots b_1 \, b_0}{g_n \quad g_{n-1} \cdots g_2 \, g_1}$$

The operations of shifting and exclusive or are inefficient in higher-level programming languages; therefore we are also interested in generating the elements of $G(n)$ successively from one codeword to the next without keeping track of the index and converting it each time to the corresponding codeword. It is sufficient to be able to generate the sequence T_n efficiently, and we can do so by using a stack operated as follows.

The stack initially contains $n, n-1, \ldots, 1$ (with the 1 on top). The algorithm pops off the top element i and puts it into the sequence; the elements $i-1, i-2, \ldots, 1$ are then pushed onto the stack. Thus Algorithm 5.7(a) generates the Gray code.

$$S \leftarrow \text{empty stack}$$

for $j = n$ **to** 1 **by** -1 **do** $\begin{cases} g_j \leftarrow 0 \\ S \Leftarrow j \end{cases}$

while $S \neq \text{empty}$ **do** $\begin{cases} \text{output } (g_n g_{n-1} \cdots g_1) \\ i \Leftarrow S \\ g_i \leftarrow \bar{g}_i \\ \textbf{for } j = i-1 \textbf{ to } 1 \textbf{ by } -1 \textbf{ do } S \Leftarrow j \end{cases}$

Algorithm 5.7(a)   Generation of the binary-reflected Gray code.

Notice that the stack in Algorithm 5.7(a) operates in a highly restricted manner; whenever $j > 1$ is placed on the stack, we know *a priori* that $j-1, \ldots, 1$ will be placed above it. Using this knowledge in a clever fashion allows us to use an array $(\tau_n, \tau_{n-1}, \ldots, \tau_0)$ as the stack as follows. τ_0 points to the top stack element, and, for $j \geqq 1$, τ_j is the element below j on the stack *if* j is on the stack. If j is not on the stack, then the value of τ_j cannot affect the computation, and its value can be *reset* to $j+1$, since we know that the next time that $j+1$ is put on the stack the element above it must be j. Moreover, since the elements $i-1, \ldots, 1$ will be put on the stack when i is removed, we need only set $\tau_{i-1} \leftarrow \tau_i$, assuming that τ_j was reset when j was removed for all j, $1 \leqq j \leqq i-2$. Finally, for $i \neq 1$, the elements $i - 1, \ldots, 1$ are put on the stack by the assignment $\tau_0 \leftarrow 1$. In this manner, the algorithm becomes as shown in Algorithm 5.7(b).

$$\textbf{for } j = 0 \textbf{ to } n + 1 \textbf{ do } \begin{cases} g_j \leftarrow 0 \\ \tau_j \leftarrow j + 1 \end{cases}$$

$$i \leftarrow 0$$

$$\textbf{while } i < n + 1 \textbf{ do} \begin{cases} \text{output } (g_n g_{n-1} \cdots g_1) \\ i \leftarrow \tau_0 \\ g_i \leftarrow \bar{g}_i \\ \tau_{i-1} \leftarrow \tau_i \\ \tau_i \leftarrow i + 1 \\ \textbf{if } i \neq 1 \textbf{ then } \tau_0 \leftarrow 1 \end{cases}$$

Algorithm 5.7(b)   Generation of the binary-reflected Gray code.

Observe that if the statement "$\tau_0 \leftarrow 1$" is inserted after "$g_i \leftarrow \bar{g}_i$" and the **if** statement is deleted, the loop will work correctly when $i \neq 1$; and when $i = 1$, the statement "$\tau_{i-1} \leftarrow \tau_i$" immediately corrects the incorrect value assigned to τ_0. Thus the final algorithm is Algorithm 5.7(c).

$$\textbf{for } j = 0 \textbf{ to } n + 1 \textbf{ do } \begin{cases} g_j \leftarrow 0 \\ \tau_j \leftarrow j + 1 \end{cases}$$

$$i \leftarrow 0$$

$$\textbf{while } i < n + 1 \textbf{ do} \begin{cases} \text{output } (g_n g_{n-1} \cdots g_1) \\ i \leftarrow \tau_0 \\ g_i \leftarrow \bar{g}_i \\ \tau_0 \leftarrow 1 \\ \tau_{i-1} \leftarrow \tau_i \\ \tau_i \leftarrow i + 1 \end{cases}$$

Algorithm 5.7(c)   Generation of the binary-reflected Gray code.

Note that implementing the stack in this manner does not reduce the number of stack operations, but it does eliminate the inner loop in which $i - 1, \ldots, 1$ are pushed onto the stack. Thus the algorithm goes from one codeword to the next in constant time and without the overhead of the inner loop.

To prove that Algorithm 5.7(c) is correct, we need only show that the sequence of values of the variable i is T_n. Consider the loop

$$\textbf{while } \tau_0 < k + 1 \textbf{ do} \begin{cases} i \leftarrow \tau_0 \\[4pt] \text{output } i \\[4pt] \tau_0 \leftarrow 1 \\[4pt] \tau_{i-1} \leftarrow \tau_i \\[4pt] \tau_i \leftarrow i + 1 \end{cases}$$

when started with $\tau_j = \alpha_j, 0 \leq j \leq k$, where $\alpha_j = j + 1$ for $0 \leq j < k$ and $\alpha_k \geq k + 1$. The loop in Algorithm 5.7 is essentially this loop, with the end test changed so that the last codeword is produced. Consequently, we are done if we demonstrate that the preceding loop produces the sequence T_k, stopping with $\tau_j = j + 1$ for $1 \leq j \leq k$ and $\tau_0 = \alpha_k \geq k + 1$. This is obviously true for $k = 1$. Suppose that it is true for $k = n - 1$ and consider what happens when $k = n$. Since the hypothesis for $k = n - 1$ is also satisfied, we know by induction that T_{n-1} is produced, leaving $\tau_j = j + 1$ for $1 \leq j \leq n - 1$ and $\tau_0 = \alpha_{n-1} = n$. The next iteration produces n, leaving $\tau_j = j + 1$ for $0 \leq j < n - 1$ and sets $\tau_{n-1} = \alpha_n \geq n + 1$. Then, again by induction, we know that T_{n-1} is produced before $\tau_0 = \alpha_n \geq n + 1$, leaving $\tau_j = j + 1$ for $1 \leq j \leq n - 1$, and hence $T_{n-1}, n, T_{n-1} = T_n$ has been produced.

5.2.2   k Subsets (Combinations)

Generally not all the subsets of $\{a_1, a_2, \ldots, a_n\}$ are required but only those that satisfy some restrictions; the most common restriction is on the size of the subsets. Of particular interest are the subsets of size k, that is, the $\binom{n}{k}$ combinations of n things taken k at a time. In this section we discuss generating them in lexicographic order, in a minimal change order based on the Gray code of the previous section, and at random.

Lexicographic Order.   As usual, we make the simplifying assumption that the underlying set is $\{1, 2, \ldots, n\}$; so we want to generate all combinations of size k of the integers $\{1, 2, \ldots, n\}$. The most natural order is the increasing lexicographic order, with the components of each combination in increasing order from left to right. Thus for example, the $\binom{6}{3} = 20$ combinations of six things taken three at a time (i.e., the three-element subsets of $\{1, 2, 3, 4, 5, 6\}$) listed in lexicographic order are

123	135	234	256
124	136	235	345
125	145	236	346
126	146	245	356
134	156	246	456

The recurrence relation for the index position of the combination $(m_1, m_2, \ldots, m_k)$ in the list of the k subsets of $\{1, 2, \ldots, n\}$, $1 \leq m_1 < m_2 < \cdots < m_k \leq n$, can be found by observing that, if $k = 1$, then the index position is m_1, whereas, if $k \geq 2$, it is the index position of the combination $(m_2 - m_1, m_3 - m_1, \ldots, m_k - m_1)$ of $n - m_1$ things taken $k - 1$ at a time, plus the number of combinations of n things taken k at a time that begin with a first component less than m_1. Thus.

$$\text{index}_{n,k}\,(m_1, m_2, \ldots, m_k)$$

$$= \text{index}_{n-m_1,k-1}\,(m_2 - m_1, \ldots, m_k - m_1) + \sum_{i=1}^{m_1-1} \binom{n-i}{k-1}, \qquad k \geq 2$$

and

$$\text{index}_{n,1}\,(m_1) = m_1.$$

Letting $m_0 = 0$, we find that

$$\text{index}_{n,k}\,(m_1, m_2, \ldots, m_k) = 1 + \sum_{j=0}^{k-1} \sum_{i=1}^{m_{j+1}-m_j-1} \binom{n - m_j - i}{k - j - 1}.$$

Using the identity in equation (3.22),

$$\sum_{I=1}^{T} \binom{N - I}{J} = \binom{N}{J + 1} - \binom{N - T}{J + 1},$$

with $T = m_{j+1} - m_j - 1$, $N = n - m_j$, and $J = k - j - 1$, the formula becomes

$$1 + \sum_{j=0}^{k-1} \left[\binom{n - m_j}{k - j} - \binom{n - m_{j+1} + 1}{k - j} \right],$$

which simplifies to

$$\binom{n}{k} - \sum_{j=1}^{k} \binom{n - m_j}{k - j + 1}.$$

The combinations in lexicographic order can be generated sequentially in a straightforward way. Starting with the combination $(1, 2, \ldots, k)$, the next combination is found by scanning the current combination from right to left so as to locate the rightmost element that has not yet attained its maximum value. This element is incremented by one, and all positions to its right are reset to the lowest values possible, as shown in Algorithm 5.8.

$$c_0 \leftarrow -1$$

for $i = 1$ **to** k **do** $c_i \leftarrow i$

$$j \leftarrow 1$$

while $j \neq 0$ **do** $\begin{cases} \text{output } (c_1, c_2, \ldots, c_k) \\ j \leftarrow k \\ \textbf{while } c_j = n - k + j \textbf{ do } j \leftarrow j - 1 \\ c_j \leftarrow c_j + 1 \\ \textbf{for } i = j + 1 \textbf{ to } k \textbf{ do } c_i \leftarrow c_{i-1} + 1 \end{cases}$$

Algorithm 5.8   Generation of combinations (k subsets) in lexicographic order.

The entire performance of Algorithm 5.8 can easily be determined once we know the number of times that the comparison "$c_j = n - k + j$" is made. This comparison is made once for each combination [i.e., $\binom{n}{k}$ times], once more for each combination with $c_n = n$ [i.e., $\binom{n-1}{k-1}$ times], once more still for each combination with $c_{n-1} = n - 1$ and $c_n = n$ [i.e., $\binom{n-2}{k-2}$ times], ..., and once more still for each combination with $c_1 = 1, c_2 = 2, \ldots, c_n = n$ [i.e., $\binom{n-k}{k-k} = 1$ time]. So it is made

$$\sum_{j=0}^{k} \binom{n-j}{k-j} = \binom{n+1}{k}$$

times. Knowing this, it is easy to see that the statement "$c_i \leftarrow c_{i-1} + 1$" is executed $\binom{n+1}{k} - \binom{n}{k} = \binom{n}{k-1}$ times. The comparison "$j \neq 0$" is obviously made $\binom{n}{k} + 1$ times, and thus the behavior of the algorithm is completely determined. Since $\binom{n+1}{k} = \binom{n}{k}[(n + 1)/(n - k + 1)]$ and $\binom{n}{k-1} = \binom{n}{k}k/(n - k + 1)$, the algorithm is linear except when $k = n - o(n)$. In this case the algorithm can be applied to generate the combinations of n things taken $n - k$ at a time; these combinations can then be "complemented" to obtain the combinations of n things taken k at a time.

Minimal Change Order.   The smallest change possible between successive combinations in a list of all the combinations of n things taken k at a time is the replacement of one element by another. In terms of bit strings, this means that we want to list all n-bit strings containing exactly k 1-bits so that successive strings differ in exactly two bits (one changes from zero to one, another from one to zero).

Let $G(n)$ be the n-bit binary-reflected Gray code of the previous section and let $G(n, k)$, $0 \leqq k \leqq n$, be the sequence of all codewords with exactly k 1-bits. This $G(n, k)$ clearly has the desired property that all codewords contain exactly k 1-bits. We must now show that successive codewords differ in exactly two bits.

Theorem 5.1

The first codeword in $G(n)$ with exactly k 1-bits is $0^{n-k} 1^k$ and the last is $10^{n-k} 1^{k-1}$ for $k \geqq 1$ and 0^n if $k = 0$.†

Proof: The proof is by induction on n for any fixed $k \geqq 1$. The theorem is certainly true for all n if $k = 0$ and for $n = k$, since only one codeword has the correct form; thus the theorem holds for $n = 1$. Suppose that the theorem is true for $G(n)$, $n \geqq 1$. Letting $G(n)^R$ be the codewords (rows) of $G(n)$ in reverse order, $G(n + 1)$ is formed recursively from $G(n)$ by

$$G(n + 1) = \begin{pmatrix} 0G(n) \\ 1G(n)^R \end{pmatrix},$$

and so the first codeword with $k \geqq 1$ 1-bits in $G(n + 1)$ is the first such codeword in $G(n)$, with a zero concatenated on the left. That is, it is $0^{n+1-k} 1^k$, as desired. Similarly, the last codeword with $k \geqq 1$ 1-bits in $G(n + 1)$ is the first codeword with $k - 1$ 1-bits in $G(n)$, with a one concatenated on the left. That is, it is $10^{n+1-k} 1^{k-1}$, as desired.

Theorem 5.2

Successive codewords in $G(n, k)$ differ in exactly two bits.

Proof: The proof is by induction on n. The theorem is obviously true for $n = 1$, since $k = 0$ or $k = 1$. Suppose that it is true for n and k, $0 \leqq k \leqq n$. We must show that it is true for $n + 1$ and k, $0 \leqq k \leqq n + 1$. If $k = 0$ or $k = n + 1$, it is trivially true, since

$$G(n + 1) = \begin{pmatrix} 0G(n) \\ 1G(n)^R \end{pmatrix}$$

contains only one codeword with no 1-bits and only one codeword with $n + 1$ 1-bits. For $1 \leqq k \leqq n$, the theorem holds by the inductive hypothesis if the successive codewords are both in either $0G(n)$ or $1G(n)^R$. Thus the only case left is to show that the last codeword with k 1-bits in $0G(n)$ differs in exactly two bits from the first codeword with k 1-bits in $1G(n)^R$. From Theorem 5.1 we know that, for $k = 1$, these two codewords are, respectively, 010^{n-1} and 100^{n-1}; for $k \geqq 2$, they are $010^{n-k}1^{k-1} = 010^{n-k}11^{k-2}$ and $110^{n-k+1}1^{k-2} = 110^{n-k}01^{k-2}$, respectively.

---

†The exponent is with regard to concatenation. Thus $0^{n-k}1^k$ is an n-bit vector consisting of $n - k$ zeros followed by k ones.

In order to generate the codewords of $G(n, k)$, $n \geq k \geq 0$, successively in an efficient manner, it is helpful to give a recursive definition analogous to the first definition given for the Gray code $G(n)$. For $n > k > 0$, we have

$$G(n, k) = \begin{pmatrix} 0G(n - 1, k) \\ 1G(n - 1, k - 1)^R \end{pmatrix}$$

and

$$G(n, 0) = 0^n$$
$$G(n, n) = 1^n.$$

A proof by induction on n shows that it indeed defines the same sequence of codewords as is obtained by deleting all codewords without exactly k 1-bits from $G(n)$.

The recursive structure of $G(n, k)$ is shown as a binary tree in Figure 5.1. A simple inductive proof shows that the leaves of the tree (not shown) must be either $1G(m, 0)^R$ or $0G(m, m)$. The problem of generating the codewords one to the next is then the problem of traversing the leaves of the tree in Figure 5.1 from top to bottom. This traversal is accomplished by a backtrack-style search in which advancing and backing up are simplified by the recursive structure of the tree.

To get from one codeword to the next, we look at the path from the current leaf to the root of the tree and follow this path in the direction of the root as long as nodes are lower sons. Upon reaching a node that is an upper son, we go to its brother (a lower son) and follow upper sons to a leaf. For example, in Figure 5.1 we would follow the boldface, solid line from right to left and then the boldface, dashed line from left to right. Because of the recursive nature of the tree, the node $G(n, k)$ in Figure 5.1 can be considered to be the root of a subtree of a larger tree. In this way, Figure 5.1 exemplifies the transformation from any codeword to the next.

As the generation progresses, we maintain the path from the root to the current leaf in the form of the current codeword $(g_n, g_{n-1}, \ldots, g_1)$. We also maintain a pushdown stack of the nodes $g_i G(i - 1, t)$ to which we must back up.[†] Since we can easily calculate the value of t, it need not be saved on the stack; only the value of i is saved. Notice that t is a count of the number of 1-bits among $g_{i-1}, g_{i-2}, \ldots, g_1$.

[†]It is obvious from Figure 5.1 that there will never be a node $g_i G(i - 1, t)^R$ on the stack, for only lower sons are reflected, and we never back up to a lower son.

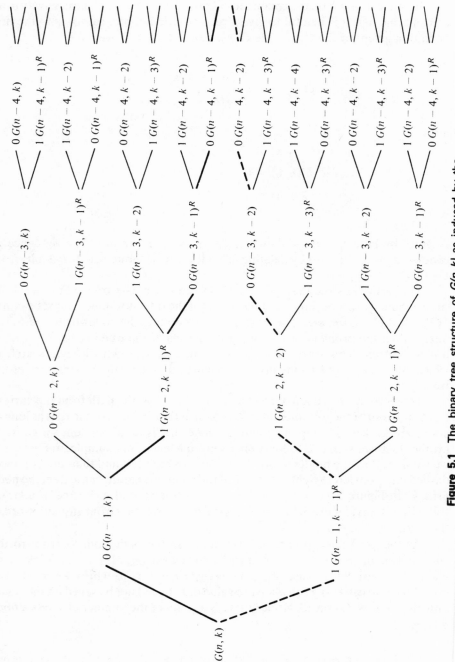

Figure 5.1 The binary tree structure of $G(n, k)$ as induced by the recursive definition. Getting from one codeword to the next means, for example, following the boldface, dashed line instead of the boldface, solid line.

Careful examination of the tree in conjunction with Theorem 5.1 shows that transforming the current codeword to the next when we have backed up into a node $g_i G(i - 1, t)$ requires the consideration of four cases. These cases depend on the values of g_i and t:

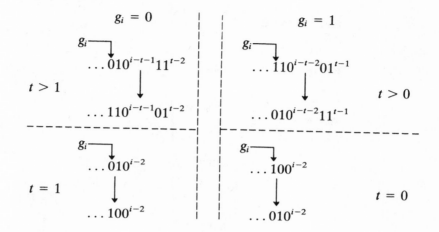

These transformations affect the number of 1-bits to the right of g_i, which means that we must decrement t by 1 if $g_i = 0$ and increment it by 1 if $g_i = 1$. The two positions complemented by the transformation are g_i and

$$g_{t-1} \text{ if } g_i = 0 \text{ and } t > 1$$

$$g_{i-1} \text{ if } g_i = 0 \text{ and } t = 1$$

$$g_t \quad \text{ if } g_i = 1 \text{ and } t > 0$$

$$g_{i-1} \text{ if } g_i = 1 \text{ and } t = 0.$$

In addition to the preceding transformation, we must add to the stack the nodes on the path between the current node to the current leaf in order to backtrack later on. Because it depends on the top node in the stack, the value of t may therefore change again.

If $t = 0$ or $t = i - 1$, then clearly no new nodes are pushed onto the stack, since the node we are at is $g_i G(i - 1, i - 1)$ or $g_i G(i - 1, 0)^R$, a leaf in either case. The value of t must be recomputed for the top stack element, and we assert that the new value of t in this case is always $t + 1$. This is true, since we are going to the left from g_i until we reach a node that is the upper son of its father; but it is easy to show that if $1 G(m, t)^R$ is a lower son, then its father must be an upper son. Similarly, if $0 G(m, t)^R$ is a lower son, its father is also a lower son.

On the other hand, if $t \neq 0$ and $t \neq i - 1$, we must push the nodes $i - 1, i - 2, \ldots, t + 1$ onto the stack unless $g_{i-2} = \cdots = g_1 = 0$; in this case, only $i - 1$ is pushed onto the stack. The value of t must be recalculated, and since the only possible 1-bit among $g_{i-1}, g_{i-2}, \ldots, g_{t+1}$ is g_{i-1}, we can set $t \leftarrow t - g_{i-1}$. If this causes t to become zero, we must have $g_{i-2} = \cdots = g_1 = 0$.

All the details are given in Algorithm 5.9.

$$\textbf{for } j = 1 \textbf{ to } k \textbf{ do} \begin{cases} g_j \leftarrow 1 \\ \tau_j \leftarrow j + 1 \end{cases}$$

$$\textbf{for } j = k + 1 \textbf{ to } n + 1 \textbf{ do} \begin{cases} g_j \leftarrow 0 \\ \tau_j \leftarrow j + 1 \end{cases}$$

$$t \leftarrow k$$

$$\tau_1 \leftarrow k + 1$$

$$i \leftarrow 0$$

$$\textbf{while } i \neq n + 1 \textbf{ do} \begin{cases} \text{output } (g_n, g_{n-1}, \ldots, g_1) \\[4pt] i \leftarrow \tau_1 \\[4pt] \tau_1 \leftarrow \tau_i \\[4pt] \tau_i \leftarrow i + 1 \\[6pt] \textbf{if } g_i = 1 \textbf{ then} \begin{cases} \textbf{if } t \neq 0 \textbf{ then } g_t \leftarrow \bar{g}_t \\ \qquad\quad \textbf{else } g_{i-1} \leftarrow \bar{g}_{i-1} \\ t \leftarrow t + 1 \end{cases} \\[10pt] \qquad\qquad\textbf{else} \begin{cases} \textbf{if } t \neq 1 \textbf{ then } g_{t-1} \leftarrow \bar{g}_{t-1} \\ \qquad\quad \textbf{else } g_{i-1} \leftarrow \bar{g}_{i-1} \\ t \leftarrow t - 1 \end{cases} \\[10pt] g_i \leftarrow \bar{g}_i \\[4pt] \textbf{if } t = i - 1 \textbf{ or } t = 0 \textbf{ then } t \leftarrow t + 1 \\[6pt] \qquad\quad\textbf{else} \begin{cases} t \leftarrow t - g_{i-1} \\ \tau_{i-1} \leftarrow \tau_1 \\ \textbf{if } t = 0 \textbf{ then } \tau_1 \leftarrow i - 1 \\ \qquad\quad \textbf{else } \tau_1 \leftarrow t + 1 \end{cases} \end{cases}$$

Algorithm 5.9 Generation of combinations (k subsets) in the minimal change order derived from the n-bit binary-reflected Gray code, $n \geq k > 0$.

Notice that in Algorithm 5.9 the stack is implemented in the same way as in Algorithm 5.7(c): τ_1 is the pointer to the top of the stack, and each element τ_j in the stack is immediately reset as soon as it is popped off the stack. Since we have set $\tau_{i-1} \leftarrow \tau_1$, the single assignment $\tau_1 \leftarrow t + 1$ serves to push $i - 1, \ldots, t + 1$

onto the stack. A proof of correctness for Algorithm 5.9, for $n \geq k > 0$, can be constructed along the lines of the proof that Algorithm 5.7(c) is correct. Algorithm 5.9 does not work correctly if $k = 0$, since the stack is improperly initialized. Only a few small changes are needed for this case (what are they?).

Algorithm 5.9 obviously requires only a bounded number of operations, independent of n and k, to generate a codeword from the previous codeword. We can determine the precise number of operations needed to generate all $\binom{n}{k}$ codewords of $G(n, k)$ once we know how many times, during the generation of $G(n, k)$, we have $g_i = 1$ and $t \neq 0$ in the statement "**if** $g_i = 1$" and how many times we have $t = i - 1$ or $t = 0$ in the statement "**if** $t = i - 1$ **or** $t = 0$." In the other cases, the number of operations needed is uniform over all possible alternatives.

Let $a_{n,k}$ be the number of times during the generation of $G(n, k)$ that $g_i = 1$ and $t \neq 0$ in the statement "**if** $g_i = 1$," not including the final pass through the loop after generating the last codeword. On that final pass we have $i = n + 1$ and $g_i = 0$, which means that the statement "$g_t \leftarrow \bar{g}_t$" will not be executed. Thus $a_{n,k}$ is exactly the number of times that it will be executed during the generation of $G(n, k)$. Let $b_{n,k}$ be the number of times that this happens during the generation of $G(n, k)^R$, not including the final pass through the loop after generating the last codeword. In going from the last codeword of $0G(n - 1, k)$ to the first codeword of $1G(n - 1, k - 1)^R$, we have $i = n$ and $g_n = 0$. In going from the last codeword of $1G(n - 1, k - 1)$ to the first codeword of $0G(n - 1, k)^R$, we have $i = n$ and $g_n = 1$, and $t \neq 0$ if and only if $k > 1$. Thus we have the recurrence relations

$$\left. \begin{aligned} a_{n,k} &= a_{n-1,k} + b_{n-1,k-1} \\ b_{n,k} &= b_{n-1,k} + a_{n-1,k-1} + 1 - \delta_{1,k} \end{aligned} \right\} \quad n > k > 0,$$

where $\delta_{i,j} = 1$ if $i = j$ and 0 otherwise, and

$$\left. \begin{aligned} a_{n,k} &= 0 \\ b_{n,k} &= 0 \end{aligned} \right\} \text{otherwise.}$$

Similarly let $c_{n,k}$ be the number of times during the generation of $G(n, k)$, other than on the final pass, that we have $t = i - 1$ or $t = 0$ in the statement "**if** $t = i - 1$ **or** $t = 0$" and let $d_{n,k}$ be the number of times that it happens during the generation of $G(n, k)^R$, other than on the final pass. As above, we can derive the recurrence relations

$$\left. \begin{aligned} c_{n,k} &= c_{n-1,k} + d_{n-1,k-1} + \delta_{1,k} \\ d_{n,k} &= d_{n-1,k} + c_{n-1,k-1} + \delta_{k,n-1} \end{aligned} \right\} \quad n > k > 0$$

and

$$\left. \begin{aligned} c_{n,k} &= 0 \\ d_{n,k} &= 0 \end{aligned} \right\} \text{otherwise.}$$

In the final pass through the loop after generating the last codeword, we have $i = n + 1$. We cannot have $t = i - 1$ at this point, and we can have $t = 0$ only when $k = 1$. Therefore $c_{n,k} + \delta_{1,k}$ is the total number of times that the **then** clause of the statement "**if** $t = i - 1$ **or** $t = 0$" is executed during the generation of $G(n, k)$.

Such recurrence relations can be solved by using generating functions. We show how to do this for $c_{n,k}$ and leave the (similar) determination of $a_{n,k}$ as an arduous exercise for the reader. Define the generating functions

$$C(x, y) = \sum_{\substack{n \geq 0 \\ k \geq 0}} c_{n,k} x^n y^k$$

$$D(x, y) = \sum_{\substack{n \geq 0 \\ k \geq 0}} d_{n,k} x^n y^k.$$

The recurrence relations for $c_{n,k}$ and $d_{n,k}$ yield

$$C(x, y) = xC(x, y) + xyD(x, y) + \frac{x^2 y}{1 - x}$$

$$D(x, y) = xD(x, y) + xyC(x, y) + \frac{x^2 y}{1 - xy},$$

and so we obtain

$$C(x, y) = \frac{x^2 y}{[1 - (1 + y)x][1 - (1 - y)x](1 - xy)},$$

which can be expressed as

$$C(x, y) = \frac{1}{2} \frac{x^2(1 + y)^2}{1 - (1 + y)x} - \frac{1}{2} \frac{x^2(1 - y^2)}{(1 - 2y)[1 - (1 - y)x]} + \frac{x^2 y^3}{(1 - 2y)(1 - xy)}.$$

For $n \geq k > 0$, the coefficient of $x^n y^k$ in $C(x, y)$ is $c_{n,k}$, and therefore

$$c_{n,k} = \frac{1}{2}\binom{n}{k} - (-1)^k \frac{1}{2}\binom{n-2}{k} - \frac{1}{2} \sum_{i=0}^{k-2} 2^{k-2-i}(-1)^i \binom{n-2}{i},$$

which can be simplified to

$$c_{n,k} = \begin{cases} \dfrac{1}{2}\binom{n}{k} - \dfrac{1}{2} \sum\limits_{i=0}^{k} \binom{n-1-i}{k-i}(-1)^{k-i} & \text{if } n > k > 0, \\ 0 & \text{if } n = k. \end{cases}$$

As noted, the **then** clause of the statement "**if** $t = i - 1$ **or** $t = 0$" is executed $c_{n,k} + \delta_{1,k}$ times during the generation of $G(n, k)$.

Random k Sets.   To generate a random subset S of $\{a_1, a_2, \ldots, a_n\}$ with k elements, we choose one of the n elements at random (each a_i having probability $1/n$ of being chosen) and then choose a random $(k - 1)$ subset of the remaining $n - 1$ elements. The probability of the k subset $\{a_{i_1}, a_{i_2}, \ldots, a_{i_k}\}$ occurring as a result of this procedure is the product of the probabilities

$$\text{Pr (first element chosen is one of the } k \text{ elements } a_{i_1}, \ldots, a_{i_k}) = \frac{k}{n}$$

$$\text{Pr (second element chosen is one of the } k - 1 \text{ remaining elements)} = \frac{k - 1}{n - 1}$$

$$\text{Pr (third element chosen is one of the } k - 2 \text{ remaining elements)} = \frac{k - 2}{n - 2}$$

$$\vdots$$

$$\text{Pr (} k\text{th element chosen is the last remaining element)} = \frac{1}{n - k + 1},$$

which is

$$\frac{k}{n} \frac{k - 1}{n - 1} \frac{k - 2}{n - 2} \cdots \frac{1}{n - k + 1} = \frac{k!(n - k)!}{n!} = \frac{1}{\binom{n}{k}}.$$

Thus each of the $\binom{n}{k}$ k subsets is equally probable.

The algorithm can be implemented by using an auxiliary array to save the indices of those elements not yet chosen. After the first $j - 1 \leq k$ elements have been chosen, $p_j, p_{j+1}, \ldots, p_n$ are the subscripts of the $n - j + 1$ elements that have not yet been chosen (see Algorithm 5.10). If it is unnecessary to preserve the order in the set, the array p can be eliminated in an algorithm similar to the one for random permutations (Algorithm 5.4) that leaves the random combination in the first k positions of the set.

$$\textbf{for } j = 1 \textbf{ to } n \textbf{ do } p_j \leftarrow j$$

$$R \leftarrow \varnothing$$

$$\textbf{for } j = 1 \textbf{ to } k \textbf{ do } \begin{cases} r \leftarrow \text{rand}\,(j, n) \\ R \leftarrow R \cup \{a_{p_r}\} \\ p_r \leftarrow p_j \end{cases}$$

Algorithm 5.10   Generation of a random combination (k subset) R of $S = \{a_1, a_2, \ldots, a_n\}$.

5.3 COMPOSITIONS AND PARTITIONS OF INTEGERS

Here we consider the problem of generating partitions of a positive integer n into a sequence of nonnegative integers $\{p_1, p_2, \ldots, p_k\}$ so that $p_1 + p_2 + \cdots + p_k = n$. In particular, we consider the problem with various restrictions on the p_i. If the order of the p_i is important, then $(p_1, p_2, \ldots, p_k)$ is called a *composition* of n; in this case, we usually restrict $p_i > 0$. Occasionally, however, the value of k is fixed and we then allow $p_i = 0$; these compositions are called the *k-part compositions* of n. Other restrictions are possible, for example, on the maximum or minimum allowable size of the p_i or on the number of times each can be repeated, but such compositions are of lesser interest. If the order of the p_i is not important and $p_i > 0$, then $\{p_1, p_2, \ldots, p_k\}$ is a multiset and is called a *partition* of n. In this case, too, additional restrictions may be imposed on the p_i.

The differences between compositions, k-part compositions, and partitions can be easily understood by considering $n = 3$ as an example. The compositions are (3), (1, 2), (2, 1), (1, 1, 1); the two-part compositions are (0, 3), (1, 2), (2, 1), (3, 0); and the partitions are (3), (1, 2), (1, 1, 1).

5.3.1 Compositions

A convenient representation of compositions arises from considering the integer n to be a straight line composed of n segments of unit length. The line is divided at $n - 1$ points, and a composition is obtained by marking some or all of the points. The elements of the composition are simply the distances between adjacent points. Figure 5.2 shows the composition (2, 1, 2, 2, 1) of $n = 8$.

Clearly, then, each composition of n corresponds to a way of selecting a subset of the $n - 1$ points. We can associate a binary digit with each of the points, and thus an $(n - 1)$-bit number corresponds to a composition of n. Furthermore, a composition with k parts corresponds to an $(n - 1)$-bit number with exactly $k - 1$ 1-bits, and so there are $\binom{n-1}{k-1}$ such compositions.

Figure 5.2 Graphical representation of the composition (2, 1, 2, 2, 1).

We can consider the k-part compositions of n as the ways of selecting $k - 1$ of the points along a line divided by $n + k - 1$ points into $n + k$ unit segments. In this case, the distance between successive selected points is one more than the corresponding component of the composition. With this interpretation, for example, Figure 5.2 is the graphical interpretation of five-part compositions (1, 0, 1, 1, 0) of 3. In general, then, there are $\binom{n+k-1}{k-1} = \binom{n+k-1}{n}$ k-part compositions of n.

It is clear that the methods of the previous section can be easily applied to the sequential and random generation of compositions. The lexicographic order is the subject of Exercise 30.

5.3.2 Partitions

The partitions of n differ from its compositions in that the order of the components is of no importance, so that, for example, no distinction is made among, say $1 + 1 + 2$, $1 + 2 + 1$, and $2 + 1 + 1$. In particular, it is usually convenient to force the partition $\{p_1, p_2, \ldots, p_k\}$ of n to have its components in some order, for example $p_1 \leqq p_2 \leqq \cdots \leqq p_k$.

The partitions of n with l components can be generated in increasing lexicographic order by starting with $p_1 = p_2 = \cdots = p_{l-1} = 1, p_l = n - l + 1$ and continuing as follows. To obtain the next partition from the current one, scan the elements from right to left, stopping at the rightmost p_i such that $p_l - p_i \geqq 2$. Replace p_j by $p_i + 1$ for $j = i, i + 1, \ldots, l - 1$ and then replace p_l by $n - \sum_{j=1}^{l-1} p_j$. For example, if we have $n = 12$, $l = 5$, and the partition $\{1, 1, 3, 3, 4\}$, we find that the 4 is greater by 2 than the rightmost 1, and so the next partition is $\{1, 2, 2, 2, 5\}$. When no element of the partition differs from the last by more than 1, we are finished. The proof that this process, the basis of Algorithm 5.11, works is left as Exercise 31.

$$l \leftarrow 1$$
$$p_1 \leftarrow n$$
$$p_0 \leftarrow -1$$

$$\textbf{while } l \leqq n \textbf{ do} \begin{cases} \text{output } \{p_1, p_2, \ldots, p_l\} \\ i \leftarrow l - 1 \\ \textbf{while } p_l - p_i < 2 \textbf{ do } i \leftarrow i - 1 \\ \textbf{if } i \neq 0 \textbf{ then for } j = l - 1 \textbf{ to } i \textbf{ by } -1 \textbf{ do } p_j \leftarrow p_i + 1 \\ \qquad\qquad \textbf{else} \begin{cases} \textbf{for } j = 1 \textbf{ to } l \textbf{ do } p_j \leftarrow 1 \\ l \leftarrow l + 1 \end{cases} \\ p_l \leftarrow n - \sum_{j=1}^{l-1} p_j \end{cases}$$

Algorithm 5.11   Generation of partitions of n in increasing length, in lexicographic order for each fixed length.

The analysis of this algorithm has not been attempted because the algorithm can be improved considerably by using the notation for a multiset introduced in

Section 2.4. A partition of n is written as

$$\{m_1 \bullet p_1, m_2 \bullet p_2, \ldots, m_l \bullet p_l\},$$

where there are m_1 copies of p_1, m_2 copies of p_2, and so on, and

$$n = \sum_{i=1}^{l} m_i p_i.$$

This representation eliminates the need for the right-to-left scan, giving a more efficient algorithm (Exercise 32). However, it is even more efficient to generate the partitions in "dictionary" order, in which each partition satisfies $p_1 > p_2 > \cdots > p_l$. For $n = 7$, the partitions will be generated as follows.

$$\{7 \bullet 1\} = \{1, 1, 1, 1, 1, 1, 1\}$$
$$\{1 \bullet 2, 5 \bullet 1\} = \{2, 1, 1, 1, 1, 1\}$$
$$\{2 \bullet 2, 3 \bullet 1\} = \{2, 2, 1, 1, 1\}$$
$$\{3 \bullet 2, 1 \bullet 1\} = \{2, 2, 2, 1\}$$
$$\{1 \bullet 3, 4 \bullet 1\} = \{3, 1, 1, 1, 1\}$$
$$\{1 \bullet 3, 1 \bullet 2, 2 \bullet 1\} = \{3, 2, 1, 1\}$$
$$\{1 \bullet 3, 2 \bullet 2\} = \{3, 2, 2\}$$
$$\{2 \bullet 3, 1 \bullet 1\} = \{3, 3, 1\}$$
$$\{1 \bullet 4, 3 \bullet 1\} = \{4, 1, 1, 1\}$$
$$\{1 \bullet 4, 1 \bullet 2, 1 \bullet 1\} = \{4, 2, 1\}$$
$$\{1 \bullet 4, 1 \bullet 3\} = \{4, 3\}$$
$$\{1 \bullet 5, 2 \bullet 1\} = \{5, 1, 1\}$$
$$\{1 \bullet 5, 1 \bullet 2\} = \{5, 2\}$$
$$\{1 \bullet 6, 1 \bullet 1\} = \{6, 1\}$$
$$\{1 \bullet 7\} = \{7\}$$

The idea used in Algorithm 5.12 is to begin with $n \bullet 1$ and get from one partition to the next by looking at the rightmost element of the partition, $m_l \bullet p_l$. If $m_l p_l$ is large enough ($m_l > 1$), we can remove two p_l's to add another $p_l + 1$ (or insert one if there is none currently). If $m_l = 1$, then $m_{l-1} p_{l-1} + m_l p_l$ is large enough to add a $p_{l-1} + 1$. In any case, what is left over is turned into the correct number of ones and the new partition is formed. Algorithm 5.12 uses this process to generate the partitions of n.

$$l \leftarrow 1$$
$$p_{-1} \leftarrow m_{-1} \leftarrow 0$$
$$p_0 \leftarrow n + 1$$
$$m_0 \leftarrow 0$$
$$p_1 \leftarrow 1$$
$$m_1 \leftarrow n$$

$$\textbf{while } l \neq 0 \textbf{ do} \begin{cases} \text{output } \{m_1 \bullet p_1, m_2 \bullet p_2, \ldots, m_l \bullet p_l\} \\ sum \leftarrow m_l p_l \\ \textbf{if } m_l = 1 \textbf{ then} \begin{cases} l \leftarrow l - 1 \\ sum \leftarrow sum + m_l p_l \end{cases} \\ \textbf{if } p_{l-1} = p_l + 1 \textbf{ then} \begin{cases} l \leftarrow l - 1 \\ m_l \leftarrow m_l + 1 \end{cases} \\ \qquad\qquad \textbf{else} \begin{cases} p_l \leftarrow p_l + 1 \\ m_l \leftarrow 1 \end{cases} \\ \textbf{if } sum > p_l \textbf{ then} \begin{cases} p_{l+1} \leftarrow 1 \\ m_{l+1} \leftarrow sum - p_l \\ l \leftarrow l + 1 \end{cases} \end{cases}$$

Algorithm 5.12 Generation of the partitions of n in dictionary order.

Algorithm 5.12 is clearly linear, for the number of operations necessary to go from one partition to the next is bounded by a constant, independent of n and l. The proof of its correctness is left to Exercise 33.

Random Partitions. In this section we discuss a very general method for generating random combinatorial objects; in particular, the method is applied to generating random partitions of n. As usual, the problem is to ensure that each of the potential objects is equally probable.

Suppose that there are a_n objects of "size" n and that a_n satisfies a recurrence relation

$$a_n = \sum_{m=0}^{n-1} \alpha_{mn} a_m, \qquad \alpha_{mn} \geqq 0.$$

Suppose further that there is a "constructive" proof of this identity; in other words, suppose that there is an explicit construction by which the a_m objects of size m are extended to $\alpha_{mn}a_m$ objects of size n. Then we can choose an object at random by first choosing a random value of m according to the distribution

$$\Pr(m) = \frac{\alpha_{mn}a_m}{a_n}, \qquad 0 \leq m \leq n - 1$$

and then recursively selecting a random object of size m and extending it to one of size n in such a way that each possible extension is equally probable.

Let $P(n)$ be the number of partitions of n, where $P(0) = 1$ by convention. An identity of Euler states that

$$nP(n) = \sum_{m=0}^{n-1} \sigma(n - m)P(m), \tag{5.3}$$

where

$$\sigma(m) = \text{sum of the positive divisors of } m.$$

We can rewrite equation (5.3) as

$$nP(n) = \sum_{m=0}^{n-1} \sum_{\substack{\text{divisors} \\ d \text{ of } n-m}} dP(m),$$

and all we need is a constructive proof of this identity.

Let Π be a fixed partition of $m < n$ and let d divide $(n - m)$. Associate with (Π, d) d copies of the partition Π' constructed by adjoining $(n - m)/d$ copies of d to Π:

$$(\Pi, d) \rightarrow \begin{cases} \Pi' = \Pi \ \uplus \ \left\{ \dfrac{n - m}{d} \bullet d \right\} \\ \qquad \vdots \\ \Pi' = \Pi \ \uplus \ \left\{ \dfrac{n - m}{d} \bullet d \right\} \end{cases}$$

As d varies over the divisors of $n - m$ and Π varies over the partitions of m, $0 \leq m \leq n - 1$, each possible partition of n appears in the list on the right exactly n times, establishing equation (5.3). To see why, let $\Pi' = \{m_1 \bullet p_1, \ldots, m_l \bullet p_l\}$ be a fixed partition of n. Then Π' is constructed by adjoining k copies of p_i to the partition of $n - kp_i$, obtained by deleting k copies of p_i from Π' for $1 \leq k \leq m_i$ and $1 \leq i \leq l$. Each such partition is repeated p_i times. Thus Π' occurs a total of

$$\sum_{i=1}^{l} \left(p_i \sum_{k=1}^{m_i} 1 \right) = \sum_{i=1}^{l} m_i p_i = n$$

times.

The details of this procedure are given in Algorithm 5.13.

$$\Pi \leftarrow \varnothing$$

$$t \leftarrow n$$

while $t \neq 0$ **do** $\begin{cases} \text{choose } m, 0 \leq m \leq t - 1, \text{ according to the} \\[4pt] \text{distribution } \Pr(m) = \dfrac{\sigma(t - m)P(m)}{tP(t)} \\[12pt] \text{choose a divisor } d \text{ of } t - m \text{ according to} \\[4pt] \text{the distribution } \Pr(d) = \dfrac{d}{\sigma(t - m)} \\[12pt] \Pi \leftarrow \Pi \uplus \left\{ \dfrac{t - m}{d} \bullet d \right\} \\[10pt] t \leftarrow m \end{cases}$

Algorithm 5.13   Generation of a random partition of n.

Proving that Algorithm 5.13 generates a fixed partition Π of n with probability $1/P(n)$ is a straightforward induction. For $n = 0$, the empty partition is generated with probability $1 = 1/P(0)$. Suppose that the algorithm works for all $n' < n$. The probability of Π is then

$$\Pr(\Pi) = \sum_{\substack{\text{all partitions} \\ \Pi' \text{ of some } m, \\ 0 \leq m \leq n-1}} \Pr(\Pi') \Pr(\Pi' \text{ is extended to } \Pi)$$

$$= \sum_{\Pi'} \frac{1}{P(m)} \Pr(\Pi' \text{ is extended to } \Pi).$$

But the probability that Π' is extended to Π is zero unless Π can result from Π' by the adjunction of k copies of m of the parts p_i of Π', and so

$$\Pr(\Pi' \text{ is extended to } \Pi) = \Pr(m = n - kp_i) \Pr(d = p_i)$$

$$= \frac{\sigma(kp_i)P(m)}{nP(n)} \cdot \frac{p_i}{\sigma(kp_i)}$$

$$= \frac{P(m)}{nP(n)} p_i.$$

Thus

$$\Pr(\Pi) = \sum_{\Pi'} \frac{1}{P(m)} \frac{P(m)}{nP(n)} p_i = \frac{1}{nP(n)} \sum_{\Pi'} p_i$$

$$= \frac{1}{nP(n)} \sum_{i=1}^{l} \sum_{k=1}^{m_i} p_i = \frac{1}{nP(n)} \sum_{i=1}^{l} m_i p_i$$

$$= \frac{n}{nP(n)} = \frac{1}{P(n)},$$

as desired.

5.4 REMARKS AND REFERENCES

Algorithms for generating all kinds of combinatorial configurations can be found in the algorithms department of the *Communications of the ACM* from February 1960 through June 1975 when that department was moved to the *ACM Transactions on Mathematical Software*. These algorithms are reprinted in *Collected Algorithms from ACM*.

The combinatorial properties of permutations are ably discussed in

> KNUTH, D. E., *The Art of Computer Programming*, Vol. 3 (*Sorting and Searching*), Addison-Wesley, Reading, Mass., 1973.

A general survey and bibliography of the generation of permutations appear in the two-part article

> ORD-SMITH, R. J., "Generation of Permutation Sequences," *Computer J.*, **13** (1970), 152–155 and **14** (1971), 136–139.

Lexicographic algorithms for permutations have been discovered independently many times. The first mention of them goes back to L. L. Fischer and K. C. Krause in 1812. The most efficient algorithm known in this case is described in Exercise 4; it is due to

> ORD-SMITH, R. J., "Algorithm 323: Generation of Permutations in Lexicographic Order," *Comm. ACM*, **11** (1968), 117.

See also the certification of this algorithm [*Comm. ACM*, **12** (1969), 512] and the remark on it [*Comm. ACM*, **16** (1973), 577–578].

The observation that the inversion vector uniquely determines the corresponding permutation is due to Marshall Hall, Jr. See

> TOMPKINS, C. B., "Machine Attacks on Problems whose Variables Are Permutations," in *Numerical Analysis* (*Proceedings of Symposia in Applied Mathematics*, Vol. 6), American Mathematical Society, Providence, R.I., 1956.

This article also outlines the generation of permutations by nested cycles and describes a number of combinatorial computing problems in which the sequential generation of permutations is required.

That the permutations could be generated by transpositions was first published in

WELLS, M. B., "Generation of Permutations by Transposition," *Math. Comp.*, **15** (1961), 192–195.

In Wells' method, however, the transpositions are not always of adjacent elements. The method of Section 5.1.4, in which the transpositions are always of adjacent elements, was discovered almost simultaneously by

TROTTER, H. F., "Algorithm 115: Perm," *Comm. ACM*, **5** (1962), 434–435

and by

JOHNSON, S. M., "Generation of Permutations by Adjacent Transpositions," *Math. Comp.*, **17** (1963), 282–285.

That particular sequence of permutations has an interesting musical application; for the details, see

PAPWORTH, D. G., "Computers and Change-Ringing," *Computer J.*, **3** (1960), 47–50.

The most efficient minimal change permutation generator now known results from applying the macro expansion techniques of Section 4.1.4 to the minimal change algorithm given in Section 5.1.4 (see Exercise 13). See

EHRLICH, G., "Algorithm 466: Four Combinatorial Algorithms," *Comm. ACM*, **16** (1973), 690–691

and

DERSHOWITZ, N., "A simplified Loop-Free Algorithm for Generating Permutations," *BIT*, **15** (1975), 158–164.

Ehrlich's paper also contains linear algorithms for generating combinations and compositions in minimal change order. The theory behind these and similar algorithms for Gray code and set partition algorithms is found in

EHRLICH, G., "Loopless Algorithms for Generating Permutations, Combinations, and Other Combinatorial Configurations," *J. ACM*, **3** (1973), 500–513.

The algorithm for random permutations was first published in

MOSES, L. E., and R. V. OAKLAND, *Tables of Random Permutations*, Stanford University Press, Stanford, Ca., 1963.

It was independently published in

DURSTENFELD, R., "Algorithm 235: Random Permutation," *Comm. ACM*, **7** (1964), 420

and in

DE BALBINE, G., "Note on Random Permutations," *Math. Comp.*, **21** (1967), 710–712.

Both de Balbine's paper and

PIKE, M. C., "Remark on Algorithm 235," *Comm. ACM*, **8** (1965), 445

observe that this method can be applied to obtain the random combination algorithm given in Section 5.2.2.

Gray codes were used in the nineteenth century in the form of a solution to a puzzle known as "The Chinese Ring." The name is derived from Frank Gray, who developed the codes for use in preventing errors in certain signal transmissions. The first publication of the code was in

GRAY, F., "Pulse Code Communication," U.S. Patent 2 632 058, March 17, 1953.

Further details of the history of the Gray code can be found in

HEATH, F. G., "Origins of the Binary Code," *Scientific American*, **227**, 2 (August 1972), 76–83.

Also, see the Mathematical Games column of that same issue of *Scientific American*, where Martin Gardner discusses the Gray code.

There are many Gray codes. However, the binary-reflected Gray code discussed in Section 5.2.1 is the most widely known, and it is frequently referred to as *the* Gray code. Algorithm 5.7(c) for generating the entire binary-reflected Gray code is from

BITNER, J. R., G. EHRLICH, and E. M. REINGOLD, "Efficient Generation of the Binary Reflected Gray Code and its Applications," *Comm. ACM*, **19** (1976), 517–521.

A similar algorithm is described in

MISRA, J., "Remark on Algorithm 246," *ACM Trans. Math. Software*, **1** (1975), 285.

The algorithm for generating combinations in lexicographic order first appeared in

MIFSUD, C. J., "Algorithm 154: Combination in Lexicographic Order," *Comm. ACM*, **6** (1963), 103,

although he gave no analysis of the time required or of the index function. The analyses given here are due to Brian Hansche. The possibility of a minimal change order algorithm for combinations based on the Gray code was demonstrated in

TANG, D. T., and C. N. LIU, "Distance-2 Cycle Chaining of Constant Weight Codes," *IEEE Trans. Computers*, **22** (1973), 176–180.

Their algorithm, which appeared in

LIU, C. N., and D. T. TANG, "Algorithm 452: Enumerating Combinations of *m* Out of *n* Objects," *Comm. ACM*, **16** (1973), 485,

is inordinately complicated. The algorithm and analysis given in Section 5.2.2 is taken from the paper by Bitner, Ehrlich, and Reingold cited above.

Compositions and partitions of integers have been areas of intensive mathematical research for over 300 years, since Leibniz asked Bernoulli if he had investigated $P(n)$, the number of partitions of n. Details of the history and the state of the art as of 1920 can be found in Chapter 3 of

> DICKSON, L. E., *History of the Theory of Numbers*, Vol. II, *Diophantine Analysis*, Chelsea Publishing Co., New York, 1971.

Additional details and later results can be found in most combinatorics texts; in particular, see

> HALL, M., *Combinatorial Theory*, Blaisdell, Waltham, Mass., 1967.

One result (see pages 39–42 of Hall's book) worth mentioning here is that

$$P(n) \sim \frac{1}{4n\sqrt{3}} e^{\pi\sqrt{2n/3}}.$$

The first algorithm for generating partitions given in Section 5.3.2 is very old, having been discovered by K. F. Hindenburg in 1778. The second, and much more efficient, partition generation is due to Gideon Ehrlich.

The general method for generating random combinatorial objects and its application to partitions is from

> NIJENHUIS, A., and H. S. WILF, "A Method and Two Algorithms on the Theory of Partitions," *J. Combinatorial Theory* (A), **18** (1975), 219–222.

5.5 EXERCISES

1. (a) Calculate the index of the permutation $(4, 3, 5, 2, 7, 1, 9, 8, 6)$ in the various orders.
   (b) Calculate the millionth permutation of 12 elements in the various orders.

2. Complete the analysis of Algorithm 5.1. Show that

$$C_n = nC_{n-1} + \frac{(n-1)(3n-2)}{2}$$

$$C_1 = 0$$

and solve this recurrence. [*Hint*: Substitute $T_n = C_n + (3n + 1)/2$.] Use the values of I_n and C_n to give a complete breakdown of the total number of operations used in the algorithm, not just those involving the elements of the permutation. Count an $\leftrightarrow$ operation as three $\leftarrow$ operations and consider each comparison as an arithmetic operation. Prove that the algorithm works correctly.

★3. Show that a permutation Π is even if and only if $\sum_{j=1}^{n-1} d_j$ is even, where $i = d_1 1! + d_2 2! + \cdots + d_{n-1}(n-1)!$ is the index of Π in the lexicographic order.

4. Find a different algorithm for generating the lexicographic sequence of permutations by keeping track of the factorial representation of the index of the current permutation and using it to find the i and j needed in the interchange $\pi_i \leftrightarrow \pi_j$. Analyze this algorithm and compare it with the analysis of Exercise 2.

5. Modify the algorithm described in Exercise 4 so that instead of interchanging $\pi_i \leftrightarrow \pi_j$ and reversing the $n - i$ rightmost elements, it simply reverses the $n - i + 1$ rightmost elements. Show that the resulting sequence of permutations still has the property that all $k!$ permutations of the rightmost k elements are produced before any of the leftmost $n - k$ elements move. Analyze the resulting algorithm in terms of the number of operations performed and compare it with the number required by the algorithm of Exercise 4.

6. (a) Find the inversion vector for the permutation $(4, 3, 6, 7, 2, 5, 8, 9, 1)$.
   (b) Find the permutation whose inversion vector is

k	1	2	3	4	5	6	7	8
d_k	0	1	2	1	2	5	0	2

7. Prove by induction that an inversion vector uniquely determines a permutation.

8. Write the permutation of Exercise 6(a) in nested cycle form.

9. What are the last two permutations of $\{1, 2, \ldots, n\}$ in the minimal-change sequence described in Section 5.1.4?

★10. Show that the minimal change sequence has the property that the permutations alternate even and odd. Is this true of the lexicographic order?

11. Show how the inner loop of Algorithm 5.3 can be eliminated (by the appropriate list-processing techniques) to make the time required to transform one permutation to the next constant, independent of n. (*Hint:* Read Section 5.2.1.) After working Exercise 12, determine whether this change results in a savings.

12. Show that the comparison "$\pi_{p_m + d_m} > m$" in Algorithm 5.3 is executed $\sum_{i=1}^{n} i!$ times. Explain why the comparison "$m \neq 1$" is executed $n! + 1$ times. Use these two facts to count, as in Exercise 2, the total number of operations required to generate the $n!$ permutations and compare the answers to the results of Exercises 2 and 11.

13. In most algorithms for the systematic generation of permutations, a relatively few elements do most of the moving. For example, in the lexicographic sequence, the rightmost element moves half the time, whereas in the minimal change sequence the largest value (n) moves almost all the time. Develop faster algorithms by writing separate code for these frequent interchanges rather than allowing them to be done in the general loop. This idea occurred in Section 4.1.4 in the use of macros to "customize" programs.

14. Suppose that we change the random permutation algorithm, Algorithm 5.4, to

   for $i = 1$ **to** n **do** $\pi_i \leftrightarrow \pi_{\text{rand}(1,n)}$;

that is, at the ith step the ith element is interchanged with an element chosen at random from all the n elements. Are the permutations still all equally probable? (*Hint:* $n!$ is not a divisor of n^n for $n \geq 3$.) Assuming that originally $\pi_i = i$, $1 \leq i \leq n$, calculate the probability of the identity permutation. [*Hint:* Show that the number of ways that the identity permutation can arise is given by the recurrence relation $a_n = a_{n-1} + (n-1)a_{n-2}$.]

15. Write a program to determine, for various values of n, the number of permutations $\Pi = (\pi_1, \pi_2, \ldots, \pi_n)$ of $\{1, 2, \ldots, n\}$ that have the property that $\pi_i - i \equiv \pi_j - j \pmod n$ implies that $i = j$. What do you conjecture in general for even n? Prove it.

16. Develop and analyze an algorithm for generating all permutations of a multiset. Can you make your algorithm linear? Minimal change?

17. Develop an algorithm to generate all permutations consistent with a given partial ordering. (*Hint:* Use backtrack.)

18. Develop and analyze an algorithm to compute the inverse of a permutation *in place*.

19. (a) A *derangement* of the set $\{1, 2, \ldots, n\}$ is a permutation $\Pi = (\pi_1, \pi_2, \ldots, \pi_n)$ for which $\pi_i \neq i$ for any i. How many derangements are there of $\{1, 2, \ldots, n\}$? [*Hint:* Establish the recurrence relation $d_n = (n-1)(d_{n-1} + d_{n-2})$.]
    (b) Develop and analyze a linear algorithm to generate all derangements of $\{1, 2, \ldots, n\}$. [*Hint:* Use the recurrence relation from part (a).]
    (c) Develop an algorithm to generate a random derangement of $\{1, 2, \ldots, n\}$ such that each derangement is equally likely. [*Hint:* Use the technique of Section 5.3.2 with the recurrence relation from part (a).]

20. (a) The *rosary* permutations are the $\frac{1}{2}(n-1)!$ inequivalent (under rotations and reflections) ways of placing the elements of the set $\{1, 2, \ldots, n\}$ around the circumference of a circle. Develop and analyze an algorithm for generating the rosary permutations. Is your algorithm linear? Minimal change?
    (b) Develop and prove correct an algorithm for the generation of a random rosary permutation, assuming that each of the $(n-1)!$ is equally probable.

21. Let γ_1 and γ_2 be successive codewords in $G(n)$ having exactly k 1-bits, $1 \leq k \leq n - 1$. Prove that
$$\text{index } (\gamma_2) - \text{index } (\gamma_1) = 2^t$$
for some $t > 0$.

22. Show that not all Gray codes have the property that successive codewords with exactly k 1-bits differ in only two bits. (*Hint:* Find a counterexample with $n = 4$ and $k = 2$.)

23. Develop and analyze an algorithm for generating combinations of a multiset. Can you make your algorithm linear? Minimal change?

24. Prove that every integer from 0 to $\binom{n}{k} - 1$ can be written uniquely as
$$\binom{a_1}{1} + \binom{a_2}{2} + \cdots + \binom{a_k}{k},$$

where $0 \leqq a_1 < a_2 < \cdots < a_k < n$. How can this be used to generate the combinations of n things taken k at a time? Is it a reasonable method?

25. How would you use Algorithm 5.9 to generate the $\binom{n+k-1}{n}$ distinguishable distributions of n indistinguishable objects into k distinguishable boxes? (*Hint:* Read Section 5.3.1.)

26. Generalize the construction of the binary reflected Gray code given in Section 5.2.1 to other bases.

27. In Algorithm 5.10 approximately $\lg[n!/(n-k)!]$ random bits are generated (why?), but in theory only $\lg[n!/(n-k)!k!]$ bits are needed to distinguish the $\binom{n}{k}$ possible outcomes. How are the $\lg k!$ bits wasted?

28. Develop and analyze an algorithm for generating compositions in increasing length, in lexicographic order for each fixed length. Is your algorithm linear?

29. Develop and analyze a minimal change algorithm for k-part compositions (the Gray code causes more than a minimal change in the associated k-part composition). What is a good definition of minimal change between k-part compositions?

30. Show that the k-part compositions of n can be generated in increasing lexicographic order by generating, in increasing lexicographic order, the sequences

$$0 = \delta_0 \leqq \delta_1 \leqq \delta_2 \leqq \cdots \leqq \delta_{k-1} \leqq \delta_k = n.$$

(*Hint:* Define $p_i = \delta_i - \delta_{i-1}, i = 1, \ldots, k$.) Show that in this order the index of a composition is given by

$$\text{index}\,[(p_1, p_2, \ldots, p_k)] = \sum_{i=p_k+1}^{n} \binom{i+k-t_i-1}{i},$$

where

$$t_i = \max_{1 \leqq t \leqq k} \left\{ t \,\middle|\, \sum_{j=t}^{k} p_j \geqq i \right\}.$$

Finally, show that, given an index I, the corresponding composition in this order can be determined directly from I.

31. Prove that Algorithm 5.11 works correctly. Three things must be established for partitions of length l.
    (a) The transformation yields another partition in which the components are in increasing order.
    (b) The transformation yields the next partition in lexicographic order.
    (c) Only the last partition (of length l) in lexicographic order has the property that $p_l - p_1 \leqq 1$.

32. Rewrite Algorithm 5.11, using the multiset representation of partitions. The resulting algorithm should require only constant time (independent of either n or the length of the partition) to transform one partition to the next.

33. Prove that Algorithm 5.12 works correctly.

34. (a) What changes need to be made in Algorithm 5.12 so that it generates only the partitions of n in which the components satisfy $p_i \leq u$ for some u, $1 \leq u \leq n$?
    (b) Design an algorithm similar to the one in part (a) to generate only the partitions of n in which the components satisfy $p_i \geq u$ for some u, $1 \leq u \leq n$. (*Hint*: Generate the partitions in dictionary order with each partition satisfying $p_1 < p_2 < \cdots < p_l$.)

★35. Develop a linear, minimal change algorithm for generating the partitions of a set into k subsets. Your algorithm should restrict itself to moving a single element from one subset to another subset.

★36. Apply the general technique for generating random combinatorial objects to set partitions. (*Hint*: Let a_n be the number of partitions of a set of n elements. Use the identity

$$a_n = \sum_{m=0}^{n-1} \binom{n-1}{m} a_m, \qquad n \geq 1$$

$$a_0 = 1.)$$

37. An n-dimensional lattice point is an n-tuple $(x_1, x_2, \ldots, x_n)$ of integers satisfying $l_i \leq x_i \leq u_i$ for some vectors of integers $(l_1, l_2, \ldots, l_n)$ and $(u_1, u_2, \ldots, u_n)$. Develop and analyze both lexicographic and minimal change algorithms for generating lattice points.

chapter 6

Fast
Search

The problem of searching is fundamental in combinatorial algorithms, for it is phrased in such generality that it encompasses many problems of practical importance. In its most general setting, "Search a set T to find an element satisfying a certain condition C," hardly anything of interest can be said about the search problem. Surprisingly, however, with just a few assumptions about the structure of the set T, the problem becomes interesting because a variety of search strategies of different efficiencies is possible.

In Chapter 4 we discussed the situation in which all or a large fraction of the set T must be examined in order to guarantee that an element meeting the specification is found, if it exists; for a set with n elements, this situation leads to an $O(n)$ search time. In this chapter we will make assumptions on the structure of the set T that allow us to search T in times $O(\log n)$ or $O(1)$. Most search algorithms fall into one of the three categories characterized by search times $O(n)$, $O(\log n)$, and $O(1)$.

6.1 SEARCHING AND OTHER OPERATIONS ON TABLES

Any search technique operates on elements, which we will call *names*, drawn from a set of names, S, called the *name space*. This name space can be finite or infinite. The most common name spaces are sets of integers, with their numeric order, and sets of character strings over some finite alphabet, with their lexicographic (i.e., dictionary) order.

Each of the search algorithms discussed in this chapter is based on *one* of the following three assumptions about the name space S.

Assumption 1

A linear order, called the *natural order* and denoted by $<$, is defined on S. Such an order has the following properties:

1. Any two elements x, y in S are comparable; that is, exactly one of three conditions $x < y, x = y, y < x$ must hold.

2. The order is transitive; that is, for any elements $x, y, z \in S$, if $x < y$ and $y < z$, then $x < z$.

We use the notations $>, \leq, \geq$ with their obvious meanings. In analyzing the efficiency of a search algorithm, we assume that the outcome ($<$, $=$, or $>$) of a comparison $x : y$ is obtained in time independent of $|S|$, the cardinality of the name space S, and also in time independent of the cardinality n of the table T to be searched. *see margin*

Assumption 2

Each name in S is a string of characters, or digits, over a finite, linearly ordered alphabet $A = \{a_1, a_2, \ldots, a_c\}$. The natural order on S is the lexicographic order induced by the linear order on A. We assume that the outcome ($<$, $=$, or $>$) of a comparison of two *characters* (not names) is obtained in time independent of $|S|$ or n.

Assumption 3

There is a function $h : S \rightarrow \{0, 1, \ldots, m - 1\}$ that maps the name space S into the set $\{0, 1, \ldots, m - 1\}$ in a uniform way; that is, each integer i, $0 \leq i \leq m - 1$, occurs approximately equally often as the image of a name in S under the function h. We assume that the function h can be evaluated in time independent of $|S|$ and n.

Assumption 1 is very strong because there is no theoretical reason why a comparison of names should be possible in time independent of the size of the name space. However, it often holds, in practice, when a name can be stored in a fixed number of memory locations. Assumption 2 is the most realistic of the three assumptions. Assumption 3, in its insistence that the time for evaluating h is independent of $|S|$, is theoretically shaky but often holds in practice.

As already noted, searching does not take place in the name space S itself but in a finite subset $T = \{x_1, x_2, \ldots, x_n\}$ of S, which we will call a *table*. On most tables that we consider, a linear order called the *table order* is defined; it is indicated by the subscripts attached to names (i.e., x_1 is the first name in the table, x_2 the second, and so on). The table order is often the same as the natural order defined on the name space, but it need not be.   *| T, table*

The cardinality of the table T is usually much smaller than the cardinality of the name space S even when S is finite. For example, the telephone directory of a large city might contain 10^6 entries, or names, which might be considered to be drawn from a name space of strings of at most 20 characters, over an alphabet of about 30 distinct characters. With these assumptions we have $n = |T| \approx 10^6$ and $|S| \approx 10^{30}$. Name spaces are normally so large, in practice, that it often makes no difference whether a name space is considered to be finite or infinite. In the example of the telephone directory, we could equally well assume that the name space consisted of all character strings of arbitrary length.

The notion of a table as an ordered set of names is sufficient for many purposes, but occasionally it is necessary to consider a table as a set of *cells*, each of which is capable of holding one name. For instance, if we consider the details of the process of inserting a new name, we find that the table may first have to be extended by adding a cell to make room for a new entry; only later is the new name entered.

Throughout this chapter we will assume that names occur at most once in a table (with the possible exception of the transition period during which a new name is being entered; at that time a table may contain two occurrences of the same name). In most cases, this assumption implies that a table with n names has exactly n cells. However, the important class of algorithms based on address computation (Section 6.5) is based on the assumption that the table contains more cells than names. These algorithms must explicitly consider the presence of empty cells.

A number of synonyms are used for what we call a table and a name. In data processing there are *files* whose elements are *records*; each record is a sequence of *fields*, one of which (the one that is involved in the search) is called the *key*. If the file itself is traversed during search, then "file" and "key" correspond to what we call "table" and "name," respectively. (This terminology is slightly ambiguous because "key" can refer either to the cell itself or to the contents of the cell.) However, searching a large file often does not involve a traversal of the file itself; instead a *directory* or *index* to the file is searched. If the search is successful, the entry found in the index provides a pointer to the proper record in the file. In this type of file organization, the index or directory corresponds to our notion of a table.

This chapter discusses many algorithms for searching a table for the occurrence of a name z. Of all operations that are performed on tables, finding a single name is by far the most common. Indeed, there are tables on which no other operation is ever performed once the table has been constructed. For such tables it is reasonable to expend a considerable effort arranging the table so as to minimize search times. Optimal binary search trees (Section 6.3.2) and certain address computation techniques (Section 6.5) illustrate this situation.

The majority of tables, however, are subject to operations other than searching for names. Operations like modifying the table, listing the names in a desired order, merging several tables into one, and sorting a table are the most important ones. When some of these operations must be performed efficiently, we are no longer free to optimize the structure of the table for the fastest possible searching. A compromise results in which search times are longer in return for a more flexible table organization—that is, an organization that permits several kinds of operations to be performed with relative ease. Almost all search algorithms in this chapter result from such a compromise.

We consider only four table operations: *search, insert, delete,* and *list.* The detailed specification of what these operations must do on a table $T = \{x_1, x_2, \ldots, x_n\}$ depends on the data structure used to implement the table.

search for z: if $z \in T$ then return a pointer to it; that is, assign a value to i such that $z = x_i$; otherwise indicate that $z \notin T$.

insert z: if $z \notin T$ then insert it in the appropriate place.

Inserting generally involves finding the appropriate place first; therefore it is sometimes convenient to separate the operation into two phases. First we use the search procedure to find where z must be inserted, and then we insert z at that place.

insert z at i: insert the name z immediately after name x_i.
Before insertion $T = \{x_1, \ldots, x_i, x_{i+1}, \ldots, x_n\}$,
After insertion $T = \{x_1, \ldots, x_i, z, x_{i+1}, \ldots, x_n\}$.

delete z: if $z \in T$, then remove it.

Like insertion, deletion is sometimes realized by using the search procedure to obtain the location of z and then using a procedure

delete at i: delete x_i from T.
Before deletion $T = \{x_1, \ldots, x_{i-1}, x_i, x_{i+1}, \ldots, x_n\}$,
After deletion $T = \{x_1, \ldots, x_{i-1}, x_{i+1}, \ldots, x_n\}$.

A table in which insertions or deletions are made is said to be *dynamic*; otherwise it is *static.*

The last operation we will consider for each table structure is

list: list all names in T in their natural order.

When the table order coincides with the natural order, listing is a simple operation; when the two are unrelated, as is the case with most address computation techniques, listing involves sorting.

Among the operations that might be performed on tables, the four considered in this chapter (search, insert, delete, and list), along with sorting (Chapter 7), are the most important ones. Sorting is discussed separately because it is a time-consuming operation compared to the other four: it requires time at least proportional to $n \log n$, whereas the other operations can all be done in time $O(n)$ with most table structures (with the exception of listing the elements in natural order with most address computation techniques).

6.2 SEQUENTIAL SEARCH

Sequential search examines names in the order in which they occur in the table. When it is used to search a table, a scan of the entire table results in the worst case; even on the average, sequential search tends to require a number of operations proportional to n. It should not be considered a fast search technique for large tables, since it is asymptotically far slower than the other search algorithms described in this chapter. Despite its poor asymptotic performance, there are several reasons for discussing it initially. First, although a simple technique, it allows us to introduce important concepts and techniques applicable to searching in general. Second, sequential search is the only search technique applicable for certain storage media or when there is no linear order on the name space. Finally, sequential search is fast for sufficiently small tables and for large tables that are arranged in hierarchical fashion: a faster search technique is used to move around near the top of the hierarchy, but a sequential search is used for the subtables at the lowest level of the hierarchy.

For sequential search of a table $T = \{x_1, x_2, \ldots, x_n\}$, we assume that we have a pointer i whose value is in the range $1 \leq i \leq n$, or perhaps $0 \leq i \leq n + 1$, and that the only operations that can be performed on this pointer are initializing it to 1 or to n (or, if appropriate, to 0 or $n + 1$), incrementing and/or decrementing it by one, and comparing it to 0, 1, n, or $n + 1$. With these conventions, the most obvious algorithm for searching a table T for the first occurrence of a given name z can be expressed as in Algorithm 6.1. Here, as in all other searching algorithms in this chapter, we assume that the algorithm stops immediately on finding z or on discovering that z is not in the table.

> **for** $i = 1$ **to** n **do if** $z = x_i$ **then** [found: i points to z]
> [not found: z does not occur in T]

Algorithm 6.1   Sequential search.

Let us consider some aspects of efficiency of sequential search, beginning with standard programming techniques. In a program whose structure is a single loop, such as Algorithm 6.1, any significant speedup must come from improving the code in the loop. In order to see which operations are performed inside the loop, it is necessary to rewrite Algorithm 6.1 in a notation closer to machine language.

> $i \leftarrow 1$
> loop: **if** $z = x_i$ **then** [found]
>   **if** $i = n$ **then** [not found]
> $i \leftarrow i + 1$
>   **goto** loop

During each iteration up to four instructions are executed: two comparisons, one incrementing operation, and one jump.

A common trick to make such an inner loop faster is to add special entries to the table that make it unnecessary to check explicitly whether the pointer has reached the boundaries of the table. This can be done in Algorithm 6.1. If before the search we add the name z that we are looking for at the end of the table, then the loop will always terminate by finding an occurrence of z; thus we need not make the test $i = n$ each time through the loop. At the end of the loop, a test for $i > n$, executed only once, tells us whether the occurrence of z that was found was a true table entry or the special entry. This technique is illustrated in Algorithm 6.2.

$$x_{n+1} \leftarrow z$$
$$i \leftarrow 1$$
while $z \neq x_i$ **do** $i \leftarrow i + 1$
if $i \leq n$ **then** [found: i points to z]
                else [not found]

Algorithm 6.2   Improved sequential search.

The improvement over Algorithm 6.1 is most apparent if we rewrite Algorithm 6.2 in the same lower-level notation used earlier:

$$x_{n+1} \leftarrow z$$
$$i \leftarrow 1$$
loop: **if** $z = x_1$ **then goto** maybe
        $i \leftarrow i + 1$
        goto loop
maybe: **if** $i \leq n$ **then** [found: i points to z]
              else [not found]

At each iteration, only three statements are executed instead of four as in Algorithm 6.1. Thus on most computers the loop of Algorithm 6.2 will be much faster than in Algorithm 6.1; and since the speed of the loop determines the speed of the entire program, the same comparison holds for the two programs.

It is worth noticing that the trick of adding z to the end of the table before searching is only possible if we have direct, immediate access to the end of the table. This is possible if the table is stored sequentially in random access memory but is not generally possible when a linked allocation or sequential access memory (such as tape) is used.

The only disadvantage of Algorithm 6.2 is that unsuccessful searches (those for names not in the table) always scan the entire table. If such searches occur frequently, then names should be stored in natural order; this allows the search to be terminated as soon as the first name larger than or equal to the search argument has been scanned. In this case, we add a dummy name ∞ at the end of the table to ensure that the termination condition is always met. ∞ is a new name that is

assumed to be larger than any name in the name space S. Thus we obtain
Algorithm 6.3.

$$x_{n+1} \leftarrow \infty$$
$$i \leftarrow 1$$
while $z > x_i$ **do** $i \leftarrow i + 1$
if $z = x_i$ **then** [found: i points to z]
    else [not found]

Algorithm 6.3   Sequential search in a table stored in natural order.

In order to analyze the time required for a sequential search in a table
containing n names, we consider the time required to examine one entry in the
table as one unit of time. Examining a table entry involves a comparison between
a pair of names, incrementing a pointer, transferring control, and possibly, in the
less-efficient algorithms, a comparison between a pair of pointers. Thus i units of
time are required to find the ith name x_i in the table. To compute the average
search time, we must make an assumption about the frequency with which
individual names are accessed. The simplest assumption, and the most natural
one in the absence of specific information about actual frequencies, is to assume
that the access frequencies are uniform—that is, that all names in the table are
accessed equally often. This assumption leads to an average search time S_n for
successful searches only of

$$S_n = \frac{1}{n} \sum_{i=1}^{n} i = \frac{n+1}{2}.$$

For unsuccessful searches, the average search time is approximately $n/2$ when the
names are stored in natural order or exactly n when they are not (see Exercise 1).

The assumption of uniform access frequencies rarely holds in practice; a
small fraction of the names usually accounts for a large fraction of all searches. If
a table is static (or if modifications occur only rarely), it is possible to collect
statistics about access frequencies and use the empirically discovered distribution
to reorganize the table so as to improve search times. If α_i denotes the access
frequency of the name x_i, that is, $\alpha_i \geq 0$ and $\sum_{i=1}^{n} \alpha_i = 1$, then the average search
time $S_n = \sum_{i=1}^{n} i\alpha_i$ is minimal when names occur in the table in order of nonincreasing access frequency—that is, when the table order is such that $\alpha_1 \geq \alpha_2 \geq$
$\cdots \geq \alpha_n$. (See Exercise 2 for a proof and Exercise 3 for a generalization.)

In practice, access frequencies are often far from uniform, so that arranging
the table in order of nonincreasing access frequencies can make a substantial
reduction in the average search time. For example, one frequency distribution
that governs access to many tables in practice is *Zipf's law*:

$$\alpha_i = \frac{c}{i} \qquad \text{for } i = 1, 2, \ldots, n,$$

where the normalization constant c is chosen so that

$$\sum_{i=1}^{n} \alpha_i = 1.$$

Thus

$$c = \frac{1}{\sum_{i=1}^{n} 1/i} = \frac{1}{H_n} \approx \frac{1}{\ln n}$$

by equation (3.13). When names are ordered in a table according to decreasing access frequencies that satisfy Zipf's law, then the average search time for successful searches is

$$S_n = \sum_{i=1}^{n} i\alpha_i = \frac{1}{H_n} \sum_{i=1}^{n} 1 = \frac{n}{H_n} \approx \frac{n}{\ln n},$$

which is approximately $\frac{1}{2} \ln n$ times faster than the corresponding average times of $\frac{1}{2}(n + 1)$ for tables stored in natural order. Exercise 4 describes another frequency distribution in which reordering the names makes a significant difference in the average successful search time.

Notice that in reducing the average successful search time, we have *increased* the time required in unsuccessful searches from $n/2$ to n (since the table order differs from the natural order).

When the frequency distribution is not known *a priori* and empirical observations are impossible or inconvenient, we can still take advantage of nonuniform access frequencies to reduce the average successful search time. In a *self-organizing file* the table order is changed as the table is used so that frequently accessed names move toward the beginning of the table. This topic is explored in Exercise 5.

We have seen that the implementation and performance of the search algorithm depend on such properties of the table design as whether names are stored in natural order. For the other common table operations, this dependence is even more pronounced. For instance, listing the names in the natural order is an $O(n)$ sequential scan when the table order coincides with the natural order but requires sorting [an $O(n \log n)$ process; see Chapter 7] when it does not. Insertion and deletion, on the other hand, depend primarily on the data structure chosen to represent the table rather than on the order of names in the table.

The two most important data structures for sequential search are linked lists, which can be used only in random access memories, and sequentially allocated lists, which can be used with almost any storage medium. As we saw in Chapter 2, insertions and deletions in a linked list only require changing a few pointers once the location is known (see Figures 2.4 and 2.5). Thus if finding the correct location is separated from the actual insertion or deletion, the latter two operations can be done in time independent of the table size.

In a sequentially allocated table, consecutive names are stored in consecutive locations (see Figure 2.1). So deletion of a name creates a gap that must be filled, and insertion of a name demands that a gap be created. In either case, many names in the table must be moved or copied, and as Algorithm 6.4 shows, these operations require time $O(n)$.

insert z at i: $n \leftarrow n + 1$
                      for $j = n - 1$ **to** i **by** -1 **do** $x_{i+1} \leftarrow x_j$
                      $x_i \leftarrow z$

delete at i: **for** $j = i$ **to** $n - 1$ **do** $x_j \leftarrow x_{j+1}$
                  $n \leftarrow n - 1$

Algorithm 6.4   Insertion and deletion in a sequentially allocated table. The location of the insertion/deletion is given by pointer i.

In this section we have seen that even the simple sequential search requires clever design choices. But, more importantly, we have seen that the use of a technique to improve an algorithm in one respect may degrade it in other respects. For example, a table cannot usually be stored both in the natural order and in order of nonincreasing access frequencies. Consequently, the design of a table structure and its corresponding algorithms must often compromise between several desirable, but mutually incompatible, features. Unfortunately, we will find that these design decisions become more difficult when we consider more sophisticated data structures and search algorithms in the rest of the chapter.

6.3. LOGARITHMIC SEARCH IN STATIC TABLES

Logarithmic search times can be obtained whenever it is consistently possible, in an amount of time c independent of n, to reduce the problem of searching a table containing n names to the problem of searching a table containing at most αn names, where $\alpha < 1$ is a constant. In this case, the time $t(n)$ required to search a table with n names satisfies the recurrence relation

$$t(n) = c + t(\alpha n)$$

whose solution is of the form

$$t(n) = c \log_{1/\alpha} n + b,$$

where b is determined by initial conditions and c, the proportionality constant of the logarithmic term, is the time required to reduce the table size from n names to αn names.

The most common assumptions that make it possible to reduce the table size from n to αn in an amount of time independent of n are that the name space S is linearly ordered and that the comparison of two names, x, y in S (to determine whether $x < y$, $x = y$, or $x > y$) is an elementary operation requiring a constant amount of time independent of n. As a result, the time requirement of most logarithmic search algorithms is naturally measured by the number of comparisons (with three outcomes) of pairs of names. For some of the algorithms, however, comparisons with a larger, but fixed, number of outcomes are more natural.

In this section we consider only static tables—that is, those in which insertions and deletions either do not occur or are so infrequent that a new table is constructed when they do occur. Dynamic table structures that allow logarithmic search times as well as efficient insertion and deletion algorithms are discussed in Section 6.4. For static tables, we need discuss only algorithms for searching and for constructing the table. The search algorithms are straightforward, but some of the construction algorithms are complex. This situation occurs because, when a table is static, it is reasonable to assume that the access frequencies are known; and it may be worthwhile to invest considerable effort in the construction of an *optimal* table—that is, one with minimal average search time (with respect to the given access frequencies). Algorithms for table construction and their analysis are the most important topics of this section.

6.3.1   Binary Search

When the cells of a table are sequentially allocated in random access memory and the names are stored in the table in their natural order, then *binary search*, one of the most widely used search techniques, is possible. The idea behind this technique is to search for a name z in an interval whose endpoints are two given pointers, l (for "low") and h (for "high"). A new pointer m (for "midpoint") is positioned somewhere near the middle of the interval, and either z is found at the cell pointed to by m or the comparison of z to the name in this cell reduces the search interval to one of the subintervals $[l, m - 1]$ or $[m + 1, h]$. If the interval becomes empty, the search terminates unsuccessfully.

In order to obtain a logarithmic search time, it is essential to position the pointer m in time independent of the length of the interval; this requirement rules out binary search on most secondary storage devices. It is not essential that m be placed exactly at the middle of the interval, although choosing the midpoint for m usually yields the most efficient algorithm. In certain special cases, however, it is advantageous to partition the interval into subintervals of length $\alpha(h - l + 1)$ and $(1 - \alpha)(h - l + 1)$, for a fixed value of α other than $\frac{1}{2}$ (see Exercise 7). When the table is not sequentially allocated but instead is stored as a tree-structured list, the fraction α is likely to vary from interval to interval (see Sections 6.3.2, 6.4.2, and 6.4.3).

Binary search is conceptually simple, but care should be taken in handling the details of how searches terminate. $h - l = 1$ and $h - l = 0$ are two special cases that must be carefully handled in any binary search program. In Algorithm 6.5 these cases are handled by the same code that treats the general case, and it is instructive to see how this is done by tracing the execution of the algorithm for $n = 2$ and $n = 1$.

$$l \leftarrow 1$$

$$h \leftarrow n$$

$$\textbf{while } l \leq h \textbf{ do} \begin{cases} \llbracket 1 \leq l \leq h \leq n \text{ and } z \notin \{x_1, \ldots, x_{l-1}\}, z \notin \{x_{h+1}, \ldots, x_n\} \rrbracket \\[2mm] m \leftarrow \left\lfloor \dfrac{l+h}{2} \right\rfloor \\[2mm] \textbf{case} \begin{cases} z = x_m: \llbracket \text{found: } m \text{ points to } z \rrbracket \\ z < x_m: h \leftarrow m - 1 \\ z > x_m: l \leftarrow m + 1 \end{cases} \end{cases}$$

$\llbracket \text{not found: } z \notin \{x_1, \ldots, x_n\} \rrbracket$

Algorithm 6.5   Binary search for a name z in a table $\{x_1, \ldots, x_n\}$ stored in natural order.

The correctness of Algorithm 6.5 follows from the assertion given in the comment at the beginning of the body of the loop. It states that if z occurs anywhere in the table, then it must occur in the interval $[l, h]$; equivalently, under our assumption that a name occurs at most once in the table, it states that z does not occur in either of the intervals $[1, l - 1]$ or $[h + 1, n]$. This assertion is trivially true the first time that the loop is entered when $l = 1$ and $h = n$, and it is straightforward to verify inductively that this assertion holds each time through the loop. When control exits from the loop, we must have $l > h$, and so the assertion becomes $z \notin \{x_1, \ldots, x_{l-1}\}$ and $z \notin \{x_{h+1}, \ldots, x_n\}$, which together imply that $z \notin \{x_1, \ldots, x_n\}$.

In analyzing Algorithm 6.5, it is useful to describe the sequence of comparisons "$z : x_i$" as an extended binary tree (see Section 2.3.3). The first comparison is the root of the tree

and the left and right subtrees are the tree structures of the succeeding comparisons when $z < x_{\lfloor (n+1)/2 \rfloor}$ and $z > x_{\lfloor (n+1)/2 \rfloor}$, respectively. The tree corresponding to a binary search of the range $x_l, x_{l+1}, \ldots, x_h$ is defined recursively by (6.1).

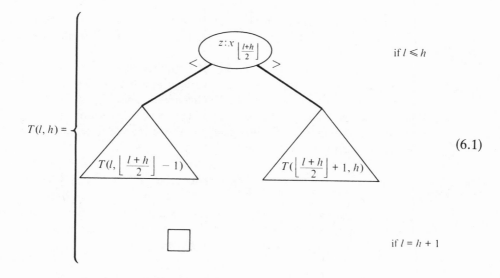

$$T(l, h) = \begin{cases} \includegraphics & \text{if } l \leqslant h \\ \\ \square & \text{if } l = h + 1 \end{cases} \tag{6.1}$$

Thus $T(1, 10)$, the tree corresponding to a binary search for z in a table containing ten names, is shown in Figure 6.1.

We can consider such a tree to be a flowchart of the binary search algorithm. The algorithm stops when it finds $z = x_i$ or when it reaches a leaf. The leaves correspond to the "gaps" between x_i and x_{i+1} in which z might lie, with the obvious interpretation for the leftmost and rightmost leaves. For example, in Figure 6.1 the left and right sons of the node $\boxed{z : x_4}$ are leaves corresponding to the conditions $x_3 < z < x_4$ and $x_4 < z < x_5$, respectively.

It is easy to prove by induction on $h - l$ that the height of the tree $T(l, h)$ defined by equation (6.1) is $\lceil \lg (h - l + 2) \rceil$. Thus the height of $T(1, n)$, the tree corresponding to binary search on a table with n names, is $\lceil \lg (n + 1) \rceil$. Since the

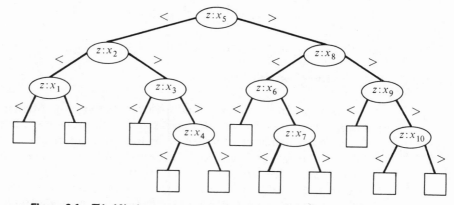

Figure 6.1 $T(1, 10)$, the tree corresponding to binary search in a table with 10 names.

height of the tree is the number of comparisons $z : x_i$ made in the worst case, we conclude that binary search requires at most $\lceil \lg(n + 1) \rceil$ comparisons for a table with n names.

To analyze the average number of comparisons required by binary search, we consider two quantities: S_n, the average number of comparisons $z : x_i$ in a successful search (z found to be in the table), and U_n, the average number of comparisons $z : x_i$ in an unsuccessful search (z found not to be in the table). Assuming that finding each of the n names is equally probable on a successful search, that is, that $\Pr(z = x_i) = 1/n$ for $1 \leq i \leq n$, we have

$$S_n = 1 + \frac{1}{n}[\text{internal path length of } T(1, n)].$$

Similarly, assuming that each gap (leaf) is equally probable on an unsuccessful search, that is, that $\Pr(x_i < z < x_{i+1}) = 1/(n + 1)$ for $1 \leq i < n$ and $\Pr(z < x_1) = \Pr(z > x_n) = 1/(n + 1)$, we have

$$U_n = \frac{1}{n + 1}[\text{external path length of } T(1, n)].$$

Using equation (2.7), we derive the interesting relation

$$S_n = \left(1 + \frac{1}{n}\right)U_n - 1. \tag{6.2}$$

Combining equation (6.2), the fact that $T(1, n)$ is a completely balanced binary tree with $2n + 1$ nodes (why?), and equation (2.10), we derive

$$U_n = \lg(n + 1) + 2 - \theta - 2^{1-\theta},$$

$$S_n = \left(1 + \frac{1}{n}\right)[\lg(n + 1) + 2 - \theta - 2^{1-\theta}] - 1,$$

where $\theta = \lg(n + 1) - \lfloor \lg(n + 1) \rfloor$.

6.3.2 Optimal Binary Search Trees

Binary search in a sequentially allocated table (Algorithm 6.5) provides very fast access to the names that occur as midpoints early in the bisection process—that is, those names near the top of the tree $T(1, n)$—and any of the names in the table can be accessed in about $\lg n$ comparisons. In practice, most tables contain some names that are accessed much more frequently than others, and it is reasonable to try to use the "privileged" table positions for the most frequently accessed names rather than for the names chosen for those positions by binary search. This is not possible with sequentially allocated tables, since the position of a name in the table is determined by its rank with respect to the natural order of the names in the table. We now introduce a data structure that has the flexibility needed to allow both a binary search technique *and* the ability to specify the points at which the table is bisected.

A *binary search tree* over names $x_1, x_2, \ldots, x_n$ is an extended binary tree (see Section 2.3.3), each of whose internal nodes is labeled with a distinct name chosen from $x_1, x_2, \ldots, x_n$ such that the symmetric order (see Section 2.3.2) of the nodes coincides with the natural order. Each of the $n + 1$ external nodes corresponds to a gap in the table. Thus the tree in Figure 6.1 is actually a binary search tree over the names $x_1, x_2, \ldots, x_{10}$. The fact that the symmetric order of the nodes is the natural order means that, for each node x_i, all names in the left subtree of x_i precede x_i in the natural order and all names in the right subtree of x_i follow x_i in the natural order.

Figure 6.2 shows four different binary search trees over the set of names $\{A, B, C, D\}$. The trees (a) and (b) are degenerate, since they are essentially linear lists that must be searched sequentially.

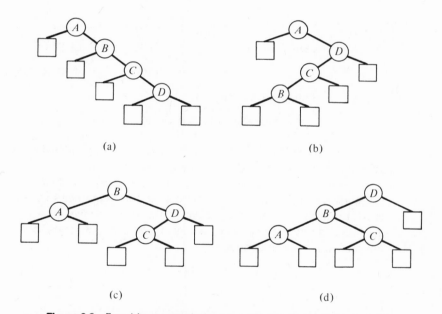

(a)　　　　　　　　　　　(b)

(c)　　　　　　　　　　　(d)

Figure 6.2　Four binary search trees over the set of names $\{A, B, C, D\}$.

Searching for a name z in a binary search tree is done by comparing z to the name at the root. Then

1.　If there is no root (the tree is empty), z is not in the table and the search terminates unsuccessfully.

2.　If z is equal to the name at the root, the search terminates successfully.

3. If z precedes the name at the root, the search continues down the left subtree of the root.

4. If z follows the name at the root, the search continues down the right subtree of the root.

Representing binary trees as in Chapter 2, each node is a triple (LEFT, NAME, RIGHT), where LEFT and RIGHT contain pointers to the left and right sons, respectively, and NAME contains the name stored in the node. The pointers can have the value Λ, meaning that the subtree to which they point is empty. If the pointer to the root of the tree is Λ, then the tree itself is empty. As expected, successful searches terminate at internal nodes of the binary search tree and unsuccessful searches terminate at external nodes.

This procedure to find a name z in a table organized as a binary search tree T is shown in Algorithm 6.6. Notice its similarity to Algorithm 6.5 (binary search).

$p \leftarrow$ root of T

$$\textbf{while } p \neq \Lambda \textbf{ do case } \begin{cases} z = \text{NAME}(p)\text{: } [\![\text{found: } p \text{ points to } z]\!] \\ z < \text{NAME}(p)\text{: } p \leftarrow \text{LEFT}(p) \\ z > \text{NAME}(p)\text{: } p \leftarrow \text{RIGHT}(p) \end{cases}$$

$[\![\text{not found: } z \text{ is not in the table}]\!]$

Algorithm 6.6   Searching in a binary search tree.

To list all the names in a binary search tree in natural order, we simply use the symmetric order traversal of binary trees (see Section 2.3.2).

Let $\beta_1, \beta_2, \ldots, \beta_n$ denote the access frequencies of the names $x_1, x_2, \ldots, x_n$, respectively, and let $\alpha_0, \alpha_1, \ldots, \alpha_n$ denote the frequencies with which the unsuccessful searches terminate at each of the $n + 1$ leaves of a binary search tree T. Then the average search time in T (over all searches, both successful and unsuccessful) can be expressed as

$$\sum_{i=1}^{n} \beta_i [1 + \text{level}(x_i)] + \sum_{\text{leaves } y_i \in T} \alpha_i \text{ level}(y_i).$$

Let

$$|T| = \sum_{i=1}^{n} \beta_i \text{ level}(x_i) + \sum_{\text{leaves } y_i \in T} \alpha_i \text{ level}(y_i).$$

be the *weighted path length* of T. The average search time in T is thus

$$|T| + \sum_{i=1}^{n} \beta_i,$$

and we can therefore restrict our attention to $|T|$, since $\sum \beta_i$ is not dependent on the structure of T.

When the frequencies α_i and β_i are interpreted as probabilities, they must be nonnegative and normalized so that

$$\sum_{i=0}^{n} \alpha_i + \sum_{i=1}^{n} \beta_i = 1.$$

In our examples we will insist on the nonnegativity but not normalization, since the normalization constant $\sum \alpha_i + \sum \beta_i$, which enters only multiplicatively in $|T|$, can be interpreted as part of the unit of measurement. The unnormalized α_i and β_i are called *weights* (leaf weights and node weights, respectively).

Continuing with the example of Figure 6.2, assume that the names A, B, C, D have access frequencies (or internal node weights) $\beta_1 = 1$, $\beta_2 = 6$, $\beta_3 = 4$, $\beta_4 = 5$, respectively, and that unsuccessful searches do not occur (all leaf weights are zero). Figure 6.3 shows the four search trees of Figure 6.2 with the external nodes deleted and the names replaced by their weights; the resulting weighted path length is also given. Notice that the average search time in the first two trees is more than twice the average search time in the third tree.

We are interested in finding an *optimal binary search tree* over a set of names with given frequencies. Given node weights $\beta_i \geq 0$, $1 \leq i \leq n$, and leaf weights $\alpha_i \geq 0$, $0 \leq i \leq n$, we want to construct a binary search tree T over the ordered set of names $x_1, \ldots, x_n$ such that the weighted path length $|T|$ is minimal.

This problem provides a striking example of the effectiveness of the dynamic programming technique of Section 4.1.7. Since there are approximately $4^n/(n\sqrt{\pi n})$ binary trees of n nodes [see formula (3.27) in Section 3.3], an exhaustive search of all trees would require an amount of time that grows exponentially in n. Dynamic programming leads to a polynomial-time algorithm because it takes advantage of the fact that an optimal binary search tree satisfies the following optimality principle: an optimal binary search tree T on weights $\alpha_0, \beta_1, \alpha_1, \ldots, \beta_n, \alpha_n$ has some weight β_i at the root, an optimal binary search tree T_l on $\alpha_0, \beta_1, \alpha_1, \ldots, \beta_{i-1}, \alpha_{i-1}$ as left subtree of the root, and an optimal binary search tree T_r on $\alpha_i, \beta_{i+1}, \ldots, \beta_n, \alpha_n$ as right subtree of the root. An optimal binary search tree also satisfies another requirement of dynamic programming—namely, that the weighted path length $|T|$ can be computed from "local" information about its two subtrees, T_l and T_r; in particular, $|T|$ can be computed from their weighted path lengths, $|T_l|$ and $|T_r|$, and the sum of all their

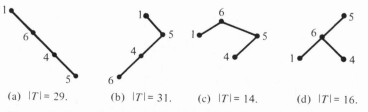

(a) $|T| = 29$.   (b) $|T| = 31$.   (c) $|T| = 14$.   (d) $|T| = 16$.

Figure 6.3 Example node weights and weighted total path lengths of the binary search trees of Figure 6.2.

weights, W_l and W_r, respectively:

$$|T| = |T_l| + |T_r| + W_l + W_r, \tag{6.3}$$

where

$$W_l = \sum_{j=0}^{i-1} \alpha_j + \sum_{j=1}^{i-1} \beta_j \quad \text{and} \quad W_r = \sum_{j=i}^{n} \alpha_j + \sum_{j=i+1}^{n} \beta_j.$$

The optimality principle says that, in order to select the root of the optimal tree, we must compute, for each node weight β_i, the optimal weighted path length when β_i is the root of T. This requires knowledge of the optimal left and right subtrees of β_i; and if the computation is organized in a recursive, top-down manner, the same optimal subtrees will be computed repeatedly (for a discussion of this problem, see Section 4.1.7). Hence the algorithm for constructing an optimal tree is best organized in a bottom-up fashion in which, for $1 \le i \le n$, we construct the $n + i - 1$ optimal trees on i consecutive node weights (and their

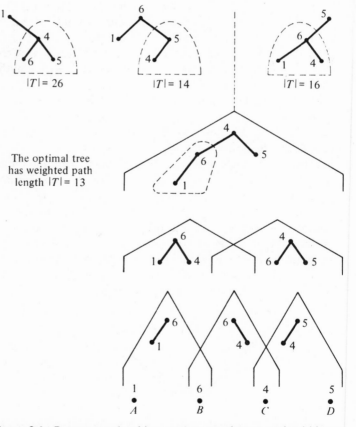

Figure 6.4 Bottom-up algorithm for constructing an optimal binary search tree.

adjacent leaf weights), using the previously constructed optimal trees on $i - 1$ or fewer consecutive node weights. Thus each optimal subtree on consecutive weights is constructed exactly once. The initialization at $i = 1$ is trivial, since there is only one tree over a single node. When the process ends at $i = n$, we have the desired optimal tree on n consecutive weights.

Figure 6.4 shows how this algorithm constructs an optimal tree over four names A, B, C, D with node weights 1, 6, 4, 5, respectively, and leaf weights of 0. The work required at each step is illustrated explicitly at the last step of the example, where the optimal tree over all four names is selected by comparing its weighted path length with that of the three other possible candidates. The subtrees surrounded by dotted lines are optimal trees constructed at earlier steps in the algorithm. Although there is a unique optimal tree in this example, uniqueness does not hold in general.

Figure 6.5 shows the same computation organized in the form of a matrix. Each entry contains all the relevant information about an optimal tree ranging over consecutive names (beginning with the name at the left of the matrix and ending with the name at the top of the matrix in the form

1. root of the optimal tree,

2. sum of all node weights in the tree,

3. weighted path length of the tree.

from \ To	A	B	C	D
A	A 1 0	B 7 1	B 11 5	C 16 13
B		B 6 0	B 10 4	C 15 11
C			C 4 0	D 9 4
D				D 5 0

Figure 6.5 Matrix representation of the bottom-up algorithm for constructing an optimal tree.

For example, the top-right entry of the matrix describes the optimal tree over all names (from A through D); it has C as its root, the sum of all node weights is 16, and the weighted path length is 13. In order to recover the entire optimal tree, it is sufficient to look up the information about its left subtree, an optimal tree over nodes A and B. Formula (6.3) is used to compute the weighted path length in each matrix entry. For example, the weighted path length of the optimal tree T_{ABCD} over the four names A, B, C, D is computed in terms of the path lengths $|T_{AB}|$ and $|T_D|$ of its optimal subtrees and the sums of their weights, W_{AB} and W_D, respectively:

$$|T_{ABCD}| = |T_{AB}| + |T_D| + W_{AB} + W_D$$
$$= 1 + 0 + 7 + 5$$
$$= 13.$$

The example in Figure 6.4 shows the case of zero leaf weights only for simplicity of presentation. When the leaf weights α_i are not zero, it is convenient to work above the diagonal in an $(n + 1) \times (n + 1)$ matrix instead of $n \times n$ as in the previous example. Each matrix entry contains three values. For $0 \leq i \leq j \leq n$, we have

R_{ij} is the index of the root of an optimal tree over the sequence of weights $\alpha_i, \beta_{i+1}, a_{i+1}, \ldots, \beta_j, \alpha_j$,

W_{ij} is the sum of weights $\alpha_i + \beta_{i+1} + \cdots + \beta_j + \alpha_j$,

P_{ij} is the weighted path length of an optimal tree over the sequence of weights $\alpha_i, \beta_{i+1}, \alpha_{i+1}, \ldots, \beta_j, \alpha_j$ (when $i = j$, the tree consists of a single leaf and this quantity is zero).

Notice that in order to compute the i, j entry of this matrix, we need the u, v entries for $u = i, v \leq j$ and $v = j, u \geq i$. The structure of Algorithm 6.7. is therefore simple. The two outer loops scan all the entries above the diagonal of the $(n + 1) \times (n + 1)$ matrix. The assignment statement

$R_{ij} \leftarrow$ any value of $k, i < k \leq j$, that minimizes the sum
$P_{i,k-1} + P_{kj} + W_{i,k-1} + W_{kj}$,

implies yet another nested loop, which, in general, is executed $j - i$ times. (Figure 6.6 shows the calculation done in this innermost loop for the case $l = 3$ and $i = 0$.) The number of operations required by Algorithm 6.7 is therefore $\frac{1}{6}n^3 + O(n^2)$ (see Exercise 10). Thus the bottom-up tree construction algorithm based on the optimality principle requires $O(n^3)$ operations and $O(n^2)$ storage.

Algorithm 6.7. uses few of the properties of optimal binary search trees: in fact, only the particular form of the expressions on the right-hand side of the assignment statements depends on properties of the trees. By analyzing the specific properties of optimal binary search trees more carefully, it is possible to improve Algorithm 6.7 so that it runs in time $O(n^2)$ instead of $O(n^3)$. The key to this improvement is a theorem whose proof is beyond the scope of this book. It allows the innermost loop (over k) to be executed fewer than $j - i$ times, as we had assumed in our analysis of Algorithm 6.7.

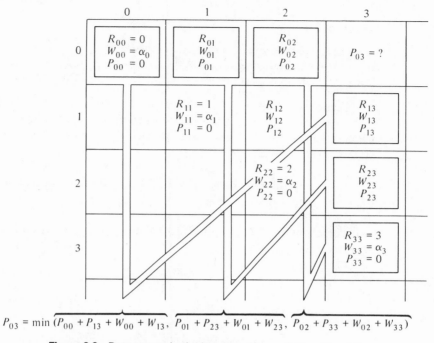

$$P_{03} = \min \overbrace{(P_{00} + P_{13} + W_{00} + W_{13}, \quad P_{01} + P_{23} + W_{01} + W_{23}, \quad P_{02} + P_{33} + W_{02} + W_{33})}$$

Figure 6.6   Data access in the innermost loop of Algorithm 6.7.

〚initialize〛

for $i = 0$ **to** n **do** $\begin{cases} R_{ii} \leftarrow i \\ W_{ii} \leftarrow \alpha_i \\ P_{ii} \leftarrow 0 \end{cases}$

〚visit each diagonal〛

for $l = 1$ **to** n **do**
    〚visit each entry
    within a diagonal〛
    for $i = 0$ **to** $n - l$ **do** $\begin{cases} j \leftarrow i + l \\[4pt] R_{ij} \leftarrow \text{any value of } k, i < k \leq j, \\ \quad\quad \text{that minimizes the sum} \\ \quad\quad P_{i,k-1} + P_{kj} + W_{i,k-1} + W_{kj} \\[4pt] W_{ij} \leftarrow W_{i,j-1} + \beta_j + \alpha_j \\[4pt] P_{ij} \leftarrow \min_{i < k \leq j} (P_{i,k-1} + P_{kj} + W_{i,k-1} + W_{kj}) \end{cases}$

Algorithm 6.7   Bottom-up algorithm for constructing an optimal search tree, given frequencies of successful and unsuccessful searches.

Theorem 6.1

Let $R_{i,j-1}$ be the root of an optimal tree over $\alpha_i, \beta_{i+1}, \ldots, \beta_{j-1}, \alpha_{j-1}$ and let $R_{i+1,j}$ be the root of an optimal tree over $\alpha_{i+1}, \beta_{i+2}, \ldots, \beta_j, \alpha_j$, where $i < j - 1$. Then there is an optimal tree over $\alpha_i, \beta_{i+1}, \ldots, \beta_j, \alpha_j$ whose root R_{ij} satisfies the inequalities $R_{i,j-1} \leq R_{ij} \leq R_{i+1,j}$.

Figure 6.7 illustrates the statement of the theorem that there is an optimal tree over names $x_i, \ldots, x_j$ whose root lies in the interval bracketed by the two roots of the smaller optimal trees shown. Theorem 6.1 allows us to refine our statement in Algorithm 6.7 to

$$R_{ij} \leftarrow \text{any value of } k, R_{i,j-1} \leq k \leq R_{i+1,j}, \text{ that minimizes the sum}$$
$$P_{i,k-1} + P_{kj} + W_{i,k-1} + W_{kj}$$

in order to obtain an $O(n^2)$-time algorithm for constructing optimal binary search trees (see Exercise 11).

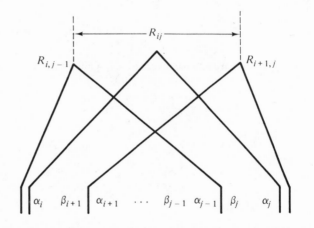

Figure 6.7 Illustration of the theorem. The root of an optimal tree is bracketed by the roots of two previously constructed trees.

6.3.3 *Near-Optimal Binary Search Trees*

Optimal binary search trees are of greater theoretical than practical interest, for the $O(n^2)$ time and storage requirements make even the refined version of Algorithm 6.7 impractical except for small tables. Moreover, since access frequencies are typically not accurately known *a priori*, it is usually not worthwhile to expend much effort in constructing an optimal tree. However, there are efficient [$O(n)$-time and storage] heuristic algorithms for constructing trees whose average search time is near that of optimal binary search trees, and such algorithms have practical value. In this section we discuss various examples of heuristic algorithms and present guidelines for avoiding poor heuristics.

There are two plausible heuristics for constructing a binary search tree, given a sequence of access frequencies. The first heuristic is to keep the tree as "balanced" as possible. This heuristic is natural because the opposite of a balanced tree, a completely skewed tree, is essentially a linear list and requires linear search times instead of the logarithmic search times that we expect from binary search trees. Among several notions of "balance," the following two make sense in the present context: choose the root (of the entire tree as well as of each subtree) so that its left and right subtrees have as nearly as possible either (a) the same number of nodes or (b) the same sum of weights. The second heuristic is to put the most frequently accessed names near the root. When applied consistently, this heuristic leads to trees whose node weights are monotonically nonincreasing along any path from the root to a leaf; such trees are called *monotonic* (compare this with the definition of a *heap* in Section 7.1.3).

To illustrate these heuristics, Figure 6.8 shows various binary search trees over six names with access frequencies $\beta_1 = 9, \beta_2 = 1, \beta_3 = 4, \beta_4 = 1, \beta_5 = 3, \beta_6 = 2$. For simplicity, we assume that the frequencies of unsuccessful searches, that is, the leaf weights α_i, are zero. (All the heuristics discussed generalize in various ways to the case of arbitrary leaf weights.) Notice that there is a ratio of $35/20 = 1.75$ between the average search times of the "worst" and the "best" among these supposedly "good" trees of Figure 6.8(a). It is easy to verify that one of the four trees that balances the number of nodes is optimal, but this need not be true in general. The example also shows that a tree that most nearly balances the sum of weights need not be optimal. The main question of interest is how close to optimal the trees constructed according to these different heuristics are likely to be.

One way to answer this question is by means of simulation experiments in which we generate many sequences of weights, construct trees according to different algorithms, and collect statistics about the results. Several such studies reported in the literature claim that, with heuristic algorithms based on a combination of the principles of balanced trees and monotonic trees, average search times within a few percent of those in optimal trees are obtained. It is worthwhile to give an intuitive explanation of this phenomenon because it occurs in many combinatorial problems. Imagine the space of all trees constructed over a given sequence of weights and imagine the weighted path length plotted as a function of the trees in this space. Apparently, for *most* sequences (but certainly not all of them), this function is very flat in the region near its minimum; that is, there are many trees whose weighted path length is close to optimal. Consequently, it is easy to find a good tree but hard to find an optimal one, since an optimal tree differs little from other nearby trees in its vicinity.

Another way to determine the effectiveness of various heuristics is to make statistical assumptions about the distribution of weight sequences and to analyze the expected weighted path length of trees constructed according to different algorithms. Here "expected" means the weighted average over all possible weight sequences. Let us summarize the results that compare the expected weighted path length of four types of trees, under the assumptions that the n

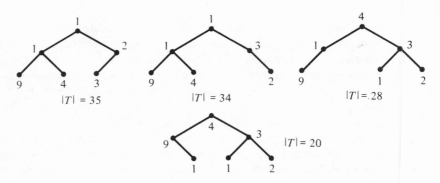

$|T| = 35$ $|T| = 34$ $|T| = .28$

$|T| = 20$

(a) Balanced by the number of nodes.

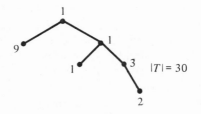

$|T| = 30$

(b) Balanced by the sum of weights.

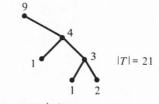

$|T| = 21$

(c) A monotonic tree.

Figure 6.8 Binary search trees constructed according to various heuristics; the weighted total path lengths are shown.

internal node weights $\beta_1 \ldots \beta_n$ are drawn in sequence as independent samples from a given probability distribution on nonnegative real numbers and that the leaf weights are zero. Let $\bar{\beta}$ be the average weight and $\bar{O}_n$, $\bar{B}_n$, and $\bar{M}_n$ be the expected weighted path length of optimal, balanced (with respect to the number of nodes), and monotonic binary search trees, respectively, constructed over these sequences of weights. Let $\bar{R}_n$ be the expected weighted path length of a "random" binary search tree on n nodes, defined by the property that every node has equal probability $1/n$ of being the root and that if node i is the root, then it has a random binary search tree of nodes $1, \ldots, i-1$ as its left subtree and a random binary search tree of nodes $i+1, \ldots, n$ as its right subtree. These random binary trees serve as a basis for evaluating the efficiency of the other algorithms; if it turns out that trees constructed according to some heuristic principle are not significantly better than the random trees, then the heuristic is suspect.

As an example of the technique used to obtain useful statistical results, consider the expected weighted path length $\bar{R}_n$ of random trees of n nodes. If such a tree T has root k, $1 \leq k \leq n$, it has a random tree T_l of $k - 1$ nodes as its left subtree and a random tree T_r of $n - k$ nodes as its right subtree. The weighted path length satisfies the equation

$$|T| = |T_l| + |T_r| + \text{(sum of weights in } T_l) + \text{(sum of weights in } T_r).$$

Considering that each value of k from 1 to n is equally likely and that the expected value of the sum of all weights is $n - 1$ times the average weight $\bar{\beta}$, and taking the expected values of both sides of this equation, we obtain

$$\bar{R}_n = (n - 1)\bar{\beta} + \frac{1}{n} \sum_{k=1}^{n} (\bar{R}_{k-1} + \bar{R}_{n-k}),$$

which can be transformed to

$$\bar{R}_n = \bar{\beta}n - \bar{\beta} + \frac{2}{n} \sum_{k=0}^{n-1} \bar{R}_k.$$

This is an instance of equation (3.20), which was solved in Section 3.2.2. Additional aspects of such derivations are pursued in Exercise 12.

The results of a comparison of the various heuristics are surprising:

$$\bar{O}_n = \bar{\beta}n \lg n + O(n)$$
$$\bar{B}_n = \bar{\beta}n \lg n + O(n)$$
$$\bar{M}_n = (2 \ln 2)\bar{\beta}n \lg n + O(n)$$
$$\bar{R}_n = (2 \ln 2)\bar{\beta}n \lg n + O(n).$$

In other words, asymptotically for large n, and averaged over the class of all weight sequences sampled as described above, balanced trees are as good as optimal trees and monotonic trees are as bad as random trees. Similar results hold in the general case when leaf weights are nonzero and are sampled independently from another given probability distribution on nonnegative real numbers.

The poor showing of the heuristic that puts heavy weights near the root can be understood as follows. Under our assumptions every node is equally likely to be assigned the largest weight—that is, to become the root of a monotonic tree. Thus the process of selecting the root of the entire tree, as well as the roots of all the subtrees, is the same for monotonic trees as it is for random trees and results in unbalancing the tree to the extent indicated by the factor $2 \ln 2 \approx 1.4$. The only compensation for the unbalancing caused by selecting the largest remaining weight as the root of each subtree is that one single weight gets multiplied by a smaller level number. However, in most weight sequences, a single weight is negligible compared to the sum of the other weights. In the preceding analysis, the effect of moving large weights up in the tree enters only in the $O(n)$ term, whereas the effect of unbalancing the tree determines the factor $2 \ln 2$ in the dominant $\bar{\beta}n \lg n$ term.

There are more precise results for binary search trees balanced by the sums of the weights. Let O_n be the weighted path length of an optimal binary search tree for a normalized frequency distribution $(\sum \beta_i + \sum \alpha_i = 1)$ and let W_n be the weighted path length of a balanced (with respect to the sums of the weights) binary search tree. Then it can be shown that

$$\mathscr{H} - \lg \mathscr{H} - \lg e + 1 \leqq O_n \leqq W_n \leqq \mathscr{H} + 2$$

where $\mathscr{H} = \mathscr{H}(\alpha_0, \beta_1, \alpha_1, \beta_2, \ldots, \alpha_{n-1}, \beta_n, \alpha_n)$ is the entropy function of Section 1.5. In other words binary search trees balanced by the sums of the weights are *never* very far from optimal.

These results indicate that a reasonable heuristic is to start with some kind of balanced tree and perturb the tree locally by placing larger weights nearer the root than smaller weights, being careful not to unbalance the tree too much in the process. (A good way to perform these local perturbations is by means of the "rotations" discussed in Section 6.4.2.) The results also indicate that if a file is static and if each record has equal memory requirements, then a sequentially allocated table has very desirable features. It can be constructed by sorting [an $O(n \log n)$ operation]; it has no storage overhead for pointers; and it allows a binary search corresponding to a complete binary tree, which is nearly as good as an optimal tree for most practical cases.

6.3.4 *Digital Search*

The search techniques discusssed so far have been based only on the assumptions that a linear order is defined on the name space S and that the time required to compare two names is independent of the cardinality of S. It has been unnecessary to consider the details of how a name is represented and how to determine whether one name precedes another. However, since names are usually represented in such a way that they can be compared by means of a simple computation on their representations, a name space will inevitably have a great deal more structure than just an abstract linear order, and certain search techniques can be designed to exploit this situation. Digital search techniques take advantage of the fact that a name, or its representation, can always be interpreted as a sequence of characters over some finite alphabet (in an extreme case, over the two-symbol alphabet $\{0, 1\}$).

As an example of digital search, consider the construction of a table to hold the following sequences of digits as approximations to well-known mathematical constants:

$$\sqrt{2} \approx 1.414$$

$$\frac{1}{\ln 2} \approx 1.44$$

$$\frac{1 + \sqrt{5}}{2} \approx 1.6$$

$$\ln 10 \approx 2.3$$

$$e \approx 2.718$$

$$\pi \approx 3.14159.$$

These six sequences of digits can be represented in the form of the tree of Figure 6.9.

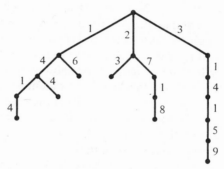

Figure 6.9   A digital search tree for six names.

Notice that the sequence of digits for each name is obtained by following some path from the root of the tree to a leaf; notice also that identical prefixes—that is, initial subsequences—of different names are merged as long as possible. Thus a search for a name z with digits $d_1 d_2 \ldots d_l$ starts at the root and proceeds by comparing the leftmost unmatched digit d_j against the labels of all branches leading out of the current node. If d_j matches the label of some branch, that branch is followed to reach a new node. If no label matches the digit d_j, then the search terminates unsuccessfully. If all the digits of the name z have been matched, the search terminates successfully.

The example in Figure 6.9 is special in that no name in the table is a prefix of any other one. Such is not the case in many applications, and so we present two methods by which digital search trees can be made to accommodate names that are prefixes of other names in the table. The first method is to add an "end-of-name" bit to each node of the tree, thereby indicating whether a name ends at that node. For example, if we want 314 to be a name as well as 314159, this bit would be set to one at the nodes containing the digits 4 and 9, and set to zero for all other nodes along the path 314519. The second method is to modify names slightly so that no name can be a prefix of another. For example, if names are English words, a trailing blank might be attached to each word; this method would assure that "A   " and "AN   " were not prefixes of "AND   ".

A digital search tree can be represented in a computer in many ways, but, for an efficient implementation, the representation must allow the simultaneous comparison of a character in the name against the labels of all branches leading out of a node. Nothing is gained by breaking a name into its constituent characters if these comparisons must be made by looking, in turn, at each branch leading out of a node; such a technique would almost certainly be slower than binary search, which compares two entire names in one elementary operation. One way to turn a multiway branch into an elementary operation (i.e., one that can be performed in an amount of time independent of the number of branches) is to associate with each node in the tree a vector with a component for each character in the alphabet (regardless of whether there is a name in the table that uses the component). Each digit in the name to be searched for is used as an index, or pointer, into such a vector, with the corresponding vector entry determining the next action of the search. Digital search trees represented in this way are called *tries*. Figure 6.10 shows the digital search tree of Figure 6.9 represented as a trie. In order to save memory space, sequences of nodes along a path where no branching occurs (as in the case of the path labeled 14159) have been collapsed into a single node drawn as an ellipse. As a result, when such a node is reached, the remaining characters of the name being searched for must be compared to the characters in the node.

Searching in a trie can be illustrated by the following example. To search the trie of Figure 6.10 for the name 2718, we use the digit 2 as an index into the vector associated with the root. Notice that we need not look at any other component of this vector, and so the time required to perform this operation is independent of the size of the alphabet. The component found is a pointer to another node, in which we use the digit 7 as an index. There we find a pointer to a special node that contains the suffix 18 of the name. By concatenating the prefix 27, which is associated with the path to this node, and the suffix 18, which is stored explicitly, we obtain the entire name. Notice that 2718 was found in two steps: we examined the smallest number of digits that could distinguish this name (in a left-to-right scan) from all the other names in the table. If we were searching for the name 23, we would encounter the null suffix at a special empty node at the end of the path labeled 23. This empty node is necessary to indicate that 23 is a name in the table. In searching for 012345, the null pointer Λ on the 0th component of the vector associated with the root terminates the search unsuccessfully, and it tells us in one step that this name is not in the table.

This example illustrates two characteristic properties of tries: they permit fast search, particularly in the case of unsuccessful searches, and they tend to waste

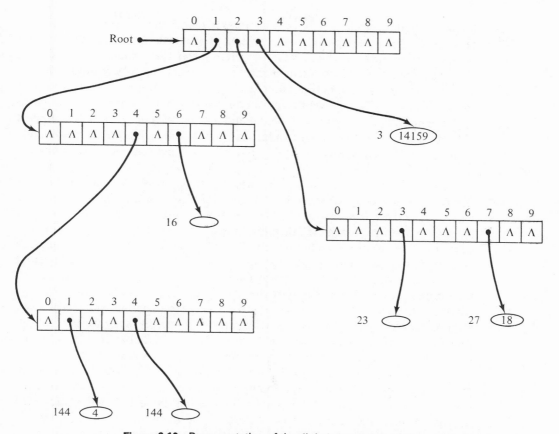

Figure 6.10 Representation of the digital search tree of Figure 6.9 as a trie.

memory. The inefficient use of memory occurs because the scheme of allocating a field for each character of the alphabet at each node is better suited to representing the entire name space rather than the contents of a particular table; thus a trie usually contains space for many names not in the table.

In general, the wasted memory space occurs at the nodes near the bottom of a trie, whereas the nodes near the root tend to be full. Attempts to exploit the speed of tries without paying the penalty in memory space usually combine trie techniques for some prefixes of the names with some other search technique for the suffixes. This is the idea behind the thumb index in a dictionary: it tells at a glance where the words beginning with any given letter start and end, but it requires some other technique to find a specific name within this narrower range.

Digital search techniques replace a comparison between two names by a sequence of comparisons between the characters in their representations. Since only one such comparison is needed per character of the representation, the search time is proportional to the (average) length of names. With an alphabet of c characters, the average name length of n distinct names cannot be less than

$\log_c n$; so digital searching yields search times that are logarithmic in the size of the table, just as binary search did. The constant of proportionality depends on many things, and, in practice, digital search is often faster than binary search, since a multiway branch based on a single character (typically, a c-way branch for an alphabet of c characters) may not take substantially longer than a two-way branch based on a comparison of two names. In this respect digital searching is similar to both the multiway trees of Section 6.4.4 and the address computation techniques of Section 6.5. It is also similar to address computation techniques in that it tends to use memory inefficiently—the table generally includes space for many nonexistent names.

Digital search differs in an important respect from all other techniques described in this chapter: there is only one digital search tree for a given set of names; thus there is no question of constructing an optimal tree. Any statistical analysis of average search time depends only on the assumptions made about the contents of the table. For techniques in which we consider comparison of names as an elementary operation, any set of n names behaves like any other (assuming that all permutations of this set are equally likely; see Section 6.4.1). In digital search trees, different sets of names lead to tries of different shapes, and, in particular, the lengths of the names have an important effect on the search times and the memory requirement of tries. Analyses and experiments merely support the natural conclusion that, in a table with n names over an alphabet of c characters, searching requires an average of approximately $\log_c n$ c-way branches.

Insertion and deletion in a trie are straightforward (Exercise 13), and, from this point of view, digital search might be considered a suitable technique for dynamic tables. Digital search has been discussed in connection with static tables because it lacks the characteristic feature of the dynamic table algorithms of Section 6.4: it does not guarantee a worst-case search time logarithmic in the table size.

6.4. LOGARITHMIC SEARCH IN DYNAMIC TABLES

Here we consider tree organizations of tables in which insertions and deletions are frequent. What happens to the search time in a tree that has been modified through insertions and deletions? If the names involved in insertion and deletion are chosen at random, it turns out that, on the average, the search time is affected little; but in the worst case, the behavior is bad—trees can degenerate into linear lists that must be searched sequentially. The problem of the tree degenerating into a linear list, which leads to $O(n)$ instead of $O(\log n)$ search times, is more pronounced in practical applications than theoretical analyses indicate. Such analyses commonly assume that insertions and deletions occur at random, but in practice they are often not random.

Three different techniques to guarantee logarithmic search times even in the worst case are presented in this section. Each technique is based on trees that are "balanced" in some sense; thus the skewed trees that lead to long search times are avoided. The major problem to be solved in any such technique is how to restore the balance of a tree after an insertion or deletion has unbalanced it. This rebalancing process must be done in $O(\log n)$ time so as to guarantee logarithmic times for search, insertion, and deletion.

6.4.1 Random Binary Search Trees

How do unconstrained binary search trees behave as dynamic trees? In other words, suppose that a binary search tree changes through random insertions and deletions; what will be the average cost of searching such a tree with Algorithm 6.6? An answer to this question gives us a basis of comparison for the sophisticated methods discussed in Sections 6.4.2 and 6.4.3. However, before considering the question, we must describe precisely the mechanics of insertions and deletions.

To insert z, we use a minor modification of Algorithm 6.6. If z has not been found, we get a new cell for z and link it to the last node visited during the unsuccessful search for z. The corresponding statement cannot simply be added to the end of Algorithm 6.6, however, because on normal termination of the **while** loop p no longer points to the last node visited but instead has the value Λ. So we must do the insertion *before* the statement $p \leftarrow$ LEFT (p) or $p \leftarrow$ RIGHT (p) is executed; we can do this by making the insertion procedure recursive, as in Algorithm 6.8. A call to *INSERT* (z, T) returns as its value a pointer to the tree in which z has been added. Thus $T \leftarrow INSERT$ (Z, T) is used to insert z into T.

$T \leftarrow INSERT (z, T)$

procedure *INSERT* (z, T)

   $p \leftarrow$ root of T

   if $p = \Lambda$ **then** $\begin{cases} p \leftarrow \textbf{get-cell} \\ \text{NAME}(p) \leftarrow z \\ \text{LEFT}(p) \leftarrow \text{RIGHT}(p) \leftarrow \Lambda \end{cases}$

   else $\begin{cases} \textbf{case} \begin{cases} z = \text{NAME}(p): [\![\text{already in the table}]\!] \\ z < \text{NAME}(p): \text{LEFT}(p) \leftarrow INSERT(z, \text{LEFT}(p)) \\ z > \text{NAME}(p): \text{RIGHT}(p) \leftarrow INSERT(z, \text{RIGHT}(p)) \end{cases} \end{cases}$

   $INSERT \leftarrow p$

return

Algorithm 6.8   Insertion into a binary search tree.

Deletion is more complex than insertion, and we present only the main idea. If the name z to be deleted has at most one son, then the deletion is done by using that son (if any) to replace z as a son of z's father [Figure 6.11(a)]. If z has two sons, it cannot be directly deleted. Instead we find either the name y_1 in the table that immediately precedes z, or the name y_2 in the table that immediately follows z, with respect to the natural order. Both names are in nodes that have at most one son (why?), and z can thus be deleted by replacing it with either the name y_1 or the name y_2 and then deleting the node that had contained y_1 or y_2, respectively [Figure 6.11(b)].

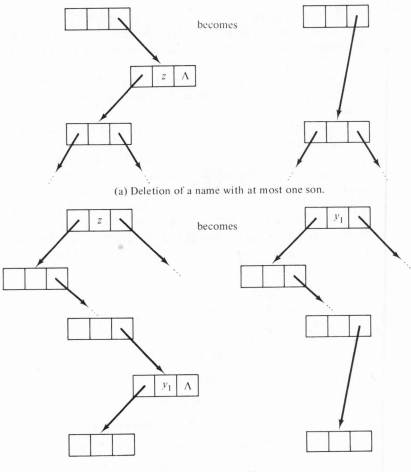

(a) Deletion of a name with at most one son.

(b) Deletion of a name with two sons.

Figure 6.11   Deletion of a name z. If z has at most one son, it is deleted as in (a). Otherwise, if z has two sons it is replaced by either its predecessor (or its successor) in the table and the original node of the predecessor (successor) is deleted.

A partial answer to the question about the effect of insertions and deletions on the shape of a binary search tree is now given by the analysis of random trees in Section 6.3.3. If n names are successively inserted by Algorithm 6.8 into an initially empty tree, and if all permutations of the input sequence are equally likely, then this "construction by repeated insertion" produces trees that are random in the sense of Section 6.3.3 with $\beta_i = 1, 1 \leq i \leq n$. Hence the average search time in such trees, measured in terms of the number of nodes visited, is $(2 \ln 2) \lg n + O(1)$, approximately 1.4 times longer than the $\lg n + O(1)$ average search time in a completely balanced binary tree of n nodes.

A search time approximately 40% longer than in completely balanced binary trees could be tolerated in most applications, but unconstrained binary search trees are unreliable in practice, since "biased" runs of insertions of names in approximate order are likely to occur and the resulting trees are not random. Moreover, the worst-case behavior is linear in the table size, not logarithmic. The balancing schemes presented in the next three sections guarantee logarithmic search times without requiring any assumptions about the sequence of insertions and deletions.

6.4.2   Height-Balanced Binary Trees

The most obvious technique to prevent a dynamic tree from becoming skewed is to keep it as balanced as possible at all times. Thus if a new name B is inserted in the tree shown at left in Figure 6.12, making it less than completely balanced, the tree is restructured as shown. Unfortunately, this transformation generally changes the entire tree: not a single father-son relationship remains unchanged in the example of Figure 6.12. So this restructuring operation, in general, requires an amount of work proportional to n.

The balancing techniques discussed here and in the next section have the property that if a transformation of a tree is required after an insertion or deletion, it can be carried out by *local changes along a single path from root to a leaf*; as we shall see, this requires $O(\log n)$ time. To achieve such flexibility, the trees must be allowed to deviate from completely balanced binary trees, but they do so to a sufficiently small extent that the average search time in them is only slightly larger than that in a completely balanced binary tree.

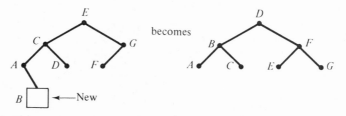

Figure 6.12   Rebalancing a tree completely requires an amount of work proportional to n.

As in Chapter 2, let $h(T)$ denote the height of a binary tree T—that is, the length of the longest path from the root to any leaf. The height of a tree with one single node is 0, and, by convention we consider the height of an empty tree to be -1.

Definition

A binary tree T is *height balanced* if and only if the two subtrees of the root, T_l and T_r, satisfy

1. $|h(T_l) - h(T_r)| \leq 1$.
2. T_l and T_r are height balanced.

The empty binary tree, having no root nor subtrees, satisfies these conditions vacuously and hence is height balanced. A tree with a single node is height balanced by our convention about the height of the empty tree. Figure 6.13 shows some height-balanced binary trees and one binary tree that is not. The height constraint prevents height-balanced binary trees from being too far away from a completely balanced binary tree. Indeed, Figure 6.13(a) shows the three "most-skewed" height-balanced binary trees among those of height 1, 2, and 3, respectively.

Despite the fact that height-balanced binary trees can look sparse, the search time that they require is only moderately longer than in completely balanced binary trees. Empirical evidence suggests that the average internal path length of a height-balanced binary tree of n nodes is $n \lg n + O(n)$; consequently, the average time is $\lg n + O(1)$, which differs asymptotically from average search time in completely balanced binary trees by only a constant.

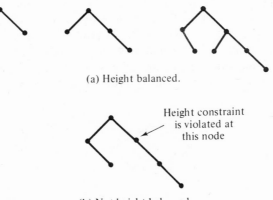

(a) Height balanced.

Height constraint
is violated at
this node

(b) Not height balanced.

Figure 6.13 Examples of height-balanced binary trees and a non-height-balanced binary tree.

There is no known mathematical analysis of the average search time in height-balanced binary trees of n nodes; it is difficult merely to count the height-balanced binary trees of n nodes (see Exercise 18), let alone analyze their shapes. However, if we consider a worst-case (most-skewed) height-balanced binary tree, we *can* analyze the maximum search time in the tree (the height) and the average successful search time in the tree (the internal path length divided by the number of nodes).

A most-skewed height-balanced binary tree of height h, T_h, has a most-skewed binary tree T_{h-1} of height $h - 1$ as one of its subtrees and a most-skewed binary tree T_{h-2} of height $h - 2$ as the other, as shown in Figure 6.14. The most-skewed height-balanced binary trees shown in Figure 6.13(a) for $h = 1, 2$, and 3 are constructed according to Fibonacci-type rule, starting with an empty tree T_{-1} and a tree T_0 with a single node.

Denoting the number of nodes of T_h by n_h and the internal path length of T_h by p_h, we obtain the recurrence relations

$$n_h = n_{h-1} + n_{h-2} + 1, \qquad n_{-1} = 0, \quad n_0 = 1, \qquad (6.4)$$

and

$$p_h = p_{h-1} + p_{h-2} + n_h - 1, \quad p_{-1} = p_0 = 0. \qquad (6.5)$$

The first is solved by the technique of Section 3.2.2 to yield the general solution

$$n_h = c\left(\frac{1 + \sqrt{5}}{2}\right)^h + c'\left(\frac{1 - \sqrt{5}}{2}\right)^h - 1.$$

Taking the initial conditions into consideration, we obtain

$$n_h = \left(1 + \frac{2}{\sqrt{5}}\right)\left(\frac{1 + \sqrt{5}}{2}\right)^h + \left(1 - \frac{2}{\sqrt{5}}\right)\left(\frac{1 - \sqrt{5}}{2}\right)^h - 1$$

$$\approx 1.9\left(\frac{1 + \sqrt{5}}{2}\right)^h. \qquad (6.6)$$

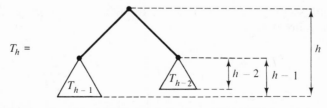

Figure 6.14 The most skewed height-balanced trees are constructed according to a Fibonacci-type rule.

Inverting equation (6.6), we find that

$$h = \frac{1}{\lg\left[(1 + \sqrt{5})/2\right]} \lg n_h + O(1)$$
$$\approx 1.44 \lg n_h.$$ (6.7)

Thus the worst-case search time (i.e., the height of the most-skewed binary tree) in height-balanced binary trees is about 44% longer than the average search time of $\lg n$ in completely balanced binary trees and about the same as the *average* search time in randomly constructed binary trees.

The analysis of the average (rather than the worst-case) successful search time $S_h = p_n/n_h$ in the most-skewed height-balanced binary tree is even more favorable. Dividing recurrence relation (6.5) by n_h gives

$$\frac{p_h}{n_h} = \frac{n_{h-1}}{n_h} \cdot \frac{p_{h-1}}{n_{h-1}} + \frac{n_{h-2}}{n_h} \cdot \frac{p_{h-2}}{n_{h-2}} + 1 - \frac{1}{n_h};$$

that is,

$$S_h = \frac{n_{h-1}}{n_h} S_{h-1} + \frac{n_{h-2}}{n_h} S_{h-2} + 1 - \frac{1}{n_h}.$$

By ignoring the term $-1/n_h$ and using expression (6.6) to approximate n_{h-1}/n_h and n_{h-2}/n_h by $\left[(1 + \sqrt{5})/2\right]^{-1}$ and $\left[(1 + \sqrt{5})/2\right]^{-2}$, respectively, we obtain

$$S_h = \left(\frac{1 + \sqrt{5}}{2}\right)^{-1} S_{h-1} + \left(\frac{1 + \sqrt{5}}{2}\right)^{-2} S_{h-2} + 1.$$ (6.8)

This recurrence relation can again be solved by the technique of Section 3.2.2. Using the identity $\left[(1 + \sqrt{5})/2\right]^{-1} + \left[(1 + \sqrt{5})/2\right]^{-2} = 1$, we find first that the general solution of the homogeneous recurrence relation corresponding to (6.8) is

$$S_h = c \cdot 1^h + c'\left(\frac{1 - \sqrt{5}}{1 + \sqrt{5}}\right)^h$$
$$\approx c \quad \text{as } h \to \infty$$

and, secondly, that (6.8) has a particular solution of the form $S_h = \text{constant} \cdot h$, or, more precisely,

$$S_h = \frac{h}{1 + \left[(1 + \sqrt{5})/2\right]^{-2}}$$
$$= \frac{5 + \sqrt{5}}{10} h \approx 0.72h.$$

Hence the solution to (6.8) is

$$S_h = \frac{5 + \sqrt{5}}{10} h + c + o(1)$$

$$\approx \frac{5 + \sqrt{5}}{10} h \quad \text{as } h \to \infty. \tag{6.9}$$

To compare this average search time for the most-skewed height-balanced binary trees to that of completely balanced binary trees, we substitute (6.7) into (6.9) to obtain

$$S_h \approx \frac{5 + \sqrt{5}}{10} \frac{1}{\lg \left[(1 + \sqrt{5})/2 \right]} \lg n_h$$

$$\approx 1.04 \lg n_h. \tag{6.10}$$

Thus the average search time in the most-skewed height-balanced binary trees is only about 4% longer than in completely balanced binary trees. This result, together with the fact that only a small fraction of height-balanced binary trees is badly skewed, supports the previously stated empirical observation that the average search time in height-balanced binary trees, averaged over all such trees, is $\lg n + O(1)$.

Our analysis establishes that height-balanced binary trees have short search times; now we must show that insertions and deletions can be performed easily while keeping the trees balanced.

Two tree transformations called *rotation* and *double rotation* allow us to restructure height-balanced trees that have been disturbed by an insertion or deletion. Figure 6.15 illustrates these two types of transformation and shows how they restore the balance of a tree disturbed by an insertion. The left parts of Figure 6.15(a) and (b) show a subtree of a height-balanced binary tree in which a new node *NEW* has just been inserted. This insertion may cause the height constraint to be violated at many nodes; among them, consider the lowest node—that is, the root of the subtree shown. (It turns out that height constraints that may have been violated at nodes higher in the tree will be automatically restored.) In the case of the rotation, two nodes with names A and B are shown explicitly in Figure 6.15(a), along with their condition codes \ and $-$, respectively, which record that before insertion the right subtree of A is of greater height than the left subtree of A, and the heights of the two subtrees of B are equal [the condition code / appears in the mirror image of the rotation shown in Figure 6.15(a)]. Similarly, three nodes labeled A, B, and C are shown explicitly in Figure 6.15(b) for the double rotation.

The reader should verify that the height conditions before and after the transformation hold, as indicated by the dotted lines, and that the *entire* tree is

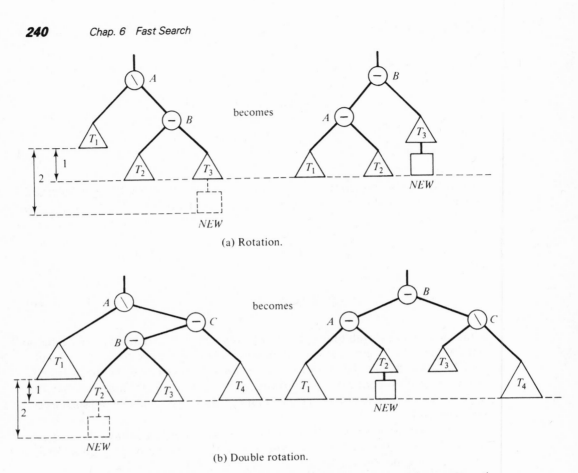

(a) Rotation.

(b) Double rotation.

Figure 6.15 Examples of a rotation and a double rotation to restore the height balance of a tree disturbed by insertion of a name *NEW*. The double rotation is equivalent to a rotation around node *A*, followed by a rotation around *C*. Both transformations have symmetrical variants.

height balanced after these local transformations. Notice that these transformations preserve the table order of all names in the tree, which is, of course, a prerequisite for performing a search by Algorithm 6.6.

The conditions under which the rotation and double rotation shown in Figure 6.15 (and their symmetrical variants) are applied remain to be specified. This is best done informally.

Insertion Algorithm. After a new name has been inserted by Algorithm 6.8, the path followed from the root to the leaf is traced backward. (Since a binary tree does not ordinarily provide explicit FATHER links, this retracing usually implies that the downward path has been stored on a stack.) The action taken, as each node is traversed along this upward path, depends first on the height condition of this node (/, −, or \) and, secondly, in case this condition is / or \, on the direction of the last step or the last two steps upward along the path. The rules are as follows.

1. If the current node has height condition −, this condition is changed to \ if the last step originated from the right son and to / if it originated from the left son. We continue to follow the path upward unless the current node is the root, in which case the procedure terminates.

2. If the current node has height condition / or \, and the last step originated from the shorter of the two subtrees of the current node, the height condition is changed to −, and the procedure terminates.

3. If the current node has height condition / or \, and the last step originated from the taller of the two subtrees of the current node, then
    (a) if the last two steps were taken in the same direction (both from left sons or both from right sons), an appropriate rotation is performed.
    (b) if the last two steps were taken in opposite directions, an appropriate double rotation is performed.
    In either case, the procedure terminates.

Deletion Algorithm.   First a name is deleted according to the algorithm of Figure 6.11. This guarantees that the node to be deleted has at least one empty subtree. If the node to be deleted has exactly one empty subtree, then the height constraint forces the other subtree to consist of a single node, as shown in Figure 6.16. The effect on the height balance of the tree is the same as if a leaf had been deleted; consequently, this is the only case we need to consider. When a leaf is deleted, the subtree consisting solely of this node has lost one unit in height (its height has been changed from 0 to −1). We now trace the path from this deleted leaf up to the root of the tree, checking at each node what effect the shortening of a subtree might have on a larger tree. The action taken at each node along this upward path depends on the height condition of the node before deletion, on the

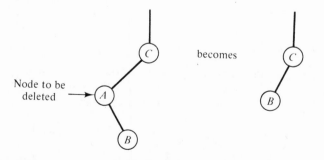

Figure 6.16   Deletion of a node with one empty subtree is reduced to the case of deleting a node with both subtrees empty.

direction of the last step, and, in certain cases, also on the height condition of a son of the node. In the following cases, dotted lines indicate a subtree that has been shortened. In certain cases, a transformation restores the height of a subtree to what it was before deletion; if so, the algorithm terminates, since the height condition of all higher nodes is then unaffected by the deletion. In other cases, the message that a subtree has been shortened must be passed up along the path to the root. In each case, symmetrical variants are *not* shown.

1. If the current node has height condition—then the shortening of the subtree does not affect the height of the tree rooted at the current node. The algorithm terminates.

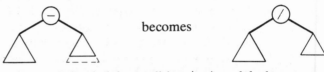

2. If the current node has height condition / or \, and the last step originated from the taller of the two subtrees of the current node, then continue to go upward, passing along the message that the current subtree has been shortened.

 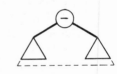

3. If the current node has height condition / or \, and the last step originated from the shorter of the two subtrees of the current node, then the height constraint is violated at the current node. Three subcases and their mirror images are distinguished according to the height condition of the other son of the current node.

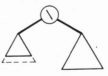

   (a) An appropriate rotation restores the height constraint at the current node without changing the height of the subtree rooted at the current node. The algorithm terminates.

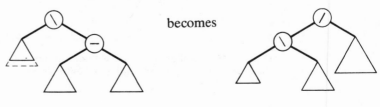

(b) An appropriate rotation restores the height constraint at the current node. Continue to go upward, passing along the message that the subtree rooted at the current node has been shortened.

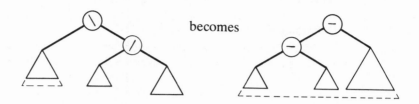

becomes

(c) An appropriate double rotation restores the height constraint at the current node. Continue to go upward, passing along the message that the subtree rooted at the current node has been shortened. (The height condition of the node labeled B can be any of $/$, $-$, or $\backslash$ as long as at least one of the two subtrees of B has the height indicated by the shaded area. The height conditions of A and C after the transformation depend on the height of the two shaded subtrees.)

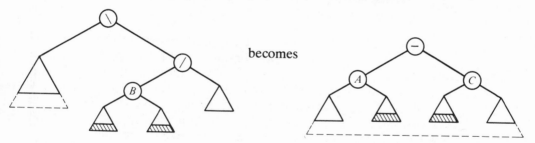

becomes

In cases 2, 3(b), and 3(c), the algorithm terminates if the current node is the root.

Notice that this deletion algorithm can require as many as $\lfloor h/2 \rfloor$ transformations, where h is the height of the entire tree (see Exercise 19); this is in contrast to the insertion algorithm, which requires at most one transformation. However, in most cases, the deletion algorithm (and the insertion algorithm as well) will not require retracing the upward path all the way to the root; usually the algorithm terminates after a number of steps that is independent of the height of the tree (see Exercise 20).

6.4.3   Weight-Balanced Binary Trees

In the previous section we considered trees that were balanced by constraining the relative heights of the subtrees of each node; here we consider trees that are balanced by constraining the relative number of nodes in the subtrees of each node. The height and the number of nodes are the two most natural measures of the size of a tree; the number of nodes has the advantage that it allows the important generalization to weighted nodes—in other words, to the case where a node, instead of being counted as a unit, enters the calculations with an arbitrary weight attached to it. Although this section considers only the case in which each node has unit weight, most results generalize easily to arbitrary weights.

Definition

Let T_n be a binary tree of $n \geq 1$ nodes in which the root has left and right subtrees T_l and T_r of $l \geq 0$ and $r \geq 0$ nodes, respectively. Then the *balance* $\beta(T_n)$ is

$$\beta(T_n) = \frac{l+1}{n+1}.$$

Notice that $\beta(T_n)$ indicates the relative number of nodes in the left subtree of T_n, and since $l + r = n - 1$, $\beta(T_n)$ always satisfies $0 < \beta(T_n) < 1$.

Definition

For $0 \leq \alpha \leq \frac{1}{2}$, a binary tree T_n of $n \geq 1$ nodes is said to be of *weight balance* α, or in the set WB[α], if the following two conditions hold:
1. $\alpha \leq \beta(T_n) \leq 1 - \alpha$.
2. The two subtrees T_l and T_r of the root of T_n are in WB[α].
The empty binary tree T_0 is in WB[α] by convention.

The class WB[α] becomes more and more restricted as α goes from 0 to $\frac{1}{2}$. Since $\alpha = 0$ imposes no restriction at all, WB[0] is the class of all binary trees. At the other extreme, since $\alpha = \frac{1}{2}$ implies that the left and right subtrees of each node contain equal numbers of nodes, only the completely balanced binary trees of $n = 2^h - 1$ nodes are in WB[$\frac{1}{2}$]. In the examples of Figure 6.17, the balance of each subtree is written next to the root of the subtree; the minimum of these balances is the maximum α such that the tree is in WB[α].

Let us make three observations.

Observation 1

For any α in the range $\frac{1}{3} < \alpha < \frac{1}{2}$, WB[$\alpha$] = WB[$\frac{1}{2}$].

Proof: If a tree T is not a completely balanced binary tree—that is, it is not in WB[$\frac{1}{2}$]—consider a minimal subtree (one of least height) T' of T not in WB[$\frac{1}{2}$]—that is, not a completely balanced binary tree. T' must have two completely balanced subtrees T_l, T_r of $l = 2^s - 1$ and $r = 2^t - 1$ nodes, respectively, with $s \neq t$ (say $s < t$). Then the balance $\beta(T') = 1/(1 + 2^{t-s}) \leq \frac{1}{3}$; hence T cannot be in WB[α] for any $\alpha > \frac{1}{3}$.

Observation 2

For any α, $0 < \alpha < \frac{1}{2}$, there are trees in WB[α] that are not height balanced.

Proof: The fourth tree in Figure 6.17 is in WB[$\frac{1}{3}$] $\subseteq$ WB[α], but it is not height balanced.

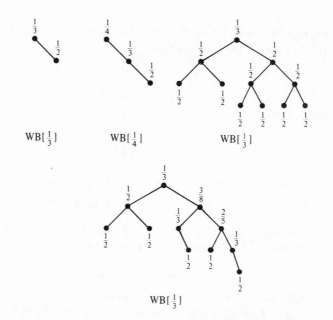

Figure 6.17   Examples of weight-balanced trees.

Observation 3

For any α, $0 < \alpha \leqq \frac{1}{2}$, there are height-balanced trees that are not in WB[α].

Proof: Consider a tree T whose left subtree is the most-skewed height-balanced tree of height h, shown in Figure 6.14, a tree with $n_h \approx 1.9[(1 + \sqrt{5})/2]^h$ nodes, and whose right subtree is a completely balanced binary tree of height h, with $2^{h+1} - 1$ nodes. T is height balanced, but $\beta(T) \to 0$ as $h \to \infty$.

We conclude that the shape of weight-balanced trees can be controlled by varying the parameter α continuously in the range $0 \leqq \alpha \leqq \frac{1}{3}$. Moreover, weight balancing and height balancing are independent balancing criteria.

In order to show that weight-balanced trees guarantee logarithmic search times in dynamic tables, we must bound the search time in WB[α] trees and show how a tree that is thrown out of WB[α] by an insertion or deletion can be rebalanced in logarithmic time. The rebalancing will be done for $0 < \alpha \leqq 1 - \sqrt{2}/2 \approx 0.29$ by using the same transformations (rotations and double rotations) as in height-balanced trees.

To bound search times, consider the most skewed tree T_n of n nodes in WB[α] ("most skewed" now takes a meaning different than in the previous section). This tree has $\alpha(n + 1) - 1 \approx \alpha n$ nodes in one of its subtrees (say the left one) and $(1 - \alpha)(n + 1) - 1 \approx (1 - \alpha)n$ in the other. Both subtrees are most skewed, hence the height $h_\alpha(n)$ of this most skewed tree T_n in WB[α] satisfies the recurrence relation

$$h_\alpha(n) = 1 + h_\alpha[(1 - \alpha)(n + 1) - 1]$$

$$\approx 1 + h_\alpha[(1 - \alpha)n],$$

with solution

$$h_\alpha(n) \approx \frac{\lg(n + 1)}{\lg[1/(1 - \alpha)]} + O(1).$$

For completely balanced binary trees ($\alpha = \frac{1}{2}$), this formula yields

$$h_{1/2}(n) \approx \lg(n + 1),$$

as it should. For $\alpha = 1 - \sqrt{2}/2$, the largest value of α for which the restructuring algorithm to be described works, we have

$$h_{1-\sqrt{2}/2} \approx 2\lg(n+1) + O(1).$$

Thus the longest search time in the most skewed WB[$1 - \sqrt{2}/2$] tree is only twice as long as in completely balanced binary trees. As in height-balanced trees, the average search time in the most skewed WB[α] trees is much better than the worst case, as the following theorem shows.

Theorem 6.2

If T_n is in WB[α], then the internal path length $I(T_n)$ satisfies the inequality

$$I(T_n) \leq \frac{1}{\mathcal{H}(\alpha)}(n + 1)\lg(n + 1) - 2n,$$

where $\mathcal{H}(\alpha) = -\alpha\lg\alpha - (1 - \alpha)\lg(1 - \alpha)$ is a simple instance of the entropy function of Section 1.5.

Proof: Exercise 22.

As a consequence of Theorem 6.2, we see that the average successful search time $I(T_n)/n$ in the most-skewed WB[α] trees is at most $[1/\mathcal{H}(\alpha)]\lg(n + 1) + O(1)$, larger by at most a factor of $1/\mathcal{H}(\alpha)$ than in completely balanced binary trees. For $\alpha = 1 - \sqrt{2}/2$, we have $1/\mathcal{H}(\alpha) \approx 1.15$, so that for trees in WB[$1 - \sqrt{2}/2$] there is at most a 15% increase in search time over completely balanced binary trees.

Only a small fraction of the trees in WB[α] are as skewed as possible, and many more are close to being completely balanced binary trees. Thus average search time, averaged over all trees, is significantly better than Theorem 6.2 indicates, perhaps as good as $\lg n + O(1)$. However, no mathematical analysis of this case exists.

Weight-balanced trees that have been disturbed by an insertion or deletion can be rebalanced in logarithmic time by using the two tree transformations introduced in the previous section. Figure 6.18 shows the effects of these transformations on the balances of the subtrees involved in them.

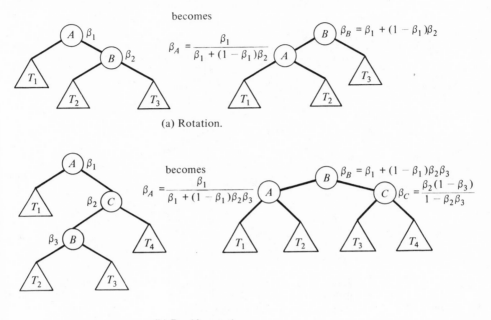

(a) Rotation.

(b) Double rotation.

Figure 6.18 The affect of rotation and double rotation on the balances of the subtrees involved.

Assume that the subtree shown at the left in Figure 6.18(a) is in WB[α], that is, β_1 and β_2 are both in the range [$\alpha, 1 - \alpha$], and that the balance β_1 has just reached its allowable limit α, where the subtree T_1 is "light" compared to the subtree consisting of T_2, B, and T_3. If we rotate around node A, we obtain the subtree shown at right, with balances $\beta_B = \beta_1 + (1 - \beta_1)\beta_2$ and $\beta_A = \beta_1/[\beta_1 + (1 - \beta_1)\beta_2]$, for the subtrees rooted at B and A, respectively; no other balances change. We hope that β_B is closer to $\frac{1}{2}$ than β_1 was and that β_A is in the range $\alpha \leqq \beta_A \leqq 1 - \alpha$. Figure 6.19 shows β_B and β_A as functions of β_2 under the assumption $\beta_1 = \alpha$, the condition that invokes this rotation.

We would like both β_A and β_B to be in the range [$\alpha, 1 - \alpha$] for $\alpha \leqq \beta_2 \leqq 1 - \alpha$. For β_A, this is guaranteed by restricting α so that when $\beta_2 = \alpha$ we have $\beta_A(\alpha) \leqq 1 - \alpha$ or, equivalently, $1/(2 - \alpha) \leqq 1 - \alpha$, that is, $\alpha^2 - 3\alpha + 1 \geqq 0$, which holds for $\alpha \leqq (3 - \sqrt{5})/2 \approx 0.38$. By Observation 1, WB[$\frac{1}{2}$] − WB[0.38] is empty, and so $\alpha \leqq 0.38$ is really no constraint at all. For β_B, Figure 6.19 shows that there is a small range $\beta_2 > (1 - 2\alpha)/(1 - \alpha)$ where $\beta_B > 1 - \alpha$ and hence where a rotation cannot be used for rebalancing the tree. However, for $\alpha \leqq \beta_2 \leqq (1 - 2\alpha)/(1 - \alpha)$, both β_A and β_B are in the range [$\alpha, 1 - \alpha$] and closer to $\frac{1}{2}$ than was β_2. Therefore our first rebalancing rule is

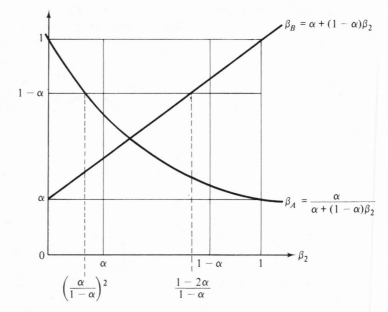

Figure 6.19 Balances of subtrees modified by the rotation of Figure 6.18(a) as a function of β_2, assuming $\beta_1 = \alpha$. β_A and β_B are then in the range $(\alpha, 1 - \alpha)$ except when $\beta_2 > (1 - 2\alpha)/(1 - \alpha)$.

Rotation Rule

A rotation rebalances a subtree when the balance β_1 of the subtree is equal to α and the balance β_2 of its right subtree is at most $(1 - 2\alpha)/(1 - \alpha)$. A symmetrical rule states that if $\beta_1 = 1 - \alpha$, and the balance β_2 of the left subtree is at least $\alpha/(1 - \alpha)$, then a rotation rebalances the subtree.

When a node is inserted or deleted, balances do not vary continuously but in discrete, small steps of size approximately $1/n$ for a tree of n nodes. Although the preceding calculation was made under the assumption $\beta_1 = \alpha$, it is clear that the result is similar when β_1 is close to α. In particular, when β_1 has just fallen slightly below α (because a node was deleted from the lighter subtree or added to the heavier subtree), the rule for rotations still holds [consider how $\beta_B = \beta_1 + (1 - \beta_1)\beta_2$ looks when $\beta_1 = \alpha - \varepsilon$]. The only restriction is that the rule must not be applied to extremely small subtrees (of less than four nodes), where the addition or deletion of a node causes a large change in β_1. These small subtrees are easily handled as special cases.

What do we do when $\beta_1 = \alpha$ and $(1 - 2\alpha)/(1 - \alpha) \leqq \beta_2 \leqq 1 - \alpha$? In this case, we use a double rotation, resulting in the balances

$$\beta_A = \beta_1[\beta_1 + (1 - \beta_1)\beta_2\beta_3],$$

$$\beta_B = \beta_1 + (1 - \beta_1)\beta_2\beta_3,$$

and
$$\beta_C = \frac{(\beta_2 - \beta_2\beta_3)}{(1-\beta_2\beta_3)},$$

as shown in Figure 6.18(b). Figure 6.20 shows β_A, β_B, and β_C as functions of the independent variable $\beta_2\beta_3$, which is constrained to lie in the range

$$\frac{\alpha - 2\alpha^2}{1 - \alpha} \leqq \beta_2\beta_3 \leqq (1 - \alpha)^2,$$

since $\alpha \leqq \beta_3 \leqq 1 - \alpha$, and $(1 - 2\alpha)/(1 - \alpha) \leqq \beta_2 \leqq 1 - \alpha$ when a double rotation is applied. With respect to this independent variable, the graphs for β_A and β_B are the same as in Figure 6.19, but the range of interest is different. In order to ensure that both β_A and β_B are in the range $[\alpha, 1 - \alpha]$ we must have

$$(1 - \alpha)^2 \leqq \frac{1 - 2\alpha}{1 - \alpha},$$

which leads to $\alpha \leqq (3 - \sqrt{5})/2 \approx 0.38$, and thus imposes no new constraint, and

$$\frac{\alpha - 2\alpha^2}{1 - \alpha} \geqq \left(\frac{\alpha}{1 - \alpha}\right)^2,$$

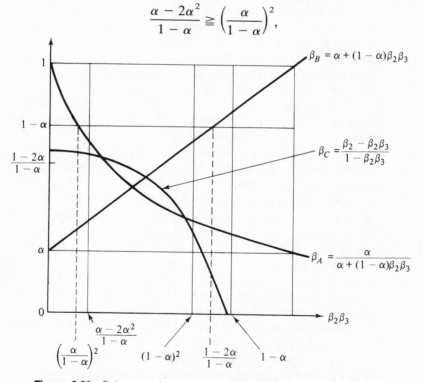

Figure 6.20 Balances of subtrees modified by the double rotation of Figure 6.18(b) as a function of $\beta_2\beta_3$, assuming $\beta_1 = \alpha$. The independent variable $\beta_2\beta_3$ is restricted to the range $(\alpha - 2\alpha^2)/(1 - \alpha) \leqq \beta_2\beta_3 \leqq (1 - \alpha)^2$; this implies that all three balances β_A, β_B, and β_C are in the range $(\alpha, 1 - \alpha)$, provided $\alpha \leqq 1 - \sqrt{2}/2$.

which leads to $\alpha \leqq \sqrt{2}/2 \approx 0.29$ and thus limits the allowable range of α to somewhat below the upper limit of $\frac{1}{3}$. Using the restriction $\alpha \leqq 1 - \sqrt{2}/2$ it is straightforward to calculate that $\alpha \leqq \beta_C \leqq 1 - \alpha$ for the allowed ranges of β_2 and β_3 (Exercise 23). Thus our second rebalancing rule is:

Double Rotation Rule

A double rotation rebalances a subtree when the root balance β_1 of any subtree equals $\alpha \leqq 1 - \sqrt{2}/2$ and the balance β_2 of its right subtree is at least $(1 - 2\alpha)/(1 - \alpha)$. A symmetrical rule states that if $\beta_1 = 1 - \alpha$, and the balance β_2 of the left subtree is at most $\alpha/(1 - \alpha)$ then a double rotation rebalances the tree.

To implement weight-balanced trees, each node must contain information sufficient to compute its balance. Instead of storing the balance directly, it is convenient to store an integer value like the size of the subtree rooted at this node (i.e., the number of nodes it contains) or the rank of the name stored at this node relative to the subtree rooted at this node (i.e., one plus the size of the left subtree of this node); the balance of a node can be computed from such information stored in the node (how?). An additional benefit of storing either quantity is that important operations, such as finding the kth name in the table, or the qth quantile, or the number of names between two given names y_1 and y_2, can all be done in time $O(\log n)$ (Exercise 24); these questions usually require time proportional to n if the size information is not explicitly stored.

The algorithm for inserting or deleting a given name z is now easily explained. Search the tree for z by starting at the root and following a path toward a leaf. As each node is traversed, the size information is updated and, if the subtree rooted at this node becomes unbalanced, the appropriate tree transformation is performed. Once z has been found (in the case of deletion) or its place has been found (in the case of insertion), the pointer manipulations described in Section 6.4.1 are applied to add or delete a node. The algorithm then terminates because all the updating and rebalancing was done during the search for z. Notice that since weight-balanced trees do not require retracing the path upward toward the root, there is no need to store the path on a stack.

What happens when there are extraneous insertions and deletions—that is, when the insertion algorithm is applied with an already present name or when the deletion algorithm is applied with a name not in the tree? In such cases, all the size information must be corrected in a second walk down the tree, but the transformations performed during the first pass need not be undone. Although not needed, they improved the balance of the tree.

If we make some assumption about the distribution of balances in a tree in WB[α], it is possible to estimate the expected number of transformations that must be performed during an insertion or deletion. Insertions and deletions have the effect of shifting the balances around in the interval [$\alpha, 1 - \alpha$], and each balance can thus be considered to be on a random walk in a continuous interval with reflecting barriers. When a step would take the balance outside the interval, a transformation is applied and the balance moves closer to $\frac{1}{2}$. If the balance were reflected the same distance that it tried to go forward, a result in probability theory states that the resulting distribution would be uniform over the interval $[\alpha, 1 - \alpha]$. However, it can be seen from Figures 6.19 and 6.20 that the situation is even more favorable; most of the time a balance that has reached the boundary of an interval is repulsed strongly toward $\frac{1}{2}$. But even under the (weak) assumption that the distribution of balances in a WB$[\alpha]$ tree is uniform over $[\alpha, 1 - \alpha]$, it can be shown that the expected number of transformations required for the insertion or deletion of a node is less than $2/(1 - 2\alpha)$, independent of n (Exercise 25). For $\alpha = \frac{1}{4}$, for example, less than four transformations are required on the average.

6.4.4 Balanced Multiway Trees

Binary search trees generalize in a natural way to m-way search trees in which each node has $k \leq m$ sons and contains $k - 1 \leq m - 1$ names. The names in a node divide a set of names into k subsets, each subset corresponding to one of the k subtrees of this node. Figure 6.21 shows a completely filled five-way tree of two levels. Notice that we cannot insist that each node of an m-way tree have exactly m sons and contain exactly $m - 1$ names; if we wish to insert Z in the tree of Figure 6.21, we must create nodes with fewer than m sons and fewer than $m - 1$ names. Thus the definition of an m-way tree states only that each node has *at most* m sons and contains *at most* $m - 1$ names. It is clear that m-way trees can be searched in a manner analogous to a search in binary trees.

To keep m-way trees balanced, we insist that all paths from the root to a leaf be of equal length and that each node (except the root) have between $\lceil \alpha m \rceil$ and m sons, for some constant $0 < \alpha < 1$. (We will discuss only the case $\alpha = \frac{1}{2}$, but other values of α are possible.) This balancing criterion guarantees logarithmic search times (why?), and, as we will see, it is easy to keep a tree balanced during insertions and deletions.

As in binary search trees, it is useful to distinguish between internal nodes and leaves. An internal node contains $k \geq 1$ names, listed in natural order, and has $k + 1$ sons, each of which may be either an internal node or a leaf. A leaf

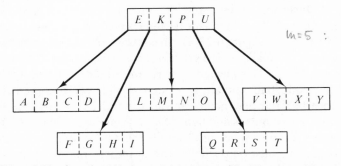

Figure 6.21   A perfectly balanced, completely filled 5-way search tree.

contains no names (except perhaps temporarily during an insertion); and, as before, the leaves are the places where unsuccessful searches terminate. For clarity, we generally omit them from figures.

Definition

A *balanced multiway tree of order m* is an *m*-way tree in which

1. All leaves are on the same level.
2. The root has k sons, $2 \leq k \leq m$.
3. Other internal nodes have k' sons, $\lceil m/2 \rceil \leq k' \leq m$.

Figure 6.22 shows a balanced five-way tree that results from the insertion the name Z into the tree of Figure 6.21 while preserving the properties of a balanced five-way tree. A step-by-step examination of the transformation of the tree of Figure 6.21 to that of Figure 6.22 will explain the process of insertion for balanced *m*-way trees.

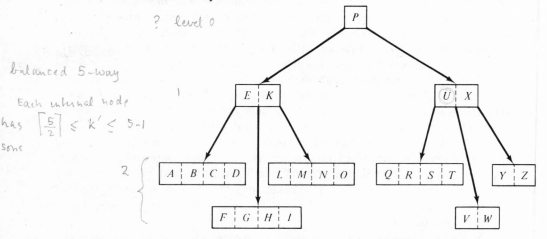

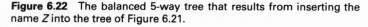

Figure 6.22 The balanced 5-way tree that results from inserting the name *Z* into the tree of Figure 6.21.

Insertion of the name Z begins by searching for Z. This search terminates unsuccessfully, which can be interpreted as Z having been found in a leaf, as illustrated in Figure 6.23(a). In contrast to binary search trees, balanced multiway trees are not allowed to grow at the leaves; instead they are forced to grow at the root. Hence Z is pushed upward, into the node occupied by V, W, X, and Y. If this node had not already been completely occupied, it would simply have absorbed the new name. In our example, however, the node already contains $m - 1 = 4$ names and hence has no room to hold another name. The five names trying to find a place in this node must be split into two groups, and the median name, X, must be pushed into the father node in order to serve as a separator between the two halves, as shown in Figure 6.23(b). If the father node had not

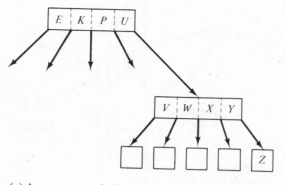

(a) A new name to be inserted is temporarily added as a leaf.

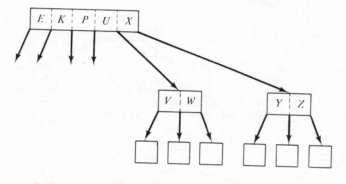

(b) It is then pushed into the father node, and if this node is already filled, the names in it are split into two nodes, pushing another name upwards.

Figure 6.23 First step of the insertion of *Z* into the tree of Figure 6.21.

been fully occupied, the insertion process would terminate, but in our example it is already filled to capacity and must be split. The median name, *P*, is used as a separator between *E, K* and *U, X* and is pushed upward. When the root splits, a new node is created that becomes the new root, and the tree grows by one level. Thus we obtain the tree of Figure 6.22.

To illustrate the process of deletion, consider the deletion of the name *U* from the tree of Figure 6.22. Since *U* serves as a separator between the two sets *Q, R, S, T* and *V, W*, it cannot simply be erased. Instead it is replaced by its immediate predecessor in the tree, *T* (which can serve as a separator between *Q, R, S* and *V, W*), or by its immediate successor in the tree, *V* (which serves as a separator between *Q, R, S, T* and *W*); both names are in nodes at the lowest level of the tree (Exercise 26). If *T* is promoted, it can simply be erased from its former

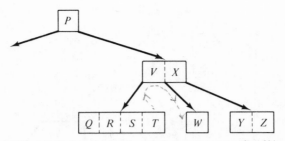

(a) Transient stage during the rebalancing process after *U* has been deleted from the tree of Fig. 6.22. The node with the single name *W* violates one of the conditions of balanced 5-way trees.

ok: This a balanced 5-way tree, with 3? levels

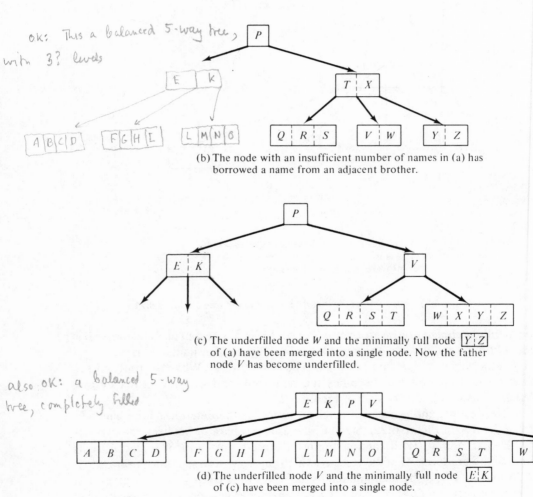

(b) The node with an insufficient number of names in (a) has borrowed a name from an adjacent brother.

(c) The underfilled node *W* and the minimally full node $\boxed{Y \vdots Z}$ of (a) have been merged into a single node. Now the father node *V* has become underfilled.

also ok: a balanced 5-way tree, completely filled

(d) The underfilled node *V* and the minimally full node $\boxed{E \vdots K}$ of (c) have been merged into a single node.

Figure 6.24   Deletion of the name *U* from the tree of Figure 6.22.

node, since that node still contains $3 \geqq \lceil m/2 \rceil - 1$ names and no constraints are violated. On the other hand, if V is promoted to the former place of U and erased from its original position, we have the situation shown in Figure 6.24(a). The node with the single name W in it violates the condition that every node other than the root must have at least $\lceil m/2 \rceil$ sons and hence must contain at least $\lceil m/2 \rceil - 1$ names.

A node that has become insufficiently full after splitting can lack only one name. Therefore we attempt to borrow a name from one of its adjacent brothers. In Figure 6.24(a) the left brother of W has a sufficient supply of names, and so we can add the separator V to the node of W and promote T to become the new separator as shown in Figure 6.24(b).

If there is no adjacent brother that can spare a name, then there must be an adjacent brother that is minimally full—that is, that contains exactly $\lceil m/2 \rceil - 1$ names [the right brother of W in Figure 6.24(a)]. Since the underfilled node contains $\lceil m/2 \rceil - 2$ names after the deletion, it, its minimally full brother, and their separator in the father node can be combined into one node with $m - 1$ or $m - 2$ names. If W and Y, Z are combined in this way in Figure 6.24(a), we obtain the tree shown in Figure 6.24(c). Now the node containing V has become underfilled, and the same process is repeated at the next higher level in the tree. In the worst case, the deletion algorithm terminates at the root; then the height of the tree may be reduced by one level, as in our example. When we combine V with its left brother E, K and the separator P, we obtain the perfectly balanced, completely filled tree of two levels shown in Figure 6.24(d).

6.5 ADDRESS COMPUTATION TECHNIQUES

Address computation techniques, also known as key to address transformations, scatter storage techniques, and hashing or hash coding, refer to any search technique that begins by computing an address from a name. Such techniques use the properties of random access memory more than any other search technique.

The assumptions made in this section are that we have a memory consisting of m cells M_j, $0 \leqq j \leqq m - 1$, each capable of holding one name, and that given a value of j we can access M_j in constant time, independent of m. These assumptions are valid on modern computers for large values of m, and so the techniques described in this section are often applicable in practice.

The main asset of address computation techniques is that their expected search times are independent of table size as long as the entire table fits into central memory. Asymptotically, and often in practice as well, they are the fastest-known search techniques, but they do have three disadvantages. First, the table order of the names is usually unrelated to their natural order. Second, the worst-case behavior can be worse than that of sequential search. Finally, tables based on address computation cannot easily be extended dynamically; this may lead to a waste of memory if the table is too large or to poor performance if the table is too small. Yet address computation techniques are valuable in certain cases: many tables, for instance, need not be listed in natural order; worst-case behavior can often be ignored if it occurs infrequently enough; and the sizes of many tables can be accurately estimated a priori.

In Section 6.5.1 the main ideas of these techniques are illustrated through simple examples. The remaining subsections discuss the most widely used techniques in more detail.

6.5.1   Hashing and its Variants

We begin with a very simple example in which the name space is atypically small. Consider the name space $S = \{A, B, \ldots, Z\}$, a *memory* or *hash table* $M = \{M_0, M_1, \ldots, M_{25}\}$ of 26 cells with an *address space* $A = \{0, 1, \ldots, 25\}$, and a *key to address transformation*, or *hash function*, $h : S \to A$ that maps letters into addresses so as to preserve the alphabetic natural order on S [i.e., $h(A) = 0$, $h(B) = 1, \ldots, h(Z) = 25$]. A table $T = \{x_1, x_2, \ldots, x_n\} \subseteq S$ is stored by putting the name x_i into cell $M_{h(x_i)}$ and using a special symbol not in S, say $-$, in the empty cells, as shown in Figure 6.25.

The four table operations are accomplished by simple and fast algorithms.

search for z: **if** $M_{h(z)} = z$ **then** ⟦found: $h(z)$ points to z⟧
                                          else ⟦not found: z is not in the table⟧

insert z: $M_{h(z)} \leftarrow z$

delete z: $M_{h(z)} \leftarrow -$

list: **for** $i \leftarrow 0$ **to** $m - 1$ **do if** $M_i \neq -$ **then** output M_i

$$S = \{A, B, \ldots, Z\}$$
$$T = \{B, C, Z\}$$

A	M
0	$-$
1	B
2	C
3	$-$
$\vdots$	$\vdots$
24	$-$
25	Z

Figure 6.25   A hash table for a small name space.

Notice the absence of an explicit loop in the search algorithm; this feature is characteristic of the ideal case in address computation techniques. It is important that any loop implicit in the computation of the hash function h be independent of the table size. This is the case in the preceding example, where h can be computed on most computers by a fixed number of simple bit operations on the binary codes of the characters. In this simple example, listing the names in natural order can be accomplished by a simple linear scan of the memory, which is most uncharacteristic of address computation techniques. It occurs here because the memory is the same size as the name space, a very unusual occurrence; more commonly, the memory is too small to hold the entire name space.

One variant of address computation applies to static tables when the hash function can be chosen *after* the contents of the table are known. By using clever ad hoc techniques, it is usually possible to find an easily computed function that maps the set of names stored in the table one to one into the address space. It is especially easy to find such a function when the memory contains more cells than names. For instance, consider storing the five words AN, AT, NO, ON, and PI in a memory $M_0, M_1, \ldots, M_6$. Assume that each letter is represented by a 5-bit code: A by 00000, B by 00001, C by 00010, and so on. Thus the five words have the representations

Ex ②

	b_9	b_8	b_7	b_6	b_5	b_4	b_3	b_2	b_1	b_0
AN	0	0	0	0	0	0	1	1	0	1
AT	0	0	0	0	0	1	0	0	1	1
NO	0	1	1	0	1	0	1	1	1	0
ON	0	1	1	1	0	0	1	1	0	1
PI	0	1	1	1	1	0	1	0	0	0

Notice that the bit string $b_6 b_5 b_4$ uniquely identifies each of these five words and that the values of these bit strings, interpreted as binary integers, lie in the address space $A = \{0, 1, 2, 3, 4, 5, 6\}$. Thus the hash function $h(x) = (b_6 b_5 b_4)_2$ maps the set of five names into the memory; moreover, the natural alphabetic order is preserved.

When the contents of the table are known before the hash function is designed, the name space itself does not matter. In this example, we could have assumed that the name space was simply the given set of five words, that it was the set of all strings of two letters, or that it was the set of all strings of arbitrary length.

In the usual application of address computation, commonly called *hashing*, a large name space S is specified along with some vague information about the contents of the table. We may know, for example, that the number of names to be stored in the table will not exceed some bound, that certain names are more likely to occur than others, or that names are likely to occur in clusters defined by common properties. On the basis of such information we choose a memory size m and design a hash function that maps S into the address space $A = \{0, 1, \ldots, m - 1\}$. The ultimate test for our choice of m and h is their performance *in practice*; theoretical analyses generally are not sufficient, since such analyses are normally based on assumptions about the contents of the table that are too simple-minded.

Ex ③

Consider, for example, the name space S of all bit strings of fixed length 8 and assume that we know that the table will never contain more than six names. Suppose that we choose a memory of eight cells $M_0, M_1, \ldots, M_7$ and that we choose the simple hashing function that maps a string into its rightmost three bits, interpreted as a binary integer. This function has two advantages: it is easy to compute and it spreads the name space uniformly over the address space.

Assume that the table consists of the five names

$$x_1 = 00010110 \qquad h(x_1) = (110)_2 = 6,$$

$$x_2 = 00100100 \qquad h(x_2) = (100)_2 = 4,$$

$$x_3 = 00111010 \qquad h(x_3) = (010)_2 = 2,$$

$$x_4 = 10101100 \qquad h(x_4) = (100)_2 = 4,$$

$$x_5 = 11110000 \qquad h(x_5) = (000)_2 = 0.$$

Since h maps x_2 and x_4 onto the same address, we have a *collision*. At least one of these two names must be stored elsewhere, and, in order to do so, we must decide on a technique for *resolving collisions*. There is still room in our eight-cell memory, and so we put x_4 in some unoccupied cell, say M_5, the first empty cell following its *home address* of $h(x_4) = 4$. Our memory now looks as shown in Figure 6.26. Notice that the table order is unrelated to the natural order of names; as a result, listing the names in their natural order requires sorting them.

P263 "linear probing"

Whenever the memory is larger than the size of the table stored, empty cells must be distinguished from occupied cells. This is normally done either by associating an extra bit with each cell, as Figure 6.26 suggests, or, if possible, by

Figure 6.26 Memory configuration showing how a collision can be handled.

using a special value not in the name space to denote an empty cell. We will simply assume that we have a test to answer the question "Is M_i empty?" in constant time; we will not worry about how it is implemented.

When collisions occur, searching becomes considerably more complex than in our earlier example and usually involves a loop that can be executed as many as m times in the worst case. With the collision resolution technique in the example above (i.e., using the first unoccupied cell beyond the home address), searching requires a sequential search of the memory, starting at the home address of z and going from cell M_{m-1} to cell M_0 in a circular fashion, as shown in Algorithm 6.9. The search terminates unsuccessfully if an empty cell is encountered or when the entire memory has been scanned. We can avoid an explicit test for the latter case if at least one cell in the memory is left empty; thus we make this assumption in Algorithm 6.9.

search (with collisions)

$i \leftarrow h(z)$

while M_i is not empty **do if** $z = M_i$ **then** [[found: i points to z]]

$\qquad\qquad\qquad\qquad\qquad\qquad$ **else** $i \leftarrow i + 1 \bmod m$

$\qquad$ [[not found: z is not in the table]]

circular search but "linear probing" p263

Algorithm 6.9   Searching in a hash table.

During an insertion we must explicitly check that the memory is not full—that is, that there is more than one empty cell. This is done in Algorithm 6.10 by comparing n, the number of names already in the table, with the memory size m. In all preceding search techniques we simply extended the table if necessary, but doing so for address computation techniques is not a simple matter, since the hash function h depends explicitly on the size of the memory. Extending the memory requires *rehashing*—that is, finding a new hash function and relocating all the previously stored names according to the new hash function.

$i \leftarrow h(z)$

while M_i is not empty **do if** $z = M_i$ **then** [[z is already in the table]]

$\qquad\qquad\qquad\qquad\qquad\qquad$ **else** $i \leftarrow i + 1 \bmod m$

[[z is not in the table; insert it if possible]]

insertion

if $n = m - 1$ **then** [[table is full; z cannot be inserted]]

n is no. of table entries
m = memory size:
address space A is $\{0, 1, \ldots, m-1\}$

$\qquad\qquad$ **else** $\begin{cases} \text{[[table is not full; insert } z\text{]]} \\ M_i \leftarrow z \\ \text{mark } M_i \text{ ``occupied''} \\ n \leftarrow n + 1 \end{cases}$

Algorithm 6.10   Insertion into a hash table in which one cell is always left empty.

Surprisingly, the obvious way of deleting the name in M_i, marking M_i as empty, is *wrong* for most hashing schemes! To understand why, consider what happens to the table in Figure 6.26 if x_2 is deleted by marking M_4 empty. If we now search for x_4, the sequential search starts at the home address 4 of x_4, finds M_4 empty, and incorrectly concludes that x_4 is not in the table; x_4 is in M_5, where it was stored because it collided with x_2. As long as M_4 was occupied, x_4 was accessible; but when M_4 became empty, x_4 became inaccessible by means of Algorithm 6.9. Ideally, deletion of x_2 should include moving x_4 from M_5 to its home address M_4. However, to ensure that the empty cell M_5 does not leave any other names inaccessible, we would have to verify that M_3 is empty and that $h(x_1) \neq 4$ and $h(x_1) \neq 5$. In more sophisticated collision resolution schemes, such a restructuring of the table requires extensive searching and verifying, and, as a rule, deletion is handled differently. If we let each entry in the column labeled "Occupied" in Figure 6.26 assume one of the three values "occupied" ($\sqrt{}$), "empty" (blank), or "deleted" (d), then deletion of a name is accomplished by marking its cell as "deleted."

In order to work correctly in the presence of cells marked deleted, Algorithm 6.9 (searching in a hash table) must be modified so that deleted cells are ignored during the search. Algorithm 6.10 (insertion in a hash table) must be modified so that a new name is entered in either an empty or a deleted cell, whichever is encountered first; in either case, the cell used is then marked as being occupied (see Exercise 30). Notice that although a deleted cell behaves like an empty cell with respect to insertion, it behaves like a full cell with respect to searching. This causes unsuccessful searches—which terminate when the first empty cell is encountered—to be very inefficient if the sequence of deletions and insertions leaves fewer and fewer cells marked "empty," even if the total number of names in the table remains roughly constant! Eventually unsuccessful searches will scan the entire memory.

The first two examples of this section illustrated two address computation techniques of limited applicability: when the name space is smaller than the available memory or when the contents of the table are known before the hash function is chosen. The third example is the most important; it illustrates a technique that is useful when the name space is large and the hashing function and collision resolution scheme must be chosen before the contents of the table are known. The remaining subsections discuss these problems in more detail.

6.5.2 Hash Functions

Ideally, a hash function $h : S \to A$ should be easy to compute and should map the name space into the address space so as to spread the names uniformly over the memory. If the second requirement is interpreted to mean that each address $i \in A$ is the image of (approximately) equally many names $x \in S$, then simple hash functions, such as using the binary integer interpretation of a substring of the name, are acceptable. However, we actually care only whether the contents of the table, not the entire name space, are spread uniformly over the memory. Thus a hash function should be designed to spread uniformly over the memory those subsets of S that are likely to occur as contents of tables. It is rarely possible to characterize these likely subsets accurately; consequently, the design of hash functions is an art that is not amenable to mathematical analysis and that relies instead on a few commonsense rules. Fortunately, as we will see, the hash function is not the most important factor in a hashing technique. Any reasonable hash function will perform satisfactorily on *most* sets of names; but, conversely, for any hash function, there are sets of names on which it performs poorly. This section describes how to avoid hash functions that are poor for certain common sets of names.

Rules stating that a certain hash function h should be avoided are always based on experimental evidence that the sets of names likely to be encountered contain many names that h maps into a small set of addresses. Such clusters of names tend to occur for a variety of reasons. If the names have been chosen by a person, there may be sequences like X1, X2, X3, or ACOST, BCOST, CCOST. If the names are originally of different lengths, they may end up padded with trailing blanks to a standard length. If the names are restricted to an alphabet smaller than the character set of the computer, certain bit patterns may occur more often than others in the representation of the names. Hash functions that tend to map such clusters of names onto clusters of addresses should be avoided. Hash functions without such glaring deficiencies are acceptable if they are easily computed and if they perform well on test data.

In order to facilitate fast computation of a hash address, most hash functions are close to the primitive operations available on a computer, and hence they are best described at the level of bit string operations on the binary representations of names and addresses. So let us assume that names and addresses are represented by bit strings of lengths l_{name} and $l_{address}$, respectively. Often l_{name} is less than or equal to the word length of the computer; and, in all cases of interest, $l_{name} > l_{address}$, for if $l_{name} \leqq l_{address}$, every name could be assigned a unique address (as in the first example of Section 6.5.1).

The simplest hash functions simply pick some particular subset of $l_{address}$ bits from the string of l_{name} bits and use it to form an address, as in the second and third examples of Section 6.5.1. From our earlier discussion it is evident that the danger in this technique is that it tends to generate an excessive number of collisions. All names whose representations happen to agree on the subset of bits chosen by the hash function are mapped into a single address. If the subset of bits chosen overlaps the set of bits that are common to some cluster of names this cluster will be mapped into a small set of addresses.

Such considerations lead us to require that each of the $l_{address}$ bits of an address depend on *all* the l_{name} bits of the name. The most efficient way to achieve this dependence is to perform arithmetic operations whose results depend on all the bits of the name and then to extract $l_{address}$ bits from this result. Most hash functions in use are based on this approach, but certain pitfalls must be avoided. For example, one of the earliest techniques for constructing hash functions, the *middle-square method*, is to consider the representation of a name as an integer, square it, and then extract $l_{address}$ bits from the middle of the representation of the square. Although often satisfactory, this technique has its dangers, since squaring does not preserve uniform distribution; if x is a uniformly distributed random variable in the range $[a, b]$, then x^2 is *not* uniformly distributed in the range $[a^2, b^2]$. A particularly troublesome manifestation of this distortion of distributions occurs when the names have representations with trailing zeros. Their squares have twice as many trailing zeros, and these zeros can propagate into the middle section of the square, thereby distorting the resulting address. Since a uniform distribution of names across memory is the goal of hashing, the middle-square method has generally been abandoned.

Multiplication by a constant does not suffer from the disadvantage of the middle-square method. If x is a uniformly distributed random variable in the range $[a, b]$, then cx is uniformly distributed in the range $[ca, cb]$, and this range can be mapped onto the address space $\{0, 1, \ldots, m - 1\}$ in a number of ways that preserve uniformity. This is the basis for *multiplicative hashing* techniques. Of course, in the design of an actual hash function, we must remember that representations of names cannot be considered uniformly distributed and that arithmetic is carried out in some discrete, finite number system whose properties (in particular, the radix and the word length) must be considered so as to avoid degenerate cases that result in an excessive number of collisions. The primary danger in multiplicative hashing concerns the many tables occurring in practice that contain subsets of names whose representations form arithmetic progressions. Multiplicative hash functions map such progressions of names into arithmetic progressions of addresses, say with an increment i. If c or m is carelessly chosen, then i may divide m, and these subsets will use only a fraction $(1/i)$ of the memory space. The effect is to create as many collisions as would be expected by using a better hash function with a smaller memory.

Division hashing techniques use a hash function of the type

$$h(x) = (\text{representation of } x \text{ interpreted as an integer}) \bmod m,$$

where m is the memory size. The choice of m is critical in this method. For example, if m is even, then $h(x)$ is even if and only if the representation of x is an even integer; this bias could be serious. If m is a power of the radix of the computer, then $h(x)$ will depend only on the rightmost characters of x, another potentially serious bias. In general, any proper divisor of m can lead to problems. Therefore it is usually best for m to be prime; the primality of m also ensures that each digit of the hash address depends on all the characters of the name.

6.5.3 Collision Resolution

As long as a hash table is not too full (say, filled up to 50% of capacity), collisions should occur rarely, and the performance of the hashing scheme is determined primarily by the time required to evaluate the hash function. As the memory becomes fuller, accessing names requires more and more time due to collisions. So when space is at a premium, the collision resolution scheme determines the performance of the hashing schemes. Hence the choice of the collision resolution scheme is usually more important than the choice of the hash function.

A collision resolution scheme assigns a sequence $\alpha_0 = h(x)$, α_1, α_2, ... of addresses to each name x. The insertion algorithm inspects the cells in the order indicated until it finds an empty one. In order to guarantee that an empty cell is encountered if it exists, each address i, $0 \le i \le m - 1$, must occur exactly once in the sequence. Essentially, this sequence can be represented in two ways, and they correspond to the sequential and linked allocations for lists (see Section 2.2).

Collision resolution

Open Addressing.   In the third example of Section 6.5.1 we used the simplest form of open addressing, called *linear probing*, which generates the address sequence $\alpha_i = (h(x) + i) \bmod m$. Unless the memory is nearly empty, the simplicity of this sequence does not compensate for a serious disadvantage that makes it generally unsatisfactory. This disadvantage is the phenomenon of *secondary clustering*. *Primary clustering* occurs when a table contains many names with the same hash address. Secondary clustering occurs when names with different hash addresses have the same (or nearly the same) sequence of addresses $\alpha_1, \alpha_2, \ldots$. When primary clustering occurs, linear probing causes a sequence of consecutive cells to become occupied, and all names that get hashed to any of these cells generate an address sequence that runs over these occupied cells.

p 258

linear probing

If the function h is a division hash function, then the severity of secondary clustering may be increased by interference with the collision resolution scheme: In practice, many tables contain subsets of names that differ only in the last characters (such as A1, A2, A3, ..., or PARTA, PARTB, PARTC, ...), and the representations of such names often form runs of consecutive integers. These names can get mapped into consecutive addresses if linear probing is used with a division hash function.

Secondary clustering and interference can be avoided by using linear probing with an increment $\Delta(x)$ that is a function of the name x. This increment gives the address sequence $\alpha_i = (h(x) + i\Delta(x)) \bmod m$, which will probe every cell if $\Delta(x)$ and m are relatively prime. Since $\Delta(x)$ is basically another hash function, this technique is called *double hashing*. As long as care is taken to ensure that the probe sequences scan the entire memory, double hashing is superior to other collision resolution techniques for open addressing.

double hashing

Chaining.   Chaining is the technique of building a sequence of pointers from the home address $h(x)$ to the cell where the name x is ultimately stored. At some cost in memory (for storing the pointers), it avoids the problem of secondary clustering during searching but not during insertion. Figure 6.27 illustrates this. Assume that x and y collide, that x is stored at its home address $\alpha_0 = h(x)$, and that y is stored at the next address α_1 of the collision sequence for α_0 [see Figure

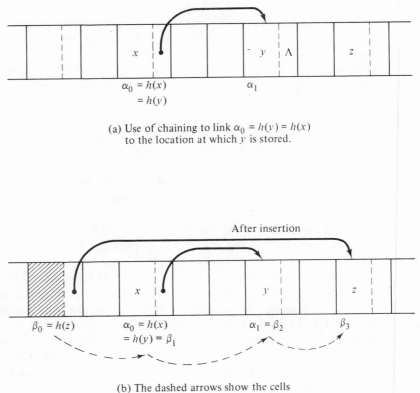

(a) Use of chaining to link $\alpha_0 = h(y) = h(x)$
to the location at which y is stored.

(b) The dashed arrows show the cells
probed during the insertion of z.

Figure 6.27   In chaining techniques the number of probes during searching may be smaller than during insertion.

6.27(a)]. Suppose now that the name z cannot be stored at its home address $h(z) = \beta_0$ and that in following its collision sequence $\beta_1, \beta_2, \ldots$ we happen to go over α_0 and α_1—for example, $\beta_1 = \alpha_0$ and $\beta_2 = \alpha_1$. In order to find an empty cell during insertion, we may have to look at α_0 and α_1; but once z has been stored in cell β_3, we link β_3 directly to $\beta_0 = h(z)$, so that fewer probes are made in searching for z than in inserting it [see Figure 6.27(b)].

In this section we have discussed collision resolution techniques that utilize only the cells in the hash table. It is possible to store colliding names in a separate overflow area. This amounts to a combination of different search techniques: hashing is used as a first step, and other techniques are used to search among all names with the same hash address.

6.5.4   *The Influence of the Load Factor*

Under the assumptions that access time to any cell in memory is constant, that the hash function spreads the name space uniformly over the address space, and that the contents of the table are an unbiased sample of the name space, the expected performance of address computation techniques is independent of n, the number of names in the table. Instead it depends primarily on the *load factor* $\lambda = n/m$, that is, the degree to which the memory is filled. In the extreme case $\lambda = 1$, when the memory is completely full, it is obvious that fast access times can no longer be achieved. However, it is important to consider the performance of address computation techniques when the memory is close to full. As the following discussion shows and experience confirms, load factors of 90% do not preclude acceptable performance; thus wasted memory need not be a significant disadvantage of address computation techniques.

What causes the good performance of address computation techniques? Possibly the secret lies in the choice of a clever hash function that avoids collisions as much as possible. However, in realistic cases, the hash function is chosen *before* the contents of the table are known, which means that there is no way to avoid collisions. Even if the hash function spreads names uniformly over the address space, a fraction $\lambda/2$ of the names in the table will collide with names previously entered (see Exercise 32). Since λ is generally close to 1, almost half the names lead to collisions; consequently, even the best hash function cannot avoid a sizable number of collisions. Fortunately, the number of collisions has little effect on the performance of an address computation technique; a collision lengthens the probe sequence by only one cell. Hence the choice of a hash function is *not* the critical part of the design of an address computation technique; the most important part concerns what happens *after* a collision has occurred—the collision resolution scheme.

Let us give an approximate analysis of double hashing. Assume, ideally, that the sequence $\alpha_0 = h(z)$, α_1, α_2, ... of addresses assigned to a name z has the property that each address α_i is equally likely to be any of the addresses $0, \ldots, m - 1$, independent of the other α_i's. Under this assumption (which is roughly true for double hashing), if the table has load factor μ and z is not in the table, then the probability of $k \geq 1$ probes being used in the unsuccessful search for z is

$$\text{Pr } (k \text{ probes are used}) = \text{Pr } (M_{\alpha_0}, M_{\alpha_1}, \ldots, M_{\alpha_{k-2}} \text{ are full, and } M_{\alpha_{k-1}} \text{ is empty})$$

$$= \mu^{k-1}(1 - \mu).$$

The expected number $U(\mu)$ of probes needed to search unsuccessfully for z in a hash table with load factor μ is thus

$$U(\mu) = \sum_{k=1}^{\infty} k\mu^{k-1}(1 - \mu)$$

$$= \frac{1}{1 - \mu}.$$

Notice that the number of probes needed to find z after it is in the hash table is the same as the number of probes on the unsuccessful search when it is inserted. Now, consider the hash table being constructed as names are inserted in sequence, with the load factor μ growing from 0 to its final value λ. It is a reasonable approximation to let μ vary continuously in the range $(0, \lambda)$ rather than in discrete steps, and we obtain for the expected number $S(\lambda)$ of probes needed to search successfully in a hash table with load factor λ:

$$S(\lambda) = \frac{1}{\lambda} \int_0^\lambda U(\mu)\, d\mu$$

$$= \frac{1}{\lambda} \ln \frac{1}{1-\lambda}.$$

Both $U(\lambda)$ and $S(\lambda)$ approach 1 as $\lambda \to 0$, thus indicating (correctly) that one probe should suffice when searching an almost empty table. Both $U(\lambda)$ and $S(\lambda)$ grow unboundedly as $\lambda \to 1$, but this is a consequence of treating μ as a continuous quantity, which amounts to assuming an infinite memory. We have

λ	0.5	0.75	0.9	0.95	0.99
$U(\lambda)$	2.0	4.0	10.0	20.0	100.0
$S(\lambda)$	1.39	1.85	2.56	3.15	4.65

Thus if addresses generated by the collision resolution technique are independent and uniformly distributed over the entire address space, as they generally are in double hashing, then the average number of probes required is small, even for load factors as high as 90%.

6.6   REMARKS AND REFERENCES

An extensive discussion of the search techniques discussed in this chapter is given in

KNUTH, D. E., *The Art of Computer Programming*, Vol. 3 (*Sorting and Searching*), Addison-Wesley, Reading, Mass., 1973.

This book also contains a brief history of the development of the various techniques.

The idea of arranging a table for sequential search so that the names are in nonincreasing order of frequency was first analyzed in

SMITH, W. E., "Various Optimizers for Single-Stage Production," *Naval Research Logistics Quart.*, **3** (1956), 59–66.

This paper also contains an interesting generalization (see Exercise 3).

The use of self-organizing files (Exercise 5) when the frequencies are not known a priori was first suggested in

MCCABE, J., "On Serial Files with Relocatable Records," *Operations Res.*, **12** (1965), 609–618.

For detailed analyses of various strategies, see

RIVEST, R. L., "On Self-Organizing Sequential Search Heuristics," *Comm. ACM*, **19** (1976), 63–67.

and

BITNER, J. R., *Heuristics that Dynamically Alter Data Structures to Decrease Their Access Time*, Ph.D. thesis, University of Illinois, Urbana, Ill., 1976.

The first published correct implementation of binary search for all table sizes (i.e., not just $2^k - 1$) is in

BOTTENBRUCH, H., "Structure and Use of Algol 60," *J. ACM*, **9** (1962), 161–221,

although it certainly must have been done much earlier. The variant in Exercise 6 (uniform binary search) is due to Ashok K. Chandra in 1971. An interesting application of a binary search technique and a clever analysis of its performance appears in

BENTLEY, J. L., and A. C.-C. YAO, "An Almost Optimal Algorithm for Unbounded Searching," *Info. Proc. Let.*, **5** (1976), 82–87.

The first description of binary search trees is given in

WINDLEY, P. F., "Trees, Forests and Rearranging," *Comp. J.*, **3** (1960), 84–88.

The idea of *optimal* binary search trees has its origins in coding theory. See

GILBERT, E. N., and E. F. MOORE, "Variable-Length Binary Encodings, *Bell System Tech. J.*, **38** (1959), 933–968.

In the special case in which all internal node weights are zero (not very useful for binary search trees, since it means that only unsuccessful searches occur, but important in the construction of alphabetic codes) there is an algorithm for constructing optimal trees in time $O(n \log n)$ and memory $O(n)$. See

HU, T. C., and A. C. TUCKER, "Optimal Computer Search Trees and Variable-Length Alphabetic Codes," *SIAM J. Appl. Math.*, **21** (1971), 514–532

and

HU, T. C., "A New Proof of the T-C Algorithm," *SIAM J. Appl. Math.*, **25** (1973), 83–94.

For the general case, Algorithm 6.7 requires $O(n^2)$ time and memory; it is due to

KNUTH, D. E., "Optimum Binary Search Trees," *Acta Informatica*, **1** (1971), 14–25.

Several generalizations of Knuth's algorithm are given in

ITAI, A., "Optimal Alphabetic Trees," *SIAM J. Comput.*, **5** (1976), 9–18.

Heuristic algorithms for constructing near optimal trees are described in

BRUNO, J., and E. G. COFFMAN, "Nearly Optimal Binary Search Trees," *Proc. IFIP Congress 71*, North-Holland Publishing Co., Amsterdam, 1972, 99–103

and

WALKER, W. A., and C. C. GOTLIEB, "A Top Down Algorithm for Constructing Nearly Optimal Lexicographic Trees," in *Graph Theory and Computing*, R. Read (Ed.), Academic Press, New York, 1972, 303–323.

The analysis of the average search time in optimal, balanced (by nodes), and random trees (described in Section 6.3.3) is from

NIEVERGELT, J., and C. K. WONG, "On Binary Search Trees," *Proc. IFIP Congress 71*, North-Holland Publishing Co., Amsterdam, 1972, 91–98.

This paper also gives a partial analysis of the average search time in monotonic trees; a complete analysis of this case is in James Bitner's University of Illinois Ph.D. thesis cited earlier. The analysis of the average search times in binary search trees balanced by sums of weights is from

BAYER, P. J., "Improved Bounds for Binary Search Trees," *Acta Informatica*, to appear.

Tries were first proposed in

FREDKIN, E., "Trie Memory," *Comm. ACM*, **3** (1960), 490–500.

Fredkin chose the name "trie" from the middle letters of the word "re*trie*val." Combinations of tries with other table organizations have been suggested in

SUSSENGUTH, E. H., "Use of Tree Structures for Processing Files," *Comm. ACM*, **6** (1963), 272–279.

and in

WALKER, W. A., *Hybrid Trees as a Data Structure*, Ph.D. thesis, Department of Computer Science, University of Toronto, 1975.

Such hybrid techniques are designed to permit fast multiway branching near the root of a tree, but to avoid the waste of memory characteristic of the unfilled nodes at lower levels of a trie.

Height-balanced trees, also known as AVL trees, were introduced in

ADELSON-VELSKII, G. M., and YE. M. LANDIS, "An Algorithm for the Organization of Information," *Dokl. Akad. Nauk SSSR 146* (1962), 263–266 (Russian). English translation in *Soviet Math. Dokl.*, **3** (1962), 1259–1262.

The obvious generalization to the case in which the difference in height between the two subtrees of each node can be at most *h* is discussed in

FOSTER, C. C., "A Generalization of AVL Trees," *Comm. ACM*, **16** (1973), 513–517

and in

LUCCIO, F., and L. PAGLI, "On the Height of Height-Balanced Trees," *IEEE Trans. Comput.*, **25** (1976), 87-90.

Various empirical results about height-balanced trees and their generalization are discussed in

KARLTON, P. L., S. H. FULLER, R. E. SCROGGS, and E. B. KAEHLER, "Performance of Height-Balanced Trees," *Comm. ACM*, **19** (1976), 23–28.

The weight-balanced trees of Section 6.4.3 were introduced in

NIEVERGELT, J., and E. M. REINGOLD, "Binary Search Trees of Bounded Balance," *SIAM J. Comput.*, **2** (1973), 33–43.

A variant of weight-balanced trees in which the balancing criterion depends on the external path length is described in

BAER, J. L., "Weight-Balanced Trees," *AFIPS Conference Proceedings* **44** (1975 National Computer Conference Proceedings), 467–472.

Balanced multiway trees (Section 6.4.4) of order m were first described in

BAYER, R., and E. MCCREIGHT, "Organization and Maintenance of Large Ordered Indexes," *Acta Informatica*, **1** (1972), 173–189.

The special case $m = 3$, called 3-2 trees, was discovered independently by John Hopcroft in 1971. This special case is discussed in detail in Chapter 4 of

AHO, A. V., J. E. HOPCROFT, and J. D. ULLMAN, *The Design and Analysis of Computer Algorithms*, Addison-Wesley, Reading, Mass., 1974.

It is sometimes useful to represent balanced multiway trees as binary trees; see

BAYER, R., "Symmetric Binary B-Trees: Data Structure and Maintenance Algorithms," *Acta Informatica*, **1** (1972), 290–306.

An important consideration in using multiway trees is the extent to which the nodes are filled to capacity; obviously, a great deal of storage is wasted if many of the nodes are only minimally full. It can be shown that, in balanced multiway trees of order m, the asymptotic storage utilization is essentially $\ln 2 \approx 69\%$ for large m. The details of this result can be found in

YAO, A. C.-C., "On Random 3-2 Trees," *Acta Informatica*, to appear.

Address computation techniques (Section 6.5) were first suggested by various people in the early 1950s, but nothing was published until

DUMEY, A. I., "Indexing for Rapid Random-Access Memory," *Computers and Automation*, **5** (1956), 6–8.

The major development of the technique occurred in three influential papers:

PETERSON, W. W., "Addressing for Random-Access Storage," *IBM J. Res. and Dev.* **1** (1957), 130–146,

BUCHHOLZ, W., "File Organization and Addressing," *IBM Syst. J.*, **2** (1963), 86–111,

and

MORRIS, R., "Scatter Storage Techniques," *Comm. ACM*, **11** (1968), 38–43.

A recent survey of hash functions is given in

KNOTT, G. D., "Hashing Functions," *Computer J.*, **18** (1975), 265–278.

A collision scheme designed to combine the advantages of hashing with those of sequential search in an ordered table (Exercise 36) is described in

AMBLE, O., and D. E. KNUTH, "Ordered Hash Tables," *Computer J.*, **18** (1975), 135–142.

In many information-retrieval applications of fast search, the object of the search is incompletely specified. For example, we may want to find all five-letter English words of the form ST*R*, where "*" is a "don't care" symbol; thus words like STARE, START, STORM, and STERN should be found. The application of various address computation and digital search techniques to such "partial-match" searches is described in

RIVEST, R. L., "Partial-Match Retrieval Algorithms," *SIAM J. Comput.*, **5** (1976), 19–50.

6.7 EXERCISES

1. It is often stated that the average number of comparisons required for a sequential search in a table of n names is approximately $n/2$. This result is based on a number of assumptions; even if we assume that all access frequencies are equal, different results can be obtained, depending on the way the search is performed.

    (a) Fill in the average number of comparisons required for a sequential search in each of the following four cases.

	successful searches only	unsuccessful searches only
table order is the same as the natural order		
table order is unrelated to the natural order		

    (b) Consider a table with names stored in natural order that can be searched sequentially backward and forward. A search is not necessarily started at the beginning of the table; it starts wherever the table pointer was left at termination of the previous search. Assuming that each name to be searched for is chosen randomly and independently of the destination of previous searches, what is the average number of comparisons per search?

2. (a) Let $\alpha_1 \geq \alpha_2 \geq \cdots \geq \alpha_n$ be the access frequencies of n names in a sequential table. Assuming that only successful searches occur, prove that among $n!$ permutations of names in the table those with access frequencies in monotonic nonincreasing order have minimal average search time.

    (b) Does the preceding statement still hold if unsuccessful searches are considered?

3. Assume that there is a cost c_i involved in looking up a name x_i *after* the position of x_i has been found. Let α_i be the access frequency of x_i and assume that only successful searches occur. Prove that among all $n!$ permutations of names in the table, those with α_i/c_i arranged in monotonic, nonincreasing order have minimal average sequential search time.

4. (a) Let $\alpha_i = 2^{-i}$, $1 \leq i \leq n - 1$, and $\alpha_n = 2^{-n+1}$ be the access frequencies of n names stored in a table in order of decreasing access frequency. Compute the average search time for successful sequential searches. Compute the same average when the names are stored in *increasing* order of access frequency.
   (b) What is the optimal binary search tree over n names with these access frequencies?
   (c) Compare the efficiency of sequential search to that of other search techniques for a table with this frequency distribution.

5. In a *self-organizing file* a name z that has just been looked up is promoted to a position nearer to the beginning of the table so that frequently accessed names will percolate to the front of the table. The idea is only practical if the work involved in the promotion is small; hence only a few simple promotion techniques are considered.
   (a) z is moved to the first position in the modified table (this is practical only if the table is stored as a linked linear list).
   (b) z is interchanged with the first name, x_1, in the table.
   (c) z is exchanged with the name that immediately precedes it in the table.
   Discuss and compare self-organizing files based on these three promotion techniques. Do they lead to shorter average search times? If so, under what conditions and by how much? Consider two cases: when successive searches are independent and when searches occur in clusters (a group of names is looked up repeatedly).

6. *Uniform binary search.* Rewrite Algorithm 6.5 so that instead of the three pointers l, h, and m, only two pointers—the current position m and its rate of change δ—are kept. Thus after an unequal comparison we would do something like $m \leftarrow m \pm \delta$ and $\delta \leftarrow \delta/2$.

7. Algorithm 6.5, binary search, requires a division by 2 at each stage to evaluate the expression $\lfloor (l + h)/2 \rfloor$. Use the recurrence relation $F_{k+1} = F_k + F_{k-1}$ of the Fibonacci numbers to replace the division by 2 by a subtraction; thus a search interval of length F_{k+1} is split into two subintervals of length F_k and F_{k-1}, respectively, leading to a bisection with $\alpha = (\sqrt{5} - 1)/2 \approx 0.62$. Work out the details of this "Fibonacci search."

8. With reasonable assumptions about the speed of various operations, estimate the table sizes for which a sequential search is faster than a binary search.

9. Construct all distinct binary search trees over four and five names and compare the numbers so obtained to the values given by the asymptotic formula (3.27) in Section 3.3.

10. Show that Algorithm 6.7 for constructing optimal binary search trees requires $n^3/6 + O(n^2)$ accesses to the elements of the matrix of Figure 6.8, assuming that in one access all three numbers stored in that element are retrieved together.

11. Show that use of Theorem 6.1, which restricts the range that must be searched to find the root of an optimal subtree, reduces the work of Algorithm 6.7 from $O(n^3)$ to $O(n^2)$.

12. Using the assumptions and the notation of Section 6.3.3, prove that
    (a) $\bar{B}_n = \bar{\beta}n \lg n + O(n)$.
        (*Hint*: In a tree balanced with respect to the number of nodes, the shape of the tree does not depend on the weights.)
    ★(b) $\bar{M}_n = (2 \ln 2)\bar{\beta}n \lg n + O(n)$.
        [*Hint*: Suppose that the weights are drawn independently according to a density function $f(x)$. Let w_i be the weight of x_i, and define A_{ji} by

    $$A_{ji} = \begin{cases} 1 \; if \; x_j \; \text{is an ancestor of } x_i, \\ 0 \; \text{otherwise.} \end{cases}$$

    The average search time is thus

    $$T_n = n\bar{\beta} + \sum_{i=1}^{n} \sum_{j \neq i} E(w_i A_{ji}),$$

    where $E(w_i A_{ji})$ is the expected value of $w_i A_{ji}$. Find an expression for T_n in terms of $f(x)$ and show that

    $$\lim_{n \to \infty} \frac{T_n}{n \ln n} = 2\bar{\beta}$$

    where $\bar{\beta}$ is the average of the density function f, i.e.,

    $$\bar{\beta} = \int_{-\infty}^{\infty} xf(x) \, dx.]$$

13. Design algorithms for insertion and deletion in a digital search tree represented as a trie.

14. Consider the name space $S = \{0, 1\}^l$ of all bit strings of length l. Construct a table T with $n \ll 2^l$ names chosen at random, assuming that each name in S is equally likely to be included in T. Organize T as a trie, using only as many nodes as are necessary to identify each name uniquely. By means of simulation experiments, investigate the average search time in terms of the number of two-way branches required.

15. Under the assumptions in Exercise 14, compare the efficiency of digital search and binary search.

16. Prove the following about binary search trees.
    (a) Let x_i, the ith name in natural order, have a left son in the tree. Then x_{i-1}, the immediate predecessor of x_i in natural order, has no right son.
    (b) Let $x_i, i \geq 2$, have no left son in the tree. Then a search for x_i must visit the node x_{i-1}.
    (c) An unsuccessful search for a name z between x_{i-1} and x_i must visit both the nodes x_{i-1} and x_i. Hence the number of comparisons required to search for z is equal to the larger of the number of comparisons required to search for x_{i-1} and the number required for x_i.

17. Let S_n be the average number of name comparisons required by a random successful search in a random binary search tree with n names and let U_n be the corresponding number for a random unsuccessful search. Prove that $S_n = 1 + (U_0 + U_1 + \cdots + U_{n-1})/n$. (*Hint*: This equation holds for any table structure in which the number of comparisons required to find a name is the same as the number of comparisons required to insert it.)

18. (a) Derive the recurrence relation for the number B_h of distinct height-balanced trees of height h.

    ★(b) Use the recurrence relation to study the asymptotic growth of B_h.

19. Prove that the deletion algorithm for height-balanced trees may require as many as $\lfloor h/2 \rfloor$ rotations and double rotations but no more.

20. Assume that there are fixed probabilities p, q, r, with $p + q + r = 1$, such that each node in a height-balanced tree has height condition $/$, $-$, or $\backslash$ with probability p, q, r respectively, independent of the height conditions of any other nodes and of the size of the tree. Why is it reasonable to assume that $p = r$? Under these assumptions, show that the average length of the upward path in the insertion algorithm is bounded by an expression independent of the tree size.

21. Assume that we change the definition of height balanced tree so as to allow a maximal difference of 2 (instead of 1) between the heights of the two subtrees of each node. Design rebalancing algorithms so as to guarantee logarithmic search times. What are the most skewed height balanced trees in this case? What are the worst case and the average search times in these most skewed trees?

22. Use induction on the number of nodes to prove the inequality

$$I(T_n) \leqq \frac{1}{\mathscr{H}(\alpha)}(n + 1)\lg(n + 1) - 2n$$

for the internal path length of a tree in WB$[\alpha]$.

23. Under the assumptions stated in Figure 6.20 and the accompanying text, prove that the root balance β_C satisfies $\alpha \leqq \beta_C \leqq 1 - \alpha$.

24. Design algorithms that work in time $O(\log n)$ for finding the kth name in a weight-balanced tree and for finding the number of names that lie between two arbitrary names x and y (in natural order). Assume that each node contains the size of the subtree rooted at the node.

25. Assuming that the root balance of each node in a tree in WB$[\alpha]$ is uniformly distributed over the interval $[\alpha, 1 - \alpha]$, independently of the root balances of other nodes, show that the expected number of rotations and double rotations required to insert or delete a node is less than $2/(1 - 2\alpha)$. (*Hint*: If T_n is a tree of n nodes in WB$[\alpha]$, with l and r nodes in its left and right subtrees, respectively, then there are only about $(1 - 2\alpha)n$ possible distinct values for l and r and all of them are equally likely. For only two of these $(1 - 2\alpha)n$ values, the largest and the smallest, can insertion or deletion cause the tree to go out of WB$[\alpha]$. Thus the probability of causing the root balance of T_n to leave the interval $[\alpha, 1 - \alpha]$ is $1/[(1 - 2\alpha)n]$, assuming that the node is inserted in (deleted from) either subtree of T_n with probability $\frac{1}{2}$.)

26. Prove the following statements about balanced multiway trees of order m.

    (a) Let x_i, the ith name in the natural order, be in a node that is not at the lowest level of the tree. Then x_{i-1}, the immediate predecessor of x_i, is either in the same node or in a node at the lowest level of the tree.

    (b) An unsuccessful search for a name z between x_{i-1} and x_i involves comparisons with both x_{i-1} and x_i.

27. Prove that if a balanced multiway tree is constructed by inserting n names one after the other and the resulting tree has N nodes and l levels, then the total number of nodes that must be split (over the n insertions) is $N - l$. Use this to bound the average number of splits per insertion.

28. Design an algorithm to list all names stored in a multiway tree in natural order.

29. It is occasionally possible to design an ad hoc address computation technique for a static table so that no collisions occur. For example, assume that the name space consists of the letters A to Z and that each letter is coded as a bit string of length 5 by using the base 2 representations of the integers 1 to 26. Assume that we must put the eight names A, E, F, J, K, Q, T, W into a memory of eight cells, with an address space $A = \{0, \ldots, 7\}$. Find a hash function h that can be evaluated with only a few instructions of the type available on most computers and that maps the binary codes of the eight names above one to one onto the binary codes of the address space.

$$
\left.\begin{array}{l}
A: 00001 \\
E: 00101 \\
F: 00110 \\
J: 01010 \\
K: 01011 \\
Q: 10001 \\
T: 10100 \\
W: 10111
\end{array}\right\}
\xrightarrow[\substack{1:1 \\ \text{onto}}]{h}
\left\{\begin{array}{l}
000 \\
001 \\
010 \\
011 \\
100 \\
101 \\
110 \\
111
\end{array}\right.
$$

    Note that h need not be order preserving; that is, the natural order and the table order may be unrelated. (*Hint*: One such hash function maps $A \to 000, E \to 001, J \to 010, K \to 011, Q \to 100, T \to 101, F \to 110, W \to 111$.)

30. Modify Algorithms 6.9 and 6.10 (searching and insertion in a hash table) so that they work correctly in the presence of cells marked "deleted."

31. Design a deletion algorithm for the hash table and collision resolution technique discussed in connection with Figure 6.26. Instead of marking a cell as "deleted," move names around in the table so that they remain accessible by Algorithm 6.9.

32. Show that out of n names entered into a hash table of m locations, $(\lambda/2)(n - 1)$ can be expected to have collided with previously entered names (λ is the load factor). p265

33. S. L. Ow, systems analyst for the Speedy Look-Up consulting firm, has the following advice.

    20% of the names in your alphabetically ordered, sequentially allocated, core-resident table of 10000 names account for 80% of the table lookups. Instead of carrying out a binary search over a single table with 10000 entries, you should have a high-frequency table, where the 2000 most frequently looked-up names are arranged alphabetically, and a low-frequency table, where the remaining 8000 names are arranged alphabetically. When you want to look up a name, you search first in the small high-frequency table. 80% of the searches will stop right there. In only 20% of the cases will you also have to search the larger low-frequency table.

Happy ending: Ow gets a salary raise for having increased significantly the throughput of Speedy Look-Up's computer center.

Sad ending: Ow gets fired for professional incompetence.

     You are the president of Speedy Look-Up, chairman of the board, and the main stockholder. Choose between the happy and the sad endings and justify your choice in a one-page statement to be presented at Speedy Look-Up's next staff meeting.

34. S. L. Ow (see Exercise 33) observes that access to many tables is governed by Zipf's law. Ow is interested in statements of the type "A fraction α of the names in the table account for a fraction β of all lookups," and hence he wants to study the function $\beta_n(\alpha), 0 \leq \alpha \leq 1$, defined as follows. In a table of n entries in which access frequencies obey Zipf's law, the $\lfloor \alpha n \rfloor$ most frequently looked-up names account for a fraction $\beta_n(\alpha)$ of all lookups.
   (a) What are $\beta_n(0)$ and $\beta_n(1)$?
   (b) Give an approximate formula for $\beta_n(\alpha)$.
   (c) Make a rough plot of $\beta_{22026}(\alpha)$ based on the results of parts (a) and (b). (*Hint*: $22026 \approx e^{10}$.)

35. S. L. Ow (see Exercises 33 and 34) develops a general theory about splitting a single table with n names into two tables: a high-frequency table containing the $\lfloor \alpha n \rfloor$ most frequently looked-up names, which together account for a fraction β of all the lookups, and a low-frequency table containing the remaining names, which account for a fraction $(1 - \beta)$ of all lookups. Ow considers three types of tables:
   (a) linearly linked lists that must be searched sequentially,
   (b) sequentially allocated lists in which binary search can be used, and
   (c) hash tables.

   In the elementary S. L. Ow theory, the original table with n names, as well as the high-frequency and the low-frequency table, must all be of the same type. For each of the three types of tables, Ow has a complete answer to the question "For what range of values of n, α, β does it pay to split a table into a high-frequency and a low-frequency table?" Independently develop the elementary S. L. Ow theory.

   In the advanced S. L. Ow theory, the types of tables can be mixed; this theory is still under development.

36. *Ordered hash tables.* An interesting variant of hash tables makes use of the numerical (or alphabetical) order of the names. This leads to faster searching at the expense of a little extra work during insertion. Suppose that there are m positions in the memory, n of which are occupied. For convenience, assume that all names have strictly positive numeric value. The table consists of memory cells M_0, $M_1, \ldots, M_{m-1}$. $M_j = 0$ if that position is empty; otherwise $M_j > 0$ is the key stored in memory cell j. Let $h(x)$ be the hash address of name x and let $\Delta(x)$ be the hash increment of name x, so that $\quad \alpha_i = (h(x) + i\Delta(x)) \bmod m, \quad 0 \leq h(x) < m,$ $1 \leq \Delta(x) < m$ and $\Delta(x)$ is relatively prime to m for all x. Assuming that the table is

never allowed to contain more than $m - 1$ names, searching an ordered hash table is done by

$$i \leftarrow h(x)$$

while $M_i > x$ **do** $i \leftarrow (i + \Delta(x)) \bmod m$

if $M_i = x$ **then** $[\![$found: i points to $x]\!]$

         else $[\![$not found: x is not in the table$]\!]$

If there are at most $m - 2$ keys in the table, insertion of a key not in the table is done by

$$i \leftarrow h(x)$$

while $M_i \neq 0$ **do** $\begin{cases} \textbf{if } M_i < x \textbf{ then } M_i \leftrightarrow x \\ i \leftarrow (i + \Delta(x)) \bmod m \end{cases}$

$$M_i \leftarrow x$$

(a) Explain how an ordered hash table differs from a conventional hash table.

(b) Explain how and why the two algorithms given above work.

(c) Prove that a set of n names $x_1, x_2, \ldots, x_n$ can be arranged in an ordered hash table in one and only one way such that the search algorithm is valid. (*Hint*: There are two things to be shown. First, at least one arrangement exists. What order of entering the names by the conventional algorithm guarantees that the hash table is ordered? Second, only one arrangement exists. Suppose that there were two and let x_j be the largest name that appears in different positions in the two arrangements. What happens?)

(d) Let S_n be the average number of probes in conventional hashing for a random successful search in a table with n names and let U_n be the corresponding number in a random unsuccessful search. Let $\hat{S}_n$ and $\hat{U}_n$ be the corresponding numbers for ordered hash tables. Explain why

$$\hat{S}_n = \frac{U_0 + U_1 + \cdots + U_{n-1}}{n}.$$

[*Hint*: Reason as in Exercise 17 and use the result in part (c).] Thus $\hat{S}_n = S_n$, so that ordered and conventional hash tables have the same expected successful search times. Explain why $\hat{U}_n = S_{n+1}$.

(e) Let C_n be the average number of times that the test of the **while** loop is made during an insertion into an ordered hash table. Show that $C_1 + C_2 + \cdots + C_n = nS_n$. (*Hint*: Each time that the loop is executed, the total number of probes to find one of the keys is increased by one.) Combine this with the results in part (d) to show that $C_n = U_{n-1}$.

chapter *7*

Sorting

The problems considered here are part of the most frequently occurring classes of combinatorial problems. In almost all computer applications, a set of items must be rearranged according to some prespecified order. For example, in business data processing it is often necessary to alphabetize records or to arrange them in increasing numerical order. In numerical computation it is sometimes necessary to know the largest of all the roots of a polynomial (see Section 3.2.1).

As in Chapter 6, we will assume that we are given a table of n names, denoted by $x_1, x_2, \ldots, x_n$. Each name x_i assumes a value from a name space on which a linear order is defined. Furthermore, we assume that no two names have equal values; that is, for any names x_i, x_j we have the property that, if $i \neq j$, either $x_i < x_j$ or $x_i > x_j$. The assumption that $x_i \neq x_j$ for $i \neq j$ simplifies the analyses without any sacrifice, since the ideas and algorithms all work correctly in the presence of equal names. Out goal is to determine something about the permutation $\Pi = (\pi_1, \pi_2, \ldots, \pi_n)$ for which $x_{\pi_1} < x_{\pi_2} < \cdots < x_{\pi_n}$. In the *general* sorting problem we must completely determine Π, although it is usually done implicitly, by rearranging the names into increasing order. In *partial* sorting problems we must either determine only partial information about Π (such as π_i for a few values of i), or we must completely determine Π, given some partial information about it (such as in merging two sorted tables).

In Section 7.1 *internal* sorting—that is, the general sorting problem in which the table is assumed to be small enough to fit into directly addressable memory—is discussed. Section 7.2 discusses *external* sorting—that is, the general sorting problem in which the table is so large that it must be accessed piecemeal on external devices. Finally, the *partial* sorting problems of selecting the ith largest name and merging two sorted tables are discussed in Section 7.3.

7.1 INTERNAL SORTING

There are at least five broad classes of internal sorting algorithms.

1. *Insertion.* At stage *i* the *i*th name is inserted into its proper place among the $i - 1$ already sorted names.

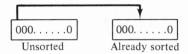

Unsorted   Already sorted

2. *Transposition.* Pairs of names that are out of order are interchanged (transposed) until no such pairs remain.

Interchange

3. *Selection.* At stage *i* the *i*th largest (smallest) name is selected from the unsorted names and put into place.

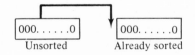

Unsorted   Already sorted

4. *Distribution.* The names are distributed into "buckets," and the contents of the buckets are then assembled in such a way as to partially sort the table; the process is repeated until the table is sorted.

5. *Merge.* The table is divided into subtables that are sorted separately and then merged.

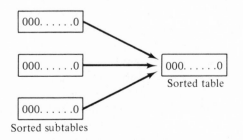

Sorted subtables   Sorted table

These classes are not mutually exclusive and they are not comprehensive: certain sorting algorithms can legitimately be put into more than one class (the bubble sort of Section 7.1.2 can be viewed as a selection sort as well as a transposition sort) and other sorting algorithms fit into none of the classes (see Exercise

1). Nevertheless, these five classes provide a useful taxonomy for discussing sorting algorithms.

In this section we concentrate on the first four classes of sorting algorithms. Algorithms based on merging, although possible for internal sorting, are more naturally discussed as external sorting methods and hence are covered in Section 7.2.

In the sorting algorithms described, the names form a sequence that we will denote by $x_1, x_2, \ldots, x_n$, regardless of data moves that occur; thus the value of x_i is whatever name is currently in the ith position of the sequence. Many sorting algorithms are best performed on an array; in this case, x_i denotes the ith component of the array. Other algorithms are more suited to working on a linked list, and here x_i denotes the ith element of the list. The following notation is used for data moves.

$x_i \leftrightarrow x_j$ means that the values of the names x_i and x_j are interchanged.

$x_i \leftarrow y$ the value of y is assigned to the name x_i.

$y \leftarrow x_j$ the value of the name x_i is assigned to y.

Thus the operation $x_i \leftarrow x_j$, which occurs in various sorting algorithms, temporarily violates the assumption that no two names have the same value. However, this condition is always ultimately restored.

In each of the sorting algorithms considered, we will assume that the names are to be sorted *in place*. In other words, the rearrangement of the names must occur entirely within the sequence $x_1, x_2, \ldots, x_n$, along with one or two additional locations in which to store the value of a name temporarily. The in-place restriction is based on the assumption that the number of names is so large as to prohibit copying them into a different storage area as they are sorted. If sufficient storage is available to allow this copying, then some of the algorithms discussed in this chapter can be speeded up considerably. These considerations force the distribution sorting and merge sorting algorithms to implement the sequence $x_1, x_2, \ldots, x_n$ as a linked list.

Lower Bounds.   Before examining various sorting algorithms, let us consider the problem of sorting from a theoretical point of view in order to have some idea of what kind of performance to expect. One method of measuring the performance of a sorting algorithm is to count the number of name comparisons "$x_i : x_j$" made during the sort. This feature is not always the determining factor in the performance of a sorting algorithm, for, as we shall see, there are sorting algorithms in which the number of interchanges of names predominates over the number of name comparisons, and there are sorting algorithms based on the representation of the names that involve no direct comparisons of the type "$x_i : x_j$." However, for many sorting algorithms the number of name comparisons is a good measure of the work being done; consequently, the minimum possible number of name comparisons required to sort n names is of interest.

In order to eliminate sorting algorithms based on the representation of the names rather than on name comparisons from this theoretical discussion, we will consider only the algorithms that are based on the abstract linear ordering of the name space: between every pair of names x_i, x_j, $i \neq j$, either $x_i < x_j$ or $x_i > x_j$. (It is not difficult to extend this discussion to the case in which equal names are allowed; see Exercise 3.) Any such sorting algorithm can be represented by an extended binary decision tree (see Sections 1.5 and 2.3.3) in which each internal node represents a name comparison and each leaf (external node) represents an outcome of the algorithm. This tree can be viewed as a flowchart of the sorting algorithm in which all loops have been "unwound" and only the name comparisons are shown. The two sons of a node thus represent the two possible outcomes of the comparison. For example, Figure 7.1 shows a binary decision tree for sorting three names. In the rest of this section we will consider *only* sorting algorithms that can be written as such decision trees.

In any such decision tree each permutation defines a unique path from the root to a leaf. Since we are considering only algorithms that work correctly on all the $n!$ permutations of the names, the leaves corresponding to different permutations must be different. Clearly, then, there must be at least $n!$ leaves in a decision tree for sorting n names.

Notice that the height of the decision tree is the number of comparisons required by the algorithm for its worst-case input. Let $S(n)$ denote the minimum

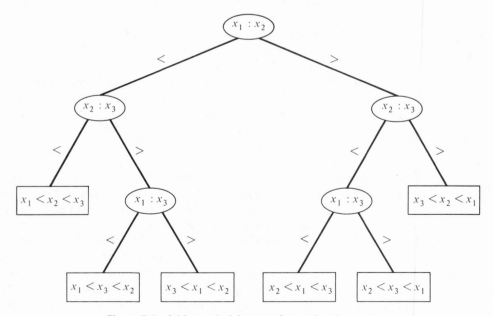

Figure 7.1   A binary decision tree for sorting three names.

number of comparisons required by any sorting algorithm in the worst case; that is,

$$S(n) = \min_{\substack{\text{sorting} \\ \text{algorithms}}} \left[\max_{\text{inputs}} \text{(number of comparisons)} \right].$$

Noticing that a binary tree of height h can have at most 2^h leaves, we conclude that

$$2^{S(n)} \geqq n!,$$

and so

$$S(n) \geqq \lg n! \approx n \lg n$$

by Stirling's formula [equation (3.12)]. Thus *any* sorting algorithm based on name comparisons will require at least $n \lg n$ name comparisons for its worst-case input.

We can also derive a lower bound for $\bar{S}(n)$, the minimum possible *average* number of name comparisons required by an algorithm that correctly sorts all the $n!$ permutations, assuming that each of these permutations is equally probable. That is,

$$\bar{S}(n) = \min_{\substack{\text{sorting} \\ \text{algorithms}}} \left[\frac{1}{n!} \sum_{\text{permutations}} \text{(number of comparisons)} \right].$$

The external path length (see Section 2.3.3) of the decision tree is the sum of all distances from the root to the leaves; dividing it by $n!$ gives the average number of comparisons for the corresponding algorithms. Theorem 2.1 shows that in an extended binary tree with N leaves the minimum possible external path length is $N \lg N + O(N)$; and setting $N = n!$, we find that

$$\bar{S}(n) \geqq \lg n! \approx n \lg n.$$

Therefore *any* sorting algorithm based on name comparisons will require an average of at least $n \lg n$ name comparisons.

The results on $S(n)$ and $\bar{S}(n)$ provide us with a crude bench mark with which to compare the performance of many sorting algorithms. Notice that these results are derived only from the properties of binary trees and so they hold for sorting algorithms based on binary decisions of any kind.

7.1.1 Insertion

The simplest insertion sort, Algorithm 7.1, goes through stages $j = 2, 3, \ldots, n$: at stage j, x_j is inserted into its correct place among $x_1, x_2, \ldots, x_{j-1}$. The insertion is done by temporarily storing x_j in X and scanning the names $x_{j-1}, x_{j-2}, \ldots, x_1$ comparing them to X and shifting them to the right if they are found greater than X. There is a dummy name, x_0, whose value is $-\infty$ to stop the scan at the left end. Figure 7.2 illustrates this algorithm on a table of five names.

$$x_0 \leftarrow -\infty$$

$$\textbf{for } j = 2 \textbf{ to } n \textbf{ do} \begin{cases} i \leftarrow j - 1 \\[4pt] X \leftarrow x_j \\[4pt] \textbf{while } X < x_i \textbf{ do} \begin{cases} x_{i+1} \leftarrow x_i \\[4pt] i \leftarrow i - 1 \end{cases} \\[4pt] x_{i+1} \leftarrow X \end{cases}$$

Algorithm 7.1   Simple insertion sort.

In this algorithm, as in most sorting algorithms, the performance depends on the number of name comparisons and the number of data moves made in the worst case, on the average (assuming that each of the $n!$ permutations of the names is equally likely), and in the best case. The key to analyzing Algorithm 7.1 is the number of times the test "$X < x_i$" is made in the **while** loop, for once that number is known, the number of data moves "$X \leftarrow x_j$," "$x_{i+1} \leftarrow x_i$," and "$x_{i+1} \leftarrow X$" can easily be deduced.

The behavior of Algorithm 7.1 thus depends completely on the number of inversions in the table to be sorted. The comparison "$X < x_i$" is made $\sum_{j=2}^{n} (1 + d_j)$ times, where d_j is the number of names larger than x_j and to its left—that is, where $(d_1, d_2, \ldots, d_n)$ is the inversion vector (see Section 5.1.2) of $x_1, x_2, \ldots, x_n$—and hence the comparison "$X < x_i$" is made $n - 1 + \sum_{j=2}^{n} d_j$ times. Since $0 \leq d_j \leq j - 1$ and the d_j's are independent, it follows that the maximum and minimum values for $n - 1 + \sum_{j=2}^{n} d_j$ are, respectively, $\frac{1}{2}(n - 1)(n + 2)$ and $n - 1$. It is not difficult to show that the average of $\sum_{j=2}^{n} d_j$ over all permutations is $\frac{1}{2} \sum_{j=1}^{n-1} j = \frac{1}{4}n(n - 1)$, and therefore the average value for

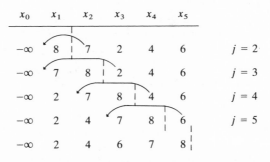

Figure 7.2 The simple insertion sort used on a table of $n = 5$ names. The dashed vertical lines separate the already sorted path of the table from the unsorted part.

$n - 1 + \sum_{j=2}^{n} d_j$ is $\frac{1}{4}(n - 1)(n + 4)$. The inside of the **while** loop is performed $\sum_{j=2}^{n} d_j$ times, so that the number of data moves is $(n - 1) + (n - 1) + \sum_{j=2}^{n} d_j$. That is, $\frac{1}{2}(n - 1)(n + 4)$ at worst, $2n - 2$ at best, and $\frac{1}{4}(n - 1)(n + 8)$ on the average.

Can insertion sorting be improved? We could do *binary insertion*, using binary search (Section 6.3.1)—and hence at most $\lceil \lg j \rceil$ name comparisons—to insert the *j*th name. A total of $\sum_{j=1}^{n} \lceil \lg j \rceil \approx n \lg n$ name comparisons would thus suffice [compare this with $S(n)$]. Unfortunately, after having found the spot in which to insert x_j, the in-place restriction causes the insertion itself to require moving $\frac{1}{2}j$ names on the average, and the algorithm would still require proportional to n^2 operations. Alternatively, we could use a linked list for the sorted portion of the table, thereby making the insertion more efficient. Of course, in this case, we could not use binary search to find the place at which x_j should be inserted, and we would be stuck with a sequential search. The resulting sorting algorithm would require proportional to n^2 operations.

If we could combine the searching ease of a sequentially allocated list with the insertion ease of a linked list, we would be able to obtain an algorithm requiring $O(n \log n)$ time. This can be done by using the balanced tree schemes of Section 6.4; but the overhead involved makes such schemes prohibitive for sorting, since we will be able to achieve $O(n \log n)$-time sorting in simpler ways. It *is* possible to use the insertion idea to obtain a $O(n(\log n)^2)$-time sorting algorithm; see Exercises 5 and 6.

7.1.2   Transposition Sorting

Sorting methods based on transpositions systematically interchange pairs of names that are out of order until no such pairs exist. In fact, Algorithm 7.1 can be considered a transposition sort in which the name x_j is interchanged with its left-hand neighbor until it is in the correct place. In this section we discuss two transposition sorts: the well-known but relatively inefficient bubble sort and quicksort, one of the best all-around internal sorting algorithms.

Bubble Sort.   The most obvious method of systematically interchanging the out-of-order pairs of names is to scan adjacent pairs of names from left to right repeatedly, interchanging those found to be out of order. This technique has come to be known as the *bubble sort*, since larger names "bubble-up" to the top (i.e., right end), of the table. Algorithm 7.2 shows how this simple idea is implemented with one slight improvement: there is clearly no point in continuing the scan into the large names (at the right end of the table) that are known to be in their final positions. Algorithm 7.2 uses the variable b, whose value at the beginning of the **while** loop is the largest index t for which x_t is *not* known to be in its final position. Figure 7.3 illustrates how the algorithm works on a table of $n = 8$ names.

$$b \leftarrow n$$

$$\textbf{while } b \neq 0 \textbf{ do} \begin{cases} t \leftarrow 0 \\[2pt] \textbf{for } j = 1 \textbf{ to } b - 1 \textbf{ do} \quad \textbf{if } x_j > x_{j+1} \textbf{ then} \begin{cases} x_j \leftrightarrow x_{j+1} \\ t \leftarrow j \end{cases} \\[2pt] b \leftarrow t \end{cases}$$

Algorithm 7.2 Bubble sort.

The analysis of the bubble sort depends on three things: the number of passes (i.e., the number of times the body of the **while** loop is executed), the number of times the comparison "$x_j > x_{j+1}$" is made, and the number of interchanges "$x_j \leftrightarrow x_{j+1}$." The number of interchanges is, as in Algorithm 7.1, the number of inversions: 0 in the best case, $\frac{1}{2}n(n-1)$ in the worst case, and $\frac{1}{4}n(n-1)$ on the average. A glance at Figure 7.3 suggests that each pass of the bubble sort except the last decreases by one each nonzero entry of the inversion vector and cyclically shifts the vector one position to the left; indeed, it is easy to prove that this is true in general, and so the number of passes is one plus the largest inversion vector entry. In the best case there is one pass, in the worst case there are n passes, and on the average there are $\sum k P_k$, where P_k is the probability that the largest entry of an inversion vector is $k - 1$. Exercise 7 shows that $\sum k P_k = n + \sqrt{\pi n/2} + \frac{5}{3} + O(n^{-1/2})$. The total number of name comparisons is difficult to determine, but it can be shown to be $n - 1$ at best, $\frac{1}{2}n(n-1)$ at worst, and $\frac{1}{2}(n^2 - n \ln n) + O(n)$ on the average.

	x_1	x_2	x_3	x_4	x_5	x_6	x_7	x_8		d_1	d_2	d_3	d_4	d_5	d_6	d_7	d_8
	4	7	3	1	5	8	2	6		0	0	2	3	1	0	5	2
pass 1																	
	4	3	1	5	7	2	6	8		0	1	2	0	0	4	1	0
pass 2																	
	3	1	4	5	2	6	7	8		0	1	0	0	3	0	0	0
pass 3																	
	1	3	4	2	5	6	7	8		0	0	0	2	0	0	0	0
pass 4																	
	1	3	2	4	5	6	7	8		0	0	1	0	0	0	0	0
pass 5																	
	1	2	3	4	5	6	7	8		0	0	0	0	0	0	0	0
pass 6																	
	1	2	3	4	5	6	7	8		0	0	0	0	0	0	0	0

Figure 7.3 The bubble sort applied to a table. The inversion vector of the table is shown after each pass.

It is possible to improve the bubble sort slightly (Exercise 8) but not enough to make it competitive with the more efficient sorting algorithms. Its only advantage is its conceptual simplicity.

In both the simple insertion sort (Algorithm 7.1) and the bubble sort (Algorithm 7.2), a major source of inefficiency is the fact that the interchanges do very little work, since names move only one position at a time. As Exercise 9 shows, such algorithms are doomed to require proportional to n^2 operations in both the average and the worst cases. Thus a promising improvement is to interchange names that are far away from each other, which is why the diminishing increment sort (Exercises 5 and 6) is asymptotically so efficient. Another way to make each interchange do more work is used in quicksort.

Quicksort.   The idea in quicksort is to select one of the names in the table and use it to partition the table into two subtables—those less than and those greater than the selected name—which are then sorted by using quicksort recursively. The partitioning can be implemented by simultaneously scanning the table from right to left and from left to right, interchanging names in the wrong parts of the table. The name used to split the table is then placed in between the two subtables and the two subtables are sorted recursively.

Algorithm 7.3 gives the details of this method for sorting the table $(x_f, x_{f+1}, \ldots, x_l)$, using x_f to partition the table into subtables. Figure 7.4 shows how Algorithm 7.3 uses the two pointers i and j to scan the table during partitioning. At the beginning of the loop "**while** $i < j$," i and j point, respectively, to the first and last names known not to be in the correct parts of the

	x_f	x_{f+1}	x_{f+2}									x_{l-1}	x_l
Start	27	99	0	8	13	64	86	16	7	10	88	25	90
First Interchange	27	99	0	8	13	64	86	16	7	10	88	25	90
Second interchange	27	25	0	8	13	64	86	16	7	10	88	99	90
Third interchange	27	25	0	8	13	10	86	16	7	64	88	99	90
Scans cross	27	25	0	8	13	10	7	16	86	64	88	99	90
x_f put into place	27	25	0	8	13	10	7	16	86	64	88	99	90
Partitioned table	16	25	0	8	13	10	7	27	86	64	88	99	90

Figure 7.4   The partitioning phase of quicksort using the first name to partition the table. The value of x_{l+1}, not shown, is assumed to be larger than the other values shown.

file. When they cross—that is, when $i \geqq j$—all names are in the correct parts of the table, and x_f is placed between the two parts by interchanging it with x_j. The algorithm assumes that x_{l+1} is defined and larger than $x_f, x_{f+1}, \ldots,$ and x_l.

procedure *QUICKSORT* (f, l)

⟦sort $x_f, x_{f+1}, \ldots, x_l$⟧

if $f \geqq l$ **then return**

⟦partition the table⟧

$i \leftarrow f + 1$

while $x_i < x_f$ **do** $i \leftarrow i + 1$

$j \leftarrow l$

while $x_j > x_f$ **do** $j \leftarrow j - 1$

while $i < j$ **do** $\begin{cases} \text{⟦ at this point we have } i < j, x_i \geqq x_f \geqq x_j \text{⟧} \\ x_i \leftrightarrow x_j \\ i \leftarrow i + 1 \\ \textbf{while } x_i < x_f \textbf{ do } i \leftarrow i + 1 \\ j \leftarrow j - 1 \\ \textbf{while } x_j > x_f \textbf{ do } j \leftarrow j - 1 \end{cases}$

$x_f \leftrightarrow x_j$

⟦sort the subtables recursively⟧

QUICKSORT $(f, j - 1)$

QUICKSORT $(j + 1, l)$

return

Algorithm 7.3   Recursive version of quicksort, using the first name to split the table. The algorithm assumes that x_{l+1} is defined and is larger than or equal to $x_f, x_{f+1}, \ldots,$ and x_l.

The proof that Algorithm 7.3 sorts correctly is left to Exercise 10.

To analyze the total number of name comparisons "$x_i < x_f$" and "$x_j > x_f$" in Algorithm 7.3, notice that at the end of the loop "**while** $i < j$" all the names $x_{f+1}, \ldots, x_l$ have been compared once with x_f except names x_s and x_{s+1} (where the scans cross), which have been compared twice with x_f. Let $\bar{C}_n$ be the average number of name comparisons to sort a table of n distinct names, assuming that each of the $n!$ permutations of the table is equally likely. We obtain

$$\bar{C}_n = \sum_{s=1}^{n} p_s(n + 1 + \bar{C}_{s-1} + \bar{C}_{n-s}), \qquad n \geqq 2$$

where

$$p_s = \text{probability that } x_1 \text{ is } s\text{th smallest name} = \frac{1}{n},$$

since the two subtables produced by the partitioning are random—that is, since each of the $(s-1)!$ permutations of the names in the left subtable is equally likely and each of the $(n-s)!$ permutations of the names in the right subtable is equally likely (Exercise 11). The recurrence simplifies to

$$\bar{C}_n = n + 1 + \frac{2}{n} \sum_{s=0}^{n-1} \bar{C}_s, \qquad n \geq 2.$$

This is similar to equation (3.20), and an analysis similar to that in Section 3.2.2 tells us that

$$\bar{C}_n = (\ln 4)n \lg n + O(n) \approx 1.386n \lg n.$$

A more exact analysis, following the technique of Exercise 12 in Chapter 3, yields

$$\bar{C}_n = 2(n+1)H_n - \frac{8}{3}n - \frac{2}{3}.$$

In the worst case, the number of name comparisons C_n is easily shown by induction to be at most $\frac{1}{2}n^2$. This result is clearly true for $n \leq 3$, since $C_1 = 0$, $C_2 = 2$, $C_3 = 4$ by inspection. Suppose that it is true for all $t < n, n \geq 4$. We have

$$C_n = n + 1 + \max_{1 \leq k \leq n} (C_{k-1} + C_{n-k});$$

so by the inductive hypothesis

$$C_n \leq n + 1 + \max_{1 \leq k \leq n} \left[\frac{1}{2}(k-1)^2 + \frac{1}{2}(n-k)^2 \right].$$

The maximum occurs for $k = \frac{1}{2}(n+1)$; therefore

$$C_n \leq n + 1 + \left(\frac{n-1}{2} \right)^2 \leq \frac{n^2}{2}$$

for $n \geq 4$. Examining what happens when quicksort is applied to an already sorted table (!), we discover that it performs $(n+1) + n + \cdots + 3$ name comparisons, and we conclude that

$$C_n = \frac{1}{2}n^2 + O(n).$$

Since the average, $\bar{C}_n$, is $O(n \log n)$, this extreme performance is rare on random tables, but it is embarrassing to have a sorting algorithm behave so poorly on an already sorted table. Moreover, in practice, the tables encountered are *not* random but often are in rough order; this makes the extreme performance more probable. The difficulty is the choice of the element to partition the table; when the table is already in rough order, the first name is unlikely to split the table evenly. We can improve quicksort (for nonrandom tables) by using a randomly chosen name to partition the table. This change is easily incorporated in Algorithm 7.3: add the statement "$x_f \leftrightarrow x_{\mathrm{rand}(f,l)}$" just before the statement "$i \leftarrow f + 1$."

Quicksort can be further improved by using the median of a small, randomly chosen sample of k names to partition the table. If $k = 3$, it can be shown that $\bar{C}_n \approx 1.188n \lg n$; of course, even in this case C_n is proportional to n^2.

Analysis of $\bar{I}_n$, the average number of interchanges "$x_i \leftrightarrow x_j$" and "$x_f \leftrightarrow x_j$," is similar to the analysis of $\bar{C}_n$ but more complicated. It can be shown (see Exercise 12) that

$$\bar{I}_n = \frac{1}{6}n + \frac{2}{3} + \frac{2}{n} \sum_{s=0}^{n-1} \bar{I}_s,$$

and so from Section 3.2.2 we know that

$$\bar{I}_n \approx 0.231n \lg n.$$

In the worst case, the number I_n of interchanges clearly satisfies

$$I_n \leq \frac{n+1}{2} + \max_{1 \leq s \leq n} (I_{s-1} + I_{n-s}),$$

and it is easy to show by induction that $I_n \leq n \lg n$. Applying quicksort to the table in which the name used to partition the table is the median at every stage, we see that $I_n \geq \frac{1}{2}n \lg n$. Thus

$$\frac{1}{2}n \lg n \leq I_n \leq n \lg n.$$

Algorithm 7.3 is elegant but not practical. The problem is that recursion is used to record subtables for later consideration, and in the worst cases (e.g., the already sorted table) the depth of the recursion can be n. Consequently, storage proportional to n is needed for the stack that implements the recursion; for large n, this is unacceptable. Furthermore, the second recursive call to quicksort in Algorithm 7.3 can easily be removed (how?). For these reasons, we give Algorithm 7.4, an iterative version of quicksort in which the stack is explicitly maintained. A stack entry is a pair (f, l); when it is on the stack, it means that the suitable $x_f, \ldots, x_l$ is to be sorted. Algorithm 7.4 puts the larger of the two subtables on the stack and applies the algorithm immediately to the smaller subtable. This reduces the worst-case stack depth to about $\lg n$ (Exercise 13). Notice that subtables of length 1 are ignored and that the partitioning of a subtable is done by using a randomly chosen name in that subtable.

$S \leftarrow$ empty stack

$S \Leftarrow (0, 0)$

$f \leftarrow 1$

$x_{n+1} \leftarrow \infty$

$l \leftarrow n$

$$
\textbf{while } f < l \textbf{ do} \begin{cases}
x_f \leftrightarrow x_{\text{rand}(f,l)} \\
i \leftarrow f + 1 \\
\textbf{while } x_i < x_f \textbf{ do } i \leftarrow i + 1 \\
j \leftarrow l \\
\textbf{while } x_j > x_f \textbf{ do } j \leftarrow j - 1 \\
\textbf{while } i < j \textbf{ do} \begin{cases}
x_i \leftrightarrow x_j \\
i \leftarrow i + 1 \\
\textbf{while } x_i < x_f \textbf{ do } i \leftarrow i + 1 \\
j \leftarrow j - 1 \\
\textbf{while } x_j > x_f \textbf{ do } j \leftarrow j - 1
\end{cases} \\
x_f \leftrightarrow x_j \\
\textbf{case} \begin{cases}
j - 1 \leqq f \textbf{ and } l \leqq j + 1: \begin{cases} \llbracket\text{both subtables are trivial}\rrbracket \\ (f, l) \Leftarrow S \end{cases} \\[2mm]
j - 1 \leqq f \textbf{ and } l > j + 1: \begin{cases} \llbracket\text{only right subtable is nontrivial}\rrbracket \\ f \leftarrow j + 1 \end{cases} \\[2mm]
j - 1 > f \textbf{ and } l \leqq j + 1: \begin{cases} \llbracket\text{only left subtable is nontrivial}\rrbracket \\ l \leftarrow j - 1 \end{cases} \\[2mm]
j - 1 > f \textbf{ and } l > j + 1 \begin{cases} \llbracket\text{neither subtable is trivial;} \\ \quad \text{put larger one on } S\rrbracket \\ \textbf{if } j - f > l - j \textbf{ then} \begin{cases} \llbracket\text{left subtable is longer}\rrbracket \\ S \Leftarrow (f, j - 1) \\ f \leftarrow j + 1 \end{cases} \\ \textbf{else} \begin{cases} \llbracket\text{right subtable is longer}\rrbracket \\ S \Leftarrow (j + 1, l) \\ l \leftarrow j - 1 \end{cases} \end{cases}
\end{cases}
\end{cases}
$$

Algorithm 7.4 Iterative version of quicksort that sorts the smaller subtable first and uses a randomly chosen name to partition a subtable.

Another significant improvement can be made to quicksort. Instead of using quicksort to sort all the subtables, we use it only for the "large" subtables, say those of length at least m, and we use the simple insertion sort for "small" subtables, those of length less than m. Since, for "small" tables, the average number of name comparisons in quicksort is more than the average number of name comparisons in the simple insertion sort, it follows that the proper choice of m will improve the performance of quicksort (Exercise 14).

7.1.3   Selection

In a selection sort the basic idea is to go through stages $i = 1, 2, \ldots, n$, finding the ith largest (smallest) name and putting it in place at the ith stage. The simplest form of selection sort is that of Algorithm 7.5: the ith largest name is found in the obvious manner by scanning the remaining $n - i + 1$ names. The number of name comparisons at the ith stage is $n - i$, leading to a total of $(n - 1) + (n - 2) + \cdots + 1 = \frac{1}{2}n(n - 1)$ name comparisons regardless of the input, and so it is clearly not a very good way to sort.

$$\mathbf{for}\ j = n\ \mathbf{to}\ 2\ \mathbf{by} - 1\ \mathbf{do} \begin{cases} i \leftarrow 1 \\ \mathbf{for}\ k = 2\ \mathbf{to}\ j\ \mathbf{do\ if}\ x_i < x_k\ \mathbf{then}\ i \leftarrow k \\ x_i \leftrightarrow x_j \end{cases}$$

Algorithm 7.5   Simple selection sort.

Despite the inefficiency of Algorithm 7.5, the idea of selection can lead to an efficient sorting algorithm. The trick is to find a more efficient method of determining the ith largest name, which can be done by using the mechanism of a *knockout tournament*. Make comparisons $x_1 : x_2, x_3 : x_4, x_5 : x_6, \ldots, x_{n-1} : x_n$ and then compare the winners (i.e., the larger names) of those comparisons in a like manner, and so on, as illustrated for $n = 16$ in Figure 7.5. Notice that this process requires $n - 1$ name comparisons to determine the largest name (see Exercise 16); but having determined the largest name, we possess a great deal of information about the second largest name: it must be one of those that lost to the largest name. Thus the second largest name can now be determined by replacing the largest name with $-\infty$ and remaking all the comparisons along the path from the largest name to the root. This is illustrated in Figure 7.6 for the tree in Figure 7.5.

Since the tree has height $\lceil \lg n \rceil$ (why?), we can find the second largest name by remaking $\lceil \lg n \rceil - 1$ comparisons instead of the $n - 2$ used in the simple selection algorithm. This process can obviously be continued. On finding the second largest name, we replace it by $-\infty$ and remake another $\lceil \lg n \rceil - 1$ comparison to find the third largest name, and so on. Thus the entire process uses at most $n - 1 + (n - 1)(\lceil \lg n \rceil - 1) \approx n \lceil \lg n \rceil$ name comparisons. This result is close to $\lg n!$, and so it *might* be a reasonable sorting method, provided that the

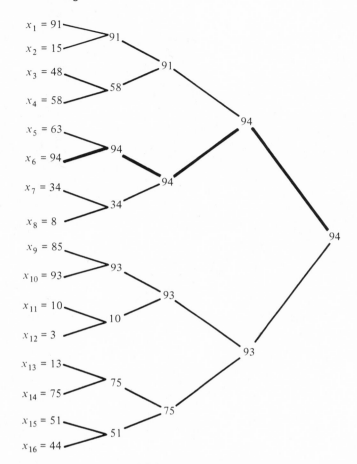

Figure 7.5   Using a knockout tournament to find the largest name. The path of the largest name is shown in boldface.

movement of names can be handled in an efficient manner (recall that binary insertion, mentioned in Section 7.1.1, uses only $n \lg n$ name comparisons but proportional to n^2 interchanges).

The idea of a *tournament selection* sort can be implemented quite neatly by maintaining the names in a *heap*. A heap is a completely balanced binary tree of height h in which all leaves are at distance h or $h - 1$ from the root (see Section 2.3.4) and all descendants of a node are smaller than it; furthermore, all leaves at level h are as far to the *left* as possible. Figure 7.7 shows a set of names arranged into a heap. A heap can be stored by levels in a one-dimensional array so as to yield a convenient linear representation of the tree: the sons of the name in the ith position are the names in positions $2i$ and $2i + 1$. Thus the heap of Figure 7.7 becomes

i:	1	2	3	4	5	6	7	8	9	10	11	12
x_i:	94	93	75	91	85	44	51	18	48	58	10	34.

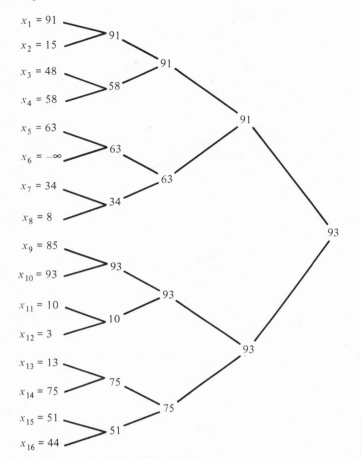

Figure 7.6 Finding the second largest name by replacing the largest with −∞ and remaking the comparisons won by the largest name.

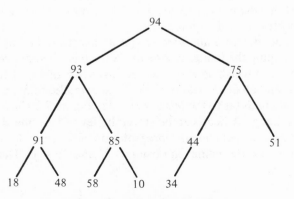

Figure 7.7 A heap containing 12 names.

Notice that in a heap the largest name must be at the root and thus always in the first position of the array representing the heap. Interchanging the first name with the nth name places the largest name into its correct position, but it destroys the heap property of the first $n - 1$ names. If we can initially build a heap and then restore it efficiently, we are finished, for we can sort as follows:

build a heap from $x_1, \ldots, x_n$

for $i = n$ **to** 2 **by** -1 **do** $\begin{cases} x_1 \leftrightarrow x_i \\ \text{restore the heap in } x_1, \ldots, x_{i-1} \end{cases}$

This is an outline of *heapsort*.

Given a binary tree, all of whose leaves are as far left as possible and both of whose subtrees are heaps, how do we transform the whole thing into a heap? We compare the root to the larger of the two sons. If the root is larger, the tree is already a heap; but if the root is smaller, we interchange it with the larger son and apply the restoration algorithm recursively to the subtree whose root has been interchanged. (See Figure 7.8.) The procedure to restore $x_j, x_{j+1}, \ldots, x_k$ to a heap, assuming that all of the subtrees are heaps, is thus

procedure *RESTORE* (j, k)

if $x_j \neq$ leaf **then** $\begin{cases} \text{let } x_m \text{ be larger of the sons of } x_j \\ \text{if } x_m > x_j \text{ then} \begin{cases} x_m \leftrightarrow x_j \\ RESTORE \ (m, k) \end{cases} \end{cases}$

return

Rewriting it in an iterative manner and filling in the details, we obtain Algorithm 7.6. Notice that x_j is a leaf if and only if $j > \lfloor \frac{1}{2} l \rfloor$ (Exercise 18).

p313   a heap w n names has $\lceil \frac{n}{2} \rceil$ leaves : $x_{\lceil \frac{n+1}{2} \rceil} > x_{\lceil \frac{n+1}{2} \rceil + 1} > \cdots x_n$. The last non-leaf is $x_{\lfloor \frac{n}{2} \rfloor}$

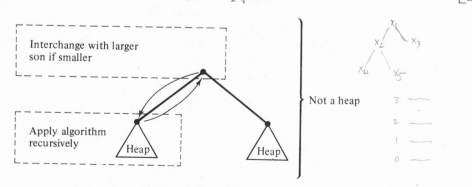

Figure 7.8 Recursive restoration of the heap.

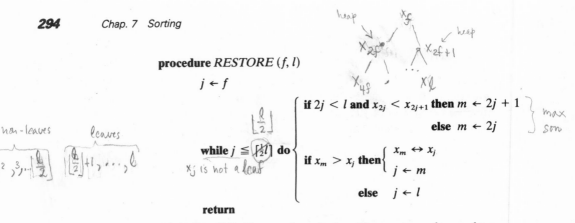

procedure *RESTORE* (f, l)

$$j \leftarrow f$$

while $j \leq \lfloor \frac{1}{2}l \rfloor$ **do** $\begin{cases} \text{if } 2j < l \text{ and } x_{2j} < x_{2j+1} \text{ then } m \leftarrow 2j+1 \\ \qquad\qquad\qquad\qquad\qquad\quad \textbf{else } m \leftarrow 2j \\ \text{if } x_m > x_j \text{ then } \begin{cases} x_m \leftrightarrow x_j \\ j \leftarrow m \end{cases} \\ \qquad\quad \textbf{else} \quad j \leftarrow l \end{cases}$

return

Algorithm 7.6   Restoration of a heap from a tree whose subtrees are heaps.

To build the heap initially, note that the heap property is already satisfied (vacuously) by each of the leaves $(x_i, i = \lfloor \frac{1}{2}n \rfloor + 1, \ldots, n)$ and that calling *RESTORE* (i, n) for $i = \lfloor \frac{1}{2}n \rfloor, \lfloor \frac{1}{2}n \rfloor - 1, \ldots, 1$ transforms the table into a heap at all higher levels. Thus heapsort is as shown in Algorithm 7.7; Exercise 17 is the proof of correctness for this algorithm. The first **for** loop builds the heap and is known as the *creation phase*; the second **for** loop is known as the *sift-up phase*.

$$\text{for } i = \lfloor \tfrac{1}{2}n \rfloor \text{ to } 1 \text{ by } -1 \text{ do } RESTORE\,(i, n)$$

$$\text{for } i = n \text{ to } 2 \text{ by } -1 \text{ do} \begin{cases} x_1 \leftrightarrow x_i \\ RESTORE\,(1, i - 1) \end{cases}$$

Algorithm 7.7   Heapsort.

The behavior of heapsort, on the average, is unknown, and so we consider only the worst case. We need to know the amount of work done during a call *RESTORE* (f, l). In particular, if h is the height of the subtree (a single node having height 0) rooted at x_f, then restore will perform at most $2h$ name comparisons and at most h interchanges; thus we need to determine h. Notice that the left son of x_f is x_{2f}, whose left son is x_{4f}, and so on. The subtree has height h, where h is the largest integer such that the leftmost descendant $x_{2^h f}$ exists, that is, such that $2^h f \leq l$, which implies that $h = \lfloor \lg l/f \rfloor$, a formula easily proved by induction (Exercise 19).

Therefore the creation phase requires at most

$$\sum_{i=1}^{\lfloor \frac{1}{2}n \rfloor} \left\lfloor \lg \frac{n}{i} \right\rfloor \leq \sum_{i=1}^{\lfloor \frac{1}{2}n \rfloor} \lg \frac{n}{i} = \left\lfloor \frac{1}{2}n \right\rfloor \lg n - \lg \left\lfloor \frac{1}{2}n \right\rfloor! = O(n)$$

interchanges and hence $O(n)$ name comparisons. Similarly, the sift-up phase requires at most

$$(n - 1) + \sum_{i=2}^{n} \left\lceil \lg \frac{i-1}{1} \right\rceil = n \lg n + O(n)$$

interchanges and hence only $2n \lg n + O(n)$ name comparisons. Thus heapsort requires $O(n \log n)$ operations for any table, in contrast to quicksort, which could require proportional to n^2 operations for some tables. Despite this fact, empirical evidence suggests that quicksort is more efficient than heapsort on the average (Exercise 20).

7.1.4   *Distribution*

The sorting algorithm discussed here differs from those considered so far in that it is based *not* on comparisons between names but on the *representation* of the names; in this respect, it is similar to the radix exchange sort described in Exercise 15. We assume that the names $x_1, x_2, \ldots, x_n$ each have the form

$$x_i = (x_{i,p}, x_{i,p-1}, \ldots, x_{i,1})$$

and that they are to be sorted into increasing *lexicographic order*; that is

$$x_i = (x_{i,p}, x_{i,p-1}, \ldots, x_{i,1}) < (x_{j,p}, x_{j,p-1}, \ldots, x_{j,1}) = x_j$$

if and only if, for some $t \leq p$, we have $x_{i,l} = x_{j,l}$ for $l > t$ and $x_{i,t} < x_{j,t}$. For simplicity, we will assume that $0 \leq x_{i,l} < r$, and so the names can be viewed as integers represented in base r, each name having p r-ary digits. If the names are of different lengths, the short names are padded with zeros to make the lengths uniform.

The *radix distribution* sort is based on the observation that if the names have been sorted with respect to the low-order positions $l, l - 1, \ldots, 1$, then they can be completely sorted by sorting them only with respect to the high-order positions $p, p - 1, \ldots, l + 1$, provided that the sorting is done in such a way as not to disturb the relative order of names with equal high-order components (for more on this type of sorting, see Exercise 25). This is the basis for mechanical card sorters: to sort cards on the field in, say, columns 76 through 80, the cards are sorted into increasing order on column 80, then on column 79, then on column 78, then on column 77, and finally on column 76. Each column sort is done by reading the column in each card and physically moving the card to the back of a pile that corresponds to the digit punched in that column of the card. Once all the cards have been placed in the proper piles, the piles are stacked together (concatenated) in increasing order; the process is then repeated for the next column to the left. This procedure is illustrated in Figure 7.9 for three-digit decimal numbers.

Notice that both the piles and the table itself are being used in a first-in, first-out manner, and so the best way to represent them is as queues. In particular, assume that a link field $LINK_i$ is associated with each key x_i; these link fields can then be used to hook all the names in the table together in an input queue, Q. The link fields can also be used to hook the names together into the queues used to represent the piles, $Q_0, Q_1, \ldots, Q_{r-1}$. After the names have been distributed into piles, the queues representing those piles are concatenated together to reform

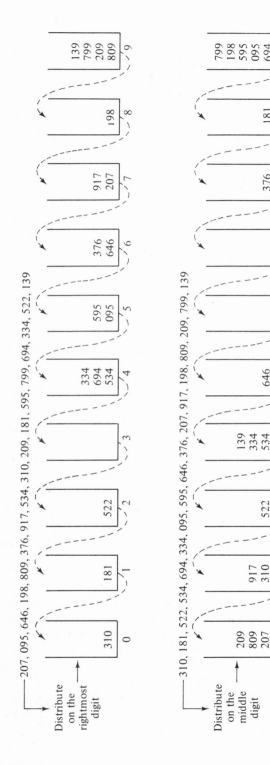
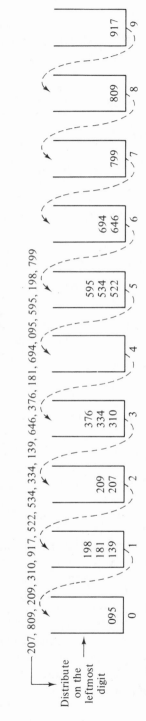

Figure 7.9 The radix distribution sort. The dashed arrows show how the piles of names are concatenated into a table.

Sorted table: 095, 139, 181, 198, 207, 209, 310, 334, 376, 522, 534, 595, 646, 694, 799, 809, 917

the table Q. The broad outline of this sorting algorithm is shown in Algorithm 7.8; the details of applying the techniques from Section 2.2.3 to build the queues is left as Exercise 23. The result of the algorithm is that the queue Q will contain the names in increasing order; that is, the names will be linked in increasing order by the link fields, starting with the front of the queue Q.

use the link fields $\text{LINK}_1, \ldots, \text{LINK}_n$ to
form $x_1, \ldots, x_n$ into an input queue, Q

$$\textbf{for } j = 1 \textbf{ to } p \textbf{ do} \left\{ \begin{array}{l} \text{initialize the queues } Q_0, \ldots, Q_{r-1} \text{ to be empty} \\[2ex] \begin{array}{l} \text{to be empty} \\ \textbf{while } Q \text{ not empty } \textbf{do} \end{array} \left\{ \begin{array}{l} X \Leftarrow Q \\ \text{let } X = (x_p, x_{p-1}, \ldots, x_1) \\ Q_{x_j} \Leftarrow X \end{array} \right. \\[3ex] \text{concatenate queues } Q_0, \ldots, Q_{r-1} \text{ together to form} \\ \text{the new queue } Q \end{array} \right.$$

Algorithm 7.8   Radix distribution sort.

The analysis of Algorithm 7.8 must be different from the analyses of the other sorting algorithms, since we cannot count the number of name comparisons and interchanges. Instead we will count the total number of queue operations. There are always p passes over the table, and each pass requires removing each name from Q and entering it on one of the Q_i. Hence there are a total of $2np$ insertion/deletion operations on the queues. Each pass also requires $r - 1$ concatenations to produce Q from $Q_0, \ldots, Q_{r-1}$, and so there are a total of $(r - 1)p$ queue concatenation operations. It is clear from Section 2.2.3 that an insertion/deletion operation or a concatenation operation can be done in constant time. Therefore Algorithm 7.8 requires time $O(np + rp)$ to sort names $x_1, \ldots, x_n$, where x_i has the form $(x_{i,p}, x_{i,p-1}, \ldots, x_{i,1})$ and $0 \leqq x_{i,l} < r$.

7.2   EXTERNAL SORTING

In the sorting methods discussed in Section 7.1 we assumed that the table fits in high-speed internal memory. Although this assumption is too strong for most real-life data processing problems, it *is* generally valid in the area of combinatorial algorithms, where sorting is usually used only to reduce slightly the cost of algorithms that are prohibitively expensive even for problems of "moderate" size. For example, it is frequently necessary to sort items during an exhaustive search (Chapter 4), but since such searches generally require exponential time, it is unlikely that the table to be sorted will be so large as to require the use of an external storage device. However, the problem of sorting a table that is too large for main memory provides a good illustration of techniques for the large-scale manipulation of data, and so in this section we discuss the important ideas of *external sorting*. More specifically, we will only consider sorting a table by using a sequentially accessible auxiliary storage medium, such as magnetic tape.

We will assume that we have a table of names $x_1, x_2, x_3, \ldots$ stored on a magnetic tape and a total of $t + 1$ tape units; that is, we have t scratch tapes and the input tape containing the table. We will also assume that the internal memory can hold only m names simultaneously, along with other data, programs, and so forth. Of course, the input table is expected to contain many more than m names.

The general strategy in such external sorting is to use the internal memory to sort the names piecemeal from the tape so as to produce *initial runs* (also known as *strings*) of names in increasing order. As they are produced, these runs are distributed onto the t scratch tapes and then merged, t runs at a time, back onto the original $(t + 1)$st tape so that it contains fewer and longer runs. Once again runs are distributed onto the other t tapes and merged, t at a time, back onto the $(t + 1)$st tape. The process continues until there is only one run—that is, the table is completely sorted. There are thus two distinct problems: how to generate the initial runs and how to do the merging.

In order to obtain the initial runs, the obvious method would be simply to read in m names, sort them in internal memory, and write them out on tape as a run, continuing in this fashion until all the names have been exhausted. The initial runs thus obtained all contain m names (except perhaps for the last one). Since the number of initial runs ultimately determines the cost of the merging, we would like to find some method of producing longer, and hence fewer, initial runs. We can do this by using the idea of a tournament sort, or heapsort, from Section 7.1.3. The m names that fit in memory are stored as a heap, but a heap in which the sons of a node are *larger* than the node (instead of smaller than it). This technique corresponds to defining the "winner" of a name comparison in the tournament sort as the *smaller* of the two names, and it allows us to keep track of the smallest name.

The generation of initial runs proceeds as follows. Read the first m names from the input tape into memory and form them into a heap as described above. Write out the smallest name as the first name in the first run and replace it in the heap by the next name from the input tape, using Algorithm 7.6 (*RESTORE*), modified to keep track of the smallest instead of the largest name, to restore the heap. The process, known as *replacement selection*, continues by always adding to the current run the smallest name in the heap that is greater than or equal to the last name added to the run, replacing it with the next name from the input tape, and restoring the heap. When there are no names in the heap larger than the last name in the current run, that run stops and the next run is begun. This process continues until all the names have been formed into runs.

A clever way of implementing this procedure is to consider each name x as a pair (r, x), where r is the number of the run that x is in. Thus the heap consists of the pairs (r_1, x_1), (r_2, x_2), $\ldots$, (r_m, x_m); comparisons between pairs are done lexicographically. When a name is read in that is *smaller* than the last name of the current run, it must be in the next run and the run number keeps it below all names in the heap that are in the current run. The details of the replacement selection algorithm are left as Exercises 28 and 29.

Does the replacement selection produce long runs? Clearly, it does at least as well as the obvious method, since all runs (except perhaps the last) contain at least m names. In fact, it can be shown that the *average* length of the runs produced by replacement selection is $2m$ (Exercise 30), quite an improvement over the obvious method in which the average length of the runs is m. Of course, in the best case, we would end up with only one initial run—that is, a sorted table.

After the initial runs have been generated, we have the problem of repeatedly distributing them onto the scratch tapes and merging them together until ultimately we obtain the final, sorted table. The simplest method of doing this is to distribute the runs evenly onto tapes $1, 2, \ldots, t$ and merge them together onto tape $t + 1$. The resulting runs are then distributed evenly onto tapes $1, 2, \ldots, t$ and merged together to form still longer runs on tape $t + 1$. The process continues until there is only one run (the sorted table) on tape $t + 1$. Notice that quite a bit of time is spent just copying the runs from tape $t + 1$ onto the other tapes. Each "pass" decreases the number of runs by a factor of $1/t$. Consequently, if there are r initial runs, then $\lceil \log_t r \rceil$ passes are needed, where each pass consists of a copy phase followed by a merge phase. Thus there are $2 \lceil \log_t r \rceil \approx (2/\lg t)\lceil \lg r \rceil$ passes over the names, half of which do nothing to reduce the number of runs. For $t + 1$ tapes (i.e., t-way merging) and initial runs, the number of passes over the data is approximately

$t + 1$	Number of passes
3	$2.000 \lg r$
4	$1.262 \lg r$
5	$1.000 \lg r$
6	$0.861 \lg r$
7	$0.774 \lg r$
8	$0.712 \lg r$
9	$0.667 \lg r$
10	$0.631 \lg r$
20	$0.471 \lg r$

The copy phases can be eliminated by being more clever. Suppose that 57 initial runs are distributed onto tapes 1, 2, 3, and 4 as shown in Figure 7.10. In this figure, $n \bullet i$ means n ith order runs, an ith order run being the result of merging i initial runs together. The runs are merged as indicated by the arrows in the figure. The idea of this *polyphase merge* is to arrange the initial runs so that after each merge except the last there is exactly one empty tape—this tape will be the recipient on the next merge. Furthermore, we want the last merge to be merging only one run from each of t non-empty tapes. A distribution of initial runs with such properties is called a *perfect* distribution.

Suppose that we had $t + 1$ tapes instead of four tapes. The polyphase merge idea generalizes easily, and we want to compute the perfect distributions for this

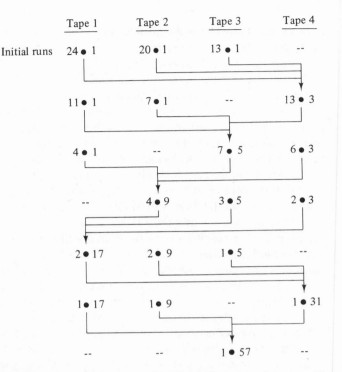

Figure 7.10 A merging pattern for 57 initial runs and 4 tapes. The notation $n \bullet i$ means n ith order runs.

case. We do this by working backward. Let $a_{k,j}$ be the number of runs on the jth tape when k merge phases remain to be done. Moreover, we will assume that tape $t + 1$ is *always* the recipient of the merge and that tape 1 contains at least as many runs as tape 2, which contains at least as many runs as tape 3, and so on. This can be done by using "logical" tape numbers and switching them around appropriately. Thus we have the distribution

$$j: \quad 1 \quad 2 \quad 3 \quad \ldots \quad t-1 \quad t \quad t+1$$

number of
runs on tape j: $\quad 1 \quad 0 \quad 0 \quad \ldots \quad 0 \quad 0 \quad 0$

when no more merge passes remain, that is, at the very end. If, when k merge passes remain, we have the distribution

$$j: \quad 1 \quad 2 \quad 3 \quad \ldots \quad t-1 \quad t \quad t+1$$

number of
runs on tape j: $\quad a_{k,1} \quad a_{k,2} \quad a_{k,3} \quad \ldots \quad a_{k,t-1} \quad a_{k,t} \quad 0 \quad ,$

then the next merge leaves the distribution

$$j: \quad 1 \quad 2 \quad \ldots \quad t-1 \quad t \quad t+1$$

number of
runs on tape j: $\quad a_{k,1} - a_{k,t} \quad a_{k,2} - a_{k,t} \quad \ldots \quad a_{k,t-1} - a_{k,t} \quad 0 \quad a_{k,t}$

It is easy to show that, for $k \geq 1$, we have $2a_{k,t} \geq a_{k,1}$, and hence switching logical tape numbers so that the numbers of runs are in decreasing order gives us

j:	1	2	3	$\ldots$	t	$t+1$
number of runs on tape j:	$a_{k,t}$	$a_{k,1} - a_{k,t}$	$a_{k,2} - a_{k,t}$	$\ldots$	$a_{k,t-1} - a_{k,t}$	0

Therefore

$$a_{k-1,1} = a_{k,t}$$

$$a_{k-1,j} = a_{k,j-1} - a_{k,t}, \qquad 2 \leq j \leq t,$$

or

$$a_{k+1,t} = a_{k,1}$$

$$a_{k+1,j} = a_{k,j+1} + a_{k,1}, \qquad 1 \leq j < t,$$

Consequently, the perfect distributions for $t + 1$ tapes are

	Tape 1	Tape 2	Tape 3		Tape $t-2$	Tape $t-1$	Tape t	Tape $t+1$
$k = 0$	1	0	0	$\ldots$	0	0	0	0
$k = 1$	1	1	1	$\ldots$	1	1	1	0
$k = 2$	2	2	2	$\ldots$	2	2	1	0
$k = 3$	4	4	4	$\ldots$	4	3	2	0
$\vdots$								
k	$a_{k,1}$	$a_{k,2}$	$a_{k,3}$	$\ldots$	$a_{k,t-2}$	$a_{k,t-1}$	$a_{k,t}$	0
$k+1$	$a_{k,2} + a_{k,1}$	$a_{k,3} + a_{k,1}$	$a_{k,4} + a_{k,1}$	$\ldots$	$a_{k,t-1} + a_{k,1}$	$a_{k,t} + a_{k,1}$	$a_{k,1}$	0

If the number of initial runs does not form a perfect distribution, we can add dummy runs to fill out the distribution. These runs are only added "virtually"; that is, we keep track of them but do not write them onto the tapes. If there are r initial runs, and dummy runs are added as needed to form a perfect distribution, it can be shown that the number of passes for $t + 1$ tapes (i.e., t-way merging) is approximately

$t + 1$	Number of passes
3	$1.042 \lg r$
4	$0.704 \lg r$
5	$0.598 \lg r$
6	$0.551 \lg r$
7	$0.528 \lg r$
8	$0.516 \lg r$
9	$0.509 \lg r$
10	$0.505 \lg r$
20	$0.500 \lg r$

As can be seen, little is gained beyond eight tapes.

The implementation of the polyphase merge is left as Exercise 31. An even better (for $t + 1 \geqq 6$) approach to merging the initial runs is contained in Exercise 33.

7.3 PARTIAL SORTING

Earlier we studied the problem of completely ordering a set of names, given no *a priori* information about the abstract order of the names. There are two obvious particularizations of this problem: instead of requiring a complete determination of the ordering, asking only for the kth largest name (selection), or instead of starting with no information about the order, starting with two sorted subtables (merging). In this section we consider both partial sorting problems.

In the algorithms discussed here we will be concerned *only* with the number of name comparisons used. This is not because the other operations are unimportant, but rather because, as we have seen in Section 7.1, it is the number of name comparisons that usually determines, to within a small constant factor, the overall performance of the algorithm. This is true for the selection and merging algorithms discussed here.

7.3.1 *Selection*

Given the names $x_1, x_2, \ldots, x_n$, how can we find the kth largest name? The problem is obviously symmetrical: finding the $(n - k + 1)$st largest (the kth smallest) can be done by using an algorithm for finding the kth largest but reversing the actions taken for the $<$ and $>$ results of name comparisons. Thus finding the largest name ($k = 1$) is equivalent to finding the smallest name ($k = n$); finding the second largest name ($k = 2$) is equivalent to finding the second smallest name ($k = n - 1$), and so on; therefore we may as well assume that $k \leq \lceil \frac{1}{2}n \rceil$. Of special interest is the problem of finding quantile values ($k = \lceil \alpha n \rceil$, $0 < \alpha < 1$) and especially the median ($\alpha = \frac{1}{2}$).

Of course, all these cases of the selection problem can be solved by using any of the methods of Section 7.1 to sort the names completely and then trivially accessing the kth largest. As we have seen, this will require proportional to $n \log n$ name comparisons *regardless of the value of k*. But we would be computing far more information than we need, and so there should be better ways. There are: first we examine the application of various sorting algorithms to the selection problem and then we describe an algorithm that requires only $O(n)$ name comparisons regardless of the value of k.

In using a sorting algorithm for selection, the most apparent choice would be one of the algorithms based on selection, either the simple selection sort (Algorithm 7.5) or heapsort (Algorithm 7.7). In each case, we can stop after the first k stages have been completed. For the simple selection sort, this means using

$$(n - 1) + (n - 2) + \cdots + (n - k) = kn - \frac{k(k + 1)}{2}$$

name comparisons, and for heapsort it means using proportional to $n + k \lg n$ name comparisons. In both cases, we are computing more information than we need because we are completely determining the order of the largest k names. This is not serious when k is a constant, independent of n, since very little extra information is being computed; but when $k = \lceil \alpha n \rceil$, $0 < \alpha \leq \frac{1}{2}$, we are sorting a table of length αn and thus computing a great deal of unneeded information.

Although it does not seem so at first glance, the quicksort idea provides a reasonable method of selecting the kth largest name in $x_1, x_2, \ldots, x_n$. The table is partitioned into two subtables, those above x_1 and those below x_1, and then the appropriate subtable of the table is examined recursively. Assume that x_1 is found to fall into position j (i.e., it is the jth largest name). If $k = j$, we are done; if $k < j$, we search for the kth largest name among $x_1, \ldots, x_{j-1}$; and if $k > j$, we search for the $(k - j)$th largest name among $x_{j+1}, \ldots, x_n$. It is easy to modify Algorithms 7.3 and 7.4 (quicksort) so that they use this technique to find the kth largest name, and we leave that to the exercises.

How efficient is this algorithm? If $\bar{C}_{n,k}$ is the number of name comparisons required on the average, we see that

$$\bar{C}_{n,k} = n + 1 + \frac{1}{n}\left(\sum_{j=1}^{k-1} \bar{C}_{n-j,k-j} + \sum_{j=k+1}^{n} \bar{C}_{j-1,k}\right),$$

and it is trivial to show by induction that $\bar{C}_{n,k} = O(n)$. Unfortunately, $C_{n,k}$, the number of name comparisons in the worst case, is proportional to n^2, and so, although good on the average, this algorithm can be *extremely* inefficient.

The inefficiency occurs because the name used to partition the table can be too close to either end of the table instead of as close to the median as possible, as we would like it (to split the table nearly in half). Thus the way to improve this algorithm is to discover a way of efficiently finding some name that is guaranteed to be near the median of the table. The following inductive technique shows how to do this. Assume that we have a selection algorithm that finds the kth largest of n names in $28n$ comparisons. This is certainly true for $n \leq 50$, since $28n \geq \frac{1}{2}n(n - 1)$ name comparisons are sufficient to sort the names completely using the bubble sort when $n \leq 50$. Suppose that $28t$ name comparisons are sufficient for $t < n$. First divide the table of n names into $\lceil n/7 \rceil \approx n/7$ subtables of 7 names each, adding some dummy $-\infty$ names, if needed, to complete the last subtable (i.e., when n is not a multiple of 7). Then completely sort each of the $n/7$ subtables. Sorting a table with 7 names requires at most $\frac{1}{2}7(7 - 1) = 21$ name comparisons when using the bubble sort, and so sorting all these subtables requires at most $21(n/7) = 3n$ name comparisons in total. Next, apply the selection algorithms recursively to the $n/7$ *medians* of the $n/7$ sorted subtables to find the *median of the medians* in $28(n/7) = 4n$ name comparisons. At this point we have information about the names, as shown in Figure 7.11.

The names in region A are thus known to be less than the median of the medians, and the names in region B are thus known to be greater than the median of the medians; the remaining names can be either less than or greater than the

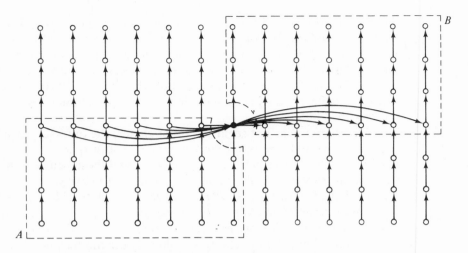

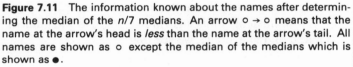

Figure 7.11 The information known about the names after determining the median of the *n*/7 medians. An arrow ○ → ○ means that the name at the arrow's head is *less* than the name at the arrow's tail. All names are shown as ○ except the median of the medians which is shown as ●.

median of the medians. Clearly, there are about $\frac{2}{7}n$ names in each of regions A and B. Therefore the median of the medians is guaranteed to be "near" the middle of the table, and we can use it to partition the table when applying the quicksort idea to selection.

How many name comparisons are needed in total? We used $3n$ to sort the subtables, $4n$ to find the median of the medians, n to partition the table, and at most $28(\frac{5}{7}n)$ to apply the algorithm recursively (since there can be as many as $\frac{5}{7}n$ names in the subtable to which the selection algorithm is recursively applied). Thus at most $28n$ name comparisons are made; and since it is clear that the total amount of work is proportional to the number of name comparisons, we have an $O(n)$ selection algorithm. Of course, the algorithm is *not* as efficient as sorting by heapsort unless $28n < 2n \lg n$, that is, $n > 2^{14} = 16384$. But we were quite sloppy; it is easy to bring the $28n$ down to $15n$ or less (Exercise 36), and, in fact, it can be reduced still further to $\frac{391}{72}n \approx 5.431n$, so that the algorithm becomes reasonable for small values of n.

7.3.2 Merging

The second aspect of partial sorting considered is the problem of merging two sorted tables $x_1 \leq x_2 \leq \cdots \leq x_n$ and $y_1 \leq y_2 \leq \cdots \leq y_m$ into a single sorted table $z_1 \leq z_2 \leq \cdots \leq z_{n+m}$. There is an obvious way to do this: scan the tables to be merged in parallel, at each stage selecting the smaller of the two names and putting it into the output table. This process is simplified a little by adding sentinel names $x_{n+1} = y_{m+1} = \infty$, as in Algorithm 7.10. In this algorithm i and j point, respectively, to the last names in the two input tables that have not yet been put into the output table.

$$x_{n+1} \leftarrow y_{m+1} \leftarrow \infty$$

$$i \leftarrow j \leftarrow 1$$

$$\textbf{for } k = 1 \textbf{ to } n + m \textbf{ do} \begin{cases} \textbf{if } x_i < y_j \textbf{ then} \begin{cases} z_k \leftarrow x_i \\ i \leftarrow i + 1 \end{cases} \\ \\ \textbf{else} \begin{cases} z_k \leftarrow y_j \\ j \leftarrow j + 1 \end{cases} \end{cases}$$

Algorithm 7.10 Straight merging.

The analysis of this algorithm is quite simple, for the comparison "$x_i < y_j$" is made exactly once for each name placed in the output table—that is, $n + m$ times. This can be reduced to $n + m - 1$ times by slightly complicating the algorithm (Exercise 41).

When $n \approx m$, this method of merging is quite good; in fact, when $n = m$, it is possible to show that in the worst case at least $n + m - 1 = 2n - 1$ name comparisons are always necessary to do the merging (Exercise 42). However, when $m = 1$, we can merge far more efficiently by using binary search (Algorithm 6.5) to find the place in which y_1 should be inserted; as a result, we want a method of merging that combines the best aspects of both binary search and Algorithm 7.10.

Figure 7.12 shows the central idea of *binary merging*, a scheme that behaves like binary search when $m = 1$ but like straight merging when $n \approx m$. It also provides a good compromise for other values of m. Assume that $n \geq m$; the idea is to divide the larger table into $m + 1$ subtables. We then compare the rightmost name of the smaller table, y_m, with the last name of the next to the rightmost subtable of the larger table, say x_l (see Figure 7.12). If $y_m < x_l$, then x_l and all the rightmost subtable of the larger table can be put into the output table. If $y_m \geq x_l$, then y_m is inserted into the rightmost subtable by using binary search; y_m and the x_i's in the subtable found to be greater than y_m can now be put into the output table, and the algorithm continues recursively. Recall, however, that binary search works most efficiently for tables of size $2^k - 1$ (why?) and so instead of having the last subtable of the larger table contain about n/m names, we do better if it has $2^{\lfloor \lg n/m \rfloor} - 1$ names. Thus, for the case shown in Figure 7.12, we have $2^{\lfloor \lg 28/6 \rfloor} - 1 = 3$, and we would compare y_6 with x_{25}, as shown in Figure 7.13. If $x_{25} > y_6$, then x_{25}, x_{26}, x_{27}, and x_{28} can be put into the output table, and we continue merging $x_1 x_2 \ldots x_{24}$ with $y_1, y_2, \ldots, y_6$. If $x_{25} < y_6$, we use binary search to find y_6's place among $x_{26} x_{27} x_{28}$ in two name comparisons and put y_6 and the x_i's larger than it into the output table. Next, we continue by merging $x_1 x_2 \ldots x_k$ with $y_1 y_2 \ldots y_5$, where k is the largest integer such that $y_6 > x_k$. An outline of this procedure is given in Algorithm 7.11. The **then** and the **else** clauses associated with "**if** $m \leq n$" are the same except that the roles of x and y (and hence n and m) are reversed.

$$x_1 x_2 x_3 x_4 \mid x_5 x_6 x_7 x_8 \mid x_9 x_{10} x_{11} x_{12} \mid x_{13} x_{14} x_{15} x_{16} \mid x_{17} x_{18} x_{19} x_{20} \mid x_{21} x_{22} x_{23} x_{24} \mid x_{25} x_{26} x_{27} x_{28}$$

$$y_1 \qquad y_2 \qquad y_3 \qquad y_4 \qquad y_5 \qquad y_6$$

Figure 7.12   The idea behind binary merging is this: assume that $n \geq m$ and divide the *x*s into $m + 1$ subtables of $n/(m + 1)$ names each. Then use binary search on the subtables. In this example, $n = 28$, $m = 6$, and $l = 24$.

$$x_1 x_2 \cdots x_{21} x_{22} \mid x_{23} x_{24} x_{25} \mid x_{26} x_{27} x_{28}$$

$$y_1 \quad \cdots \quad y_5 \qquad y_6$$

Figure 7.13   The first comparison in binary merging.

How many name comparisons are made? Let $C_{m,n}$ be the number of name comparisons made in the worst case by Algorithm 7.11 in merging $x_1, \ldots, x_n$ and $y_1, \ldots, y_m$. It is not hard to show that

$$C_{m,n} = n + m - 1, \qquad m \leq n < 2m \tag{7.1}$$

(Exercise 43). Moreover, by Exercise 44 we have

$$C_{m,n} = C_{m,\lfloor n/2 \rfloor} + m, \qquad 2m \leq n. \tag{7.2}$$

Iterating equation (7.2) and combining this with equation (7.1) yields

$$C_{m,n} = m + \left\lfloor \frac{n}{2^t} \right\rfloor - 1 + tm \qquad \text{for } m \leq n, \quad t = \left\lfloor \lg \frac{n}{m} \right\rfloor.$$

When $m = n$, this result gives

$$C_{n,n} = 2n - 1,$$

whereas when $m = 1$, it gives

$$C_{1,n} = 1 + \left\lfloor \frac{n}{2^{\lfloor \lg n \rfloor}} \right\rfloor - 1 + \lfloor \lg n \rfloor = 1 + \lfloor \lg n \rfloor;$$

that is, Algorithm 7.11 performs like straight merging at one extreme and like binary search at the other.

Furthermore, since, in analogy with the discussion of $S(n)$ in the introduction to Section 7.1, we have $\binom{m+n}{m}$ possible outcomes of a merging algorithm (why?),

$$\text{while } n \neq 0 \text{ and } m \neq 0 \text{ do} \begin{cases} \text{if } m \leq n \text{ then} \begin{cases} t \leftarrow \left\lfloor \lg \dfrac{n}{m} \right\rfloor \\[2mm] \text{if } y_m < x_{n+1-2^t} \text{ then} \begin{cases} \text{Put } x_{n+1-2^t}, \ldots, x_n \\ \text{into the output table.} \\ n \leftarrow n - 2^t \end{cases} \\[8mm] \text{else} \begin{cases} \text{Using } t \text{ name comparisons,} \\ \text{insert } y_m \text{ into} \\ x_{n+2-2^t}, \ldots, x_n \text{ by} \\ \text{binary search. Let } k \\ \text{be largest integer} \\ \text{such that } y_m > x_k. \text{ Put} \\ y_m, x_{k+1}, x_{k+2}, \ldots, x_n \\ \text{into the output table.} \\ m \leftarrow m - 1 \\ n \leftarrow k \end{cases} \end{cases} \\[30mm] \text{else} \begin{cases} t \leftarrow \left\lfloor \lg \dfrac{m}{n} \right\rfloor \\[2mm] \text{if } x_n < y_{m+1-2^t} \text{ then} \begin{cases} \text{Put } y_{m+1-2^t}, \ldots, y_m \text{ into} \\ \text{the output table.} \\ m \leftarrow m - 2^t \end{cases} \\[8mm] \text{else} \begin{cases} \text{Using } t \text{ name comparisons,} \\ \text{insert } x_n \text{ into} \\ y_{m+2-2^t}, \ldots, y_m \text{ by} \\ \text{binary search. Let } k \\ \text{be largest integer} \\ \text{such that } x_n > y_k. \text{ Put} \\ x_n, y_{k+1}, y_{k+2}, \ldots, y_m \\ \text{into the output table.} \\ n \leftarrow n - 1 \\ m \leftarrow k \end{cases} \end{cases} \end{cases}$$

if $n = 0$ **then** Put $y_1, \ldots, y_m$ into the output table.

else Put $x_1, \ldots, x_n$ into the output table.

Algorithm 7.11   An outline of binary merging.

we must have at least $\lceil \lg \binom{m+n}{m} \rceil$ name comparisons in the worst case. But as Exercise 45 shows,

$$C_{m,n} < \left\lceil \lg \binom{m+n}{m} \right\rceil + m, \qquad n \geq m,$$

and so binary merging is reasonably efficient for intermediate values of m as well as at the extremes.

7.4   REMARKS AND REFERENCES

Just about everything one might need to know about the sorting problems discussed in this chapter can be found in

KNUTH, D. E., *The Art of Computer Programming*, Vol. 3 (*Sorting and Searching*), Addison-Wesley, Reading, Mass., 1973.

This is an encyclopedic treatment of the entire subject. A lengthy bibliography of all the papers examined by Knuth in writing the section on sorting is to be found in *Computing Reviews*, **13** (1972), 283–289. Section 5.5 of Knuth's book contains an excellent comparison of sorting methods and a short history of the subject of sorting.

One of the first attempts at computer sorting was by John von Neumann, who programmed an internal merge sort (see Exercise 27) on the EDVAC computer. For details of this work, see

KNUTH, D. E., "Von Neumann's First Computer Program," *Computing Surveys*, **2** (1970), 247–260.

The bubble sort was first analyzed by Howard B. Demuth in his 1956 Stanford Ph.D. thesis on sorting, although the method had been known previously. The bubble sort was, for example, partially analyzed in an early survey paper

FRIEND, E. H., "Sorting on Electronic Computer Systems," *J. ACM*, **3** (1956), 134–168.

This paper also introduced tournament selection (the forebear of heapsort) and other new sorting techniques.

Quicksort was first described in

HOARE, C. A. R., "Quicksort," *Comp. J.*, **5** (1962), 10–15.

and the companion Algorithms 63 and 64 in the *Communications of the ACM*. The modification of using the median of a sample of names to partition the table is in Hoare's paper, but the analysis of its effect (with a three-name sample) is due to

VAN EMDEN, M. H., "Increasing the Efficiency of Quicksort," *Comm. ACM*, **13** (1970), 563–567.

Quicksort is based on a relatively simple idea, but there are many possible modifications and embellishments. All of them are ably discussed and analyzed in

SEDGEWICK, R., "Quicksort," Report Number STAN-CS-75-492, Department of Computer Science, Stanford University, Stanford, Ca., May 1975.

This report is an excellent example of the methodology of analyzing a combinatorial algorithm. Our presentation of quicksort is based on Sedgewick's report.

The behavior of quicksort in the presence of equal names is partially analyzed by Sedgewick but is more fully treated in

BURGE, W. H., "An Analysis of Binary Search Trees Formed from Sequences of Nondistinct Keys," *J. ACM*, **23** (1976), 451–454.

Heapsort was described in

WILLIAMS, J. W. J., "Algorithm 232 (Heapsort)," *Comm. ACM.*, **7** (1964), 347–348.

and the efficient approach to the creation phase was suggested in

FLOYD, R. W., "Algorithm 245 (Treesort)," *Comm. ACM*, **7** (1964), 701.

It is worth describing briefly some of what is known about $S(n)$, the smallest possible number of name comparisons needed in the worst case to sort n names. The "information-theoretic" approach given in the introduction of Section 7.1 says, essentially, that if each of the $n!$ permutations can occur with equal probability, then we can use the entropy function of Section 1.5 to obtain

$$S(n) \geq \left\lceil \mathscr{H}\left(\underbrace{\frac{1}{n!}, \frac{1}{n!}, \ldots, \frac{1}{n!}}_{n!\ \text{times}}\right)\right\rceil = \left\lceil \sum_{i=1}^{n!} \frac{1}{n!} \lg n! \right\rceil = \lceil \lg n! \rceil.$$

Thus at least $\lg n!$ binary decisions (e.g., name comparisons) are needed both in the worst case and on the average to determine the order of the names. The following table summarizes what is known about $S(n)$, $n \leq 13$.

n:	1	2	3	4	5	6	7	8	9	10	11	12	13
$\lceil \lg n! \rceil$:	0	1	3	5	7	10	13	16	19	22	26	29	33
$S(n)$:	0	1	3	5	7	10	13	16	19	22	26	30	?

For $n \leq 11$, notice that it *is* possible to sort n names in $\lceil \lg n! \rceil$ name comparisons. The method that does so, known as *merge insertion*, was presented in

FORD, L., JR., and S. JOHNSON, "A Tournament Problem," *Amer. Math. Monthly*, **66** (1959), 387–389

and was first analyzed by A. Hadian in his 1969 University of Minnesota Ph.D. thesis. The fact that $S(12) = 30$ and *not* $\lceil \lg 12! \rceil = 29$ is due to

WELLS, M., "Applications of a Language for Computing in Combinatorics," *Information Processing*, **65** (Proceedings of the 1965 IFIP Congress) Spartan Books, Washington, D.C., 1966, 497–498.

His result is based on brute-force calculation; it required 60 hours of time on the Maniac II computer.

Replacement selection (for forming the initial runs in external sorting) was first proposed by Harold H. Seward in his 1954 MIT Master's thesis. The first analysis of it is due to E. F. Moore in U.S. Patent 2983904 (1961), in which he showed that the expected run length is $2m$. The details of his fascinating proof can be found in Knuth's book on sorting cited above, along with a complete analysis of the expected length of runs (Exercise 30).

The general polyphase merge was proposed in

GILSTAD, R. L., "Polyphase Merge Sorting—An Advanced Technique," *Proc. AFIPS Eastern Joint Computer Conf.*, **18** (1960), 143–148

and the cascade merge (Exercise 33) is due to

BETZ, B. K., and W. C. CARTER, New Merge Sorting Techniques, *ACM National Conference*, **14** (1959), paper 14.

Hoare, in his paper on quicksort, suggested using it to determine the kth largest of n names. The first complete analysis of that technique is given in

KNUTH, D. E., "Mathematical Analysis of Algorithms," *Information Processing*, **71** (Proceedings of the 1971 IFIP Congress), North-Holland Publishing Co., Amsterdam, 1972, 19–27.

The $O(n)$ worst-case selection algorithm of Section 7.3.1 is due to

BLUM, M., R. W. FLOYD, V. PRATT, R. L. RIVEST, and R. E. TARJAN, "Time Bounds for Selection," *J. Comput. Sys. Sci.*, **7** (1973), 448–461;

the details of its implementation and a comparison with the quicksort approach are contained in

FLOYD, R. W., and R. L. RIVEST, "Algorithm 489 (Select)," *Comm. ACM*, **18** (1975), 173.

A refinement of the algorithm is used in

SCHONHAGE, A., M. S. PATERSON, and N. PIPPENGER, "Finding the Median," *J. Comput. Sys. Sci.*, to appear

to prove that the median of n names can be found in $3n + O((n \log n)^{3/4})$ name comparisons; this refinement can be modified to find the kth largest in a similar number of name comparisons.

The binary merging algorithm of Section 7.3.2 is due to

HWANG, F. K., and S. LIN, "A Simple Algorithm for Merging Two Disjoint Linearly Ordered Sets," *SIAM J. Comput.*, **1** (1972), 31–39.

7.5 EXERCISES

1. Implement and analyze the following *enumeration sort*. At stage i, $2 \leq i \leq n$, compare x_i with x_j for $1 \leq j < i$, increase $count_j$ by one if $x_i < x_j$, and increase $count_i$ by one if $x_i \geq x_j$. Use the values of $count_i$, $1 \leq i \leq n$, to sort the table. Does the algorithm work if names can be equal? If not, fix it so that it does.

2. Extend the argument given in the introduction of Section 7.1 to show that even if equal names are allowed, at least $\lg n!$ name comparisons are needed in the worst and average case for sorting n names. (*Note*: Let o_n be the number of outcomes when n not necessarily distinct names are sorted. The obvious extension would show that $\log_3 o_n$ are needed, not $\lg n!$.)

★3. Suppose that all the names are known to be zero or one. Prove that $n - 1$ name comparisons are necessary and sufficient in the worst case to sort n names. Prove that the minimum average number of name comparisons for such a sorting algorithm is $\frac{2}{3}n + O(1)$.

★4. Add the test "$i > 0$" to the **while** loop of Algorithm 7.1, the simple insertion sort, so that it does not need the sentinel name $x_0 = -\infty$. Analyze the modified algorithm. (*Hint*: The analysis is complicated by the fact that the number of times the comparison "$X < x_i$" is made is diminished by l, the number of times that $i = 0$, that is, the number of values j, $1 < j \leq n$, such that $x_j < x_i$ for all i, $1 \leq i < j$. Determine the minimum and maximum values of l. Show that the average value of l is $H_n - 1$, where $H_n = \sum_{i=1}^{n} 1/i$.)

5. Rewrite Algorithm 7.1 so that in a table $x_1, x_2, \ldots, x_n$ it sorts only the subtables $x_k, x_{k+\delta}, x_{k+2\delta}, \ldots$, for $1 \leq k \leq \delta$. Call this δ-*sorting* and show that, given any sequence of positive integers, $\delta_t > \delta_{t-1} > \cdots > \delta_1 = 1$, successively δ_t-sorting, δ_{t-1}-sorting, $\ldots$, δ_1-sorting, leaves the table correctly sorted. This sorting algorithm is known as the *diminishing increment sort*. Why should we expect it to be better than Algorithm 7.1?

★6. (a) Show that, for any positive integers ε and δ, if a table is ε-sorted after being δ-sorted, it remains δ-sorted.
   (b) Show that if l and m are relatively prime positive integers, then the largest integer *not* representable in the form $ul + vm, u, v \geq 0$, is $(l - 1)(m - 1) - 1$.
   (c) Let $t \geq s_2 > s_1 > s_0 \geq 1$. Use the results of parts (a) and (b) to show that if the values of δ_{s_2} and δ_{s_1} in Exercise 5 are relatively prime, then the δ_{s_0}-sorting portion of the diminishing increment sort has running time $O(n\delta_{s_1}\delta_{s_2}/\delta_{s_0})$.
   (d) Show that if the values of $\delta_t, \delta_{t-1}, \ldots$ in Exercise 5 are chosen as the set of all integers of the form $2^p 3^q < n$, then the diminishing increment sort described there requires time $O(n(\log n)^2)$. [*Hint*: How many inversions are there in a table that has been both 2-sorted and 3-sorted? How many integral points (p, q) are there in the triangular region $p \ln 2 + q \ln 3 < \ln n, p, q \geq 0$?]

★7. Let P_k be the probability that the largest entry of an inversion vector for a permutation of $\{1, 2, \ldots, n\}$ is $k - 1$. Show that

$$P_k = \frac{k^{n-k}k! - (k - 1)^{n-k+1}(k - 1)!}{n!}.$$

Show that

$$\sum_{k=1}^{n} kP_k = n + 1 - F(n),$$

where

$$F(n) = \sum_{k=1}^{n} \frac{k^{n-k}k!}{n!} = \sqrt{\frac{\pi n}{2}} - \frac{2}{3} - O(n^{-1/2}).$$

8. Implement the *cocktail-shaker sort*: alternate passes in the bubble sort to go in opposite directions. Show that if x_j and x_{j+1} are not interchanged on two consecutive passes, then they are in their final positions.

9. Let $\Pi = (\pi_1, \pi_2, \ldots, \pi_n)$ be a random permutation of $\{1, 2, \ldots, n\}$. What is the average value of

$$\frac{1}{n} \sum_{i=1}^{n} |\pi_i - i|,$$

the expected distance that a name will travel during sorting? What can be concluded about sorting algorithms that perform only adjacent interchanges?

10. Prove by induction that Algorithm 7.3 (and hence Algorithm 7.4) correctly sorts the table $x_1, x_2, \ldots, x_n$. (*Hint*: First prove by induction that the partitioning is correctly done.)

11. Prove that the partitioning method of Algorithm 7.3 (and hence Algorithm 7.4) produces random subtables; that is, prove that if each of the $n!$ permutations of the table is equally likely, then each of the permutations of the left and right subtables is equally likely.

12. Show that the average number of interchanges in quicksort, $\bar{I}_n$, is

$$\bar{I}_n = \sum_{s=1}^{n} p_s \left(1 + \sum_{t=0}^{s-1} q_{s,t} t + \bar{I}_{s-1} + \bar{I}_{n-s} \right), \quad n \geq 2,$$

where p_s is the probability that x_1 is the sth smallest name, $p_s = 1/n$, and $q_{s,t}$ is the probability that when x_1 is the sth smallest name, then among $x_2, \ldots, x_s$ there will be t names greater than x_1,

$$q_{s,t} = \frac{\binom{s-1}{t} \binom{n-s}{t}}{\binom{n-1}{s-1}}.$$

Establish the identity

$$\sum_{t=0}^{s-1} \binom{s-1}{t} \binom{n-s}{t} t = \binom{n-2}{s-2}(n-s)$$

and use it to show that for $n \geq 2$

$$\bar{I}_n = \frac{1}{6}n + \frac{2}{3} + \frac{2}{n} \sum_{i=0}^{n-1} \bar{I}_i$$

and thus that

$$\bar{I}_n = \frac{1}{3}(n+1)H_n - \frac{1}{9}n - \frac{5}{18},$$

where

$$H_n = \sum_{i=1}^{n} \frac{1}{i}.$$

13. Analyze the worst-case stack depth for Algorithm 7.4. [*Answer*: $\lceil \lg (n + 1)/3 \rceil, n \geq 2$.]

14. Modify Algorithm 7.4 (nonrecursive quicksort) so that Algorithm 7.1 (simple insertion) is used to sort subtables of length less than or equal to m; use either the modified simple insertion of Exercise 4 or add the sentinel name $x_0 = -\infty$ (in the latter case, prove that the resulting algorithm works!). Analyze the performance of the algorithm and, assuming that the cost of the algorithm is the sum of the average number of comparisons and the average number of interchanges, determine the optimum value of m.

15. Devise and implement a sorting algorithm analogous to quicksort in which the partitioning is done on the basis of the most significant bit of the names: if that bit is 1, the name goes into the right-hand part of the table; if that bit is 0, the name goes into the left-hand part of the table. Subsequent partitions are based on the less significant bits of the names. This method is known as *radix exchange* sorting. Analyze the best and worst cases for
    (a) the number of interchanges,
    (b) the number of bit inspections,
    (c) the maximum stack depth.
    Analyze the average cases for (a), (b), and (c), assuming that the names to be sorted are $0, 1, 2, \ldots, 2^t - 1$ in random order.

16. Show that *any* method of determining the largest name in a table of n names requires $n - 1$ name comparisons. (*Hint*: Each name comparison has only one loser.)

heaps

17. (a) Prove that Algorithm 7.6, the algorithm for the restoration of a heap, works correctly.
    (b) Using part (a), prove that the first **for** loop of Algorithm 7.7, heapsort, creates a heap from the table $x_1, x_2, \ldots, x_n$.
    (c) Using parts (a) and (b), prove that Algorithm 7.7 correctly sorts the table $x_1, x_2, \ldots, x_n$.

18. Prove that a heap with n names has $\lceil \frac{1}{2}n \rceil$ leaves and that they are $x_{\lceil (n+1)/2 \rceil}$, $x_{\lceil (n+1)/2 \rceil + 1}, \ldots, x_n$; thus the last nonleaf is $x_{\lfloor n/2 \rfloor}$.

19. Prove that in a heap with n names the subtree rooted at x_f has height $\lfloor \lg n/f \rfloor$, where the height of the tree consisting of a single name is 0. (*Hint*: Use "backward" induction. As the basis, show that the result is true for the leaves, $\lceil (n + 1)/2 \rceil \leq f \leq n$, and then show that it follows by induction for a nonleaf, $f \leq \lfloor n/2 \rfloor$.)

20. Perform experiments to support or refute the claim that, on the average, quicksort is more efficient than heapsort.

21. (a) Design and analyze an algorithm for adding a name to a heap of n names to form a heap of $n + 1$ names.
    (b) Design and analyze an algorithm for deleting a given name x_i from a heap of n names to form a heap of $n - 1$ names.

22. Explore the idea of a ternary heap and a sorting algorithm based on it.

23. Using the techniques of Section 2.2.3 for implementing a queue with a linked list, fill in the details of Algorithm 7.8 and implement it.

24. Study the properties of the following sorting algorithm. Use a radix distribution only on a few high-order positions and then apply Algorithm 7.1, the simple insertion sort. The purpose of using radix distribution on the high-order names is to greatly reduce the number of inversions in the file so that the simple insertion sort becomes efficient.

25. A sorting Algorithm is called *stable* if it preserves the original order of equal names. Which of the sorting algorithms discussed in Section 7.1. are stable?

★26. Given a matrix $A = (a_{ij})$, suppose that each row is sorted into increasing order and then each column is sorted into increasing order. Do the rows remain sorted into increasing order? Prove your answer. Compare this with Exercise 6(a).

27. The idea of merging can be used not only for external sorting but for internal sorting as well. Apply the ideas of Section 7.2 to obtain an internal algorithm based on merging. Analyze and implement your algorithm.

28. Modify Algorithm 7.6 (heap restoration) so that the names it compares are pairs (r, x), where $(r_1, x_1) < (r_2, x_2)$ if $r_1 < r_2$ or if $r_1 = r_2$ and $x_1 < x_2$ and so that the smallest name (not the largest) is kept at the top of the heap.

29. Use your solution to Exercise 28 to implement replacement selection.

★30. Given an infinite "random" sequence of names $x_1, x_2, \ldots$, where random means that each of the $n!$ possible relative orderings of the first n names is equally likely, compute the expected length of the first run. (*Answer*: $e - 1 \approx 1.718$.) Compute L_k, the expected length of the kth run. (*Hint*: Show that the generating function of L_k is

$$\frac{x(1 - x)}{e^{x-1} - x} - x$$

and use this to show that the answer is

$$L_k = k \sum_{j=0}^{k} e^j \frac{(-1)^{k-j} j^{k-j-1}}{(k - j)!}.\Big)$$

Show that $\lim_{k \to \infty} L_k = 2$. How fast does the sequence converge? Show that the expected length of the kth run produced by replacement selection is mL_k, where m is the number of names in the heap.

31. Implement the polyphase merge described in Section 7.2. Keep track of the dummy runs on tape j by a counter D_j. The dummy runs needed to form a perfect distribution should be distributed as evenly as possible over the t tapes.

32. What is the relationship between the Fibonacci numbers $F_0 = 0$, $F_1 = 1$, $F_{i+2} = F_{i+1} + F_i$ and the polyphase merge?

33. The *cascade merge* is best explained by means of an example. Suppose that we have 190 initial runs and 6 tapes; we merge as shown in Figure 7.14. Examine and implement this merge. It can be shown that with $t + 1$ tapes (i.e., t-way merging)

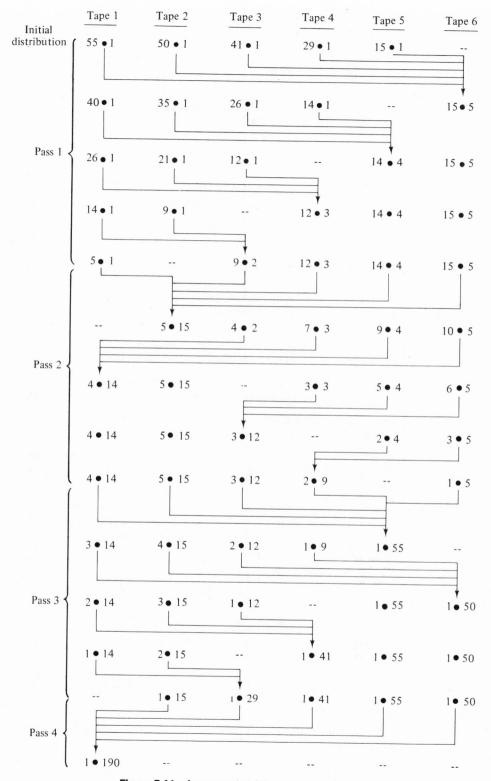

Figure 7.14 An example of the cascade merge.

and r initial runs the number of passes over the data is approximately

$t + 1$	Number of passes
3	$1.042 \lg r$
4	$0.764 \lg r$
5	$0.622 \lg r$
6	$0.536 \lg r$
7	$0.479 \lg r$
8	$0.438 \lg r$
9	$0.407 \lg r$
10	$0.383 \lg r$
20	$0.275 \lg r$

Why is the number of passes the same as the polyphase merge for $t + 1 = 3$?

★34. To use either the polyphase or cascade merging schemes in tape sorting, we need at least three tapes. Suppose that there are only two tapes available: find a method for tape sorting. (*Hint*: Combine the ideas of bubble sort and replacement selection.)

35. (a) Implement the quicksort approach to selection.
     ★(b) Analyze the expected number of comparisons needed.

36. Show that the kth largest of n names can be determined in at most $15n$ name comparisons in the worst case. Try to do it in even fewer name comparisons.

37. Is anything gained in quicksort if we use the $O(n)$ selection algorithm to determine the median of the table and use it to partition the table?

38. Suppose that instead of selecting the kth largest name, we are interested in determining the k largest names but not their relative order. Can this be done in time $O(n)$?

39. Prove that at the termination of an algorithm based on name comparisons (i.e., one represented by a decision tree) for determining the second largest name, there is enough information about the names *already known* to determine the largest name. Generalize this result.

40. Use Exercises 16 and 39 to prove that $n - k + \lceil \lg \binom{n}{k-1} \rceil$ name comparisons are necessary to determine the kth largest of n names. [*Hint*: Use an argument similar to the one at the end of Section 7.1 which shows that $S(n) \geq \lg n!$.]

41. Modify Algorithm 7.10 (straight merging) so that the names $x_{n+1} = y_{n+1} = \infty$ are not needed and thus produce an algorithm that uses at most $n + m - 1$ name comparisons to merge $x_1 \leq x_2 \leq \cdots \leq x_n$ and $y_1 \leq y_2 \leq \cdots \leq y_m$.

42. Prove that $2n - 1$ name comparisons are necessary and sufficient to merge $x_1 \leq x_2 \leq \cdots \leq x_n$ and $y_1 \leq y_2 \leq \cdots \leq y_n$. (*Hint*: What if $x_i < y_i$ if $i < j$ and $x_i > x_j$ otherwise?)

43. Prove that in the worst case Algorithm 7.11 (binary merging) uses $n + m - 1$ name comparisons when $m \leq n < 2m$.

★44.   Prove that for binary merging we have

$$C_{m,n} = C_{m,\lfloor n/2 \rfloor} + m$$

for $2m \leq n$. [*Hint*: Prove by induction on m that $C_{m,n} \leq C_{m,n+1}$ for $n \geq m$ by observing that $C_{m,n} = \max(C_{m,n-2^t} + 1, C_{m-1,n} + t + 1)$, where $t = \lfloor \lg \frac{n}{m} \rfloor$.]

★45.   Prove that

$$C_{m,n} < \left\lceil \lg \binom{m+n}{m} \right\rceil + m, \quad n \geq m.$$

(*Hint*: First prove that $m! \leq m^m/2^{m-1}$.)

chapter *8*

Graph Algorithms

Many diverse problems are naturally formulated in terms of points and connections between the points (i.e., in terms of graphs). For example, scheduling problems in operations research, network analysis in electrical engineering, identification of molecular structure in organic chemistry, program segmentation in computer science, and analysis of Markov chains in probability theory can all be so formulated. In such real-life problems we find that the resulting graphs are often so large that analysis without the aid of the computer is impossible. Thus our ability to solve practical problems by graph-theoretic techniques is only as good as our ability to manipulate large graphs on computers, and so efficient algorithms for graph-theoretic problems are of much practical importance. In this chapter we develop a number of efficient graph algorithms and use them to illustrate some general techniques for solving graph problems on the computer.

A finite *graph* $G = (V, E)$ consists of a finite set of *vertices* $V = \{v_1, v_2, \ldots\}$ and a finite set E of *edges* $E = \{e_1, e_2, \ldots\}$. To each edge there corresponds a pair of vertices: if the edge (v, w) corresponds to the edge e, then e is said to be *incident* on vertices v and w. When we draw a graph G, each vertex is represented by a dot and each edge is represented by a line segment connecting its two end vertices. A graph is *directed* if the vertex pair (v, w) associated with each edge e is an ordered pair. Edge e is then said to be *directed from* vertex v to vertex w, and the direction is indicated by an arrowhead on the edge. We will call directed graphs *digraphs*. In an *undirected* graph the end vertices of all edges are unordered and the edges do not have directions. Undirected and directed graphs are illustrated in Figures 8.1 and 8.2, respectively.

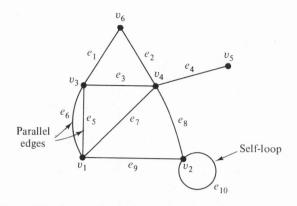

Figure 8.1 An undirected graph of 6 vertices and 10 edges.

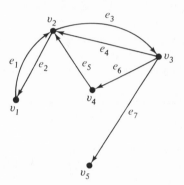

Figure 8.2 A simple directed graph of 5 vertices and 7 edges.

An edge is called a *self-loop* if it begins and ends at the same vertex. Two edges are said to be *parallel* if they have the same pair of end vertices (and if they have the same direction, in the case of directed edges). A graph is *simple* if it has neither self-loops nor parallel edges. Unless stated otherwise we will assume that the graphs under consideration are simple.

Throughout this chapter we will use $|V|$ and $|E|$, respectively, to denote the number of vertices and the number of edges in a graph $G = (V, E)$.

8.1 REPRESENTATIONS

The most familiar way of representing a graph on paper is by drawing dots and line segments. In a computer the graph must be represented in a discrete way, and a number of different representations are possible. The ease of implementation, as well as the efficiency of a graph algorithm, depends on the proper choice of the graph representation. In this section we discuss various data structures for representing graphs.

Adjacency Matrix.   One of the most popular computer representations of a simple graph is the *adjacency* or *connection matrix*. The adjacency matrix of a graph $G = (V, E)$ is a $|V| \times |V|$ matrix $A = [a_{ij}]$ in which $a_{ij} = 1$ if there is an edge from vertex i to vertex j in G; otherwise $a_{ij} = 0$. A digraph and its adjacency matrix are shown in Figure 8.3. Note that in an adjacency matrix a self-loop can be represented by making the corresponding diagonal entry 1. Parallel edges could be represented by allowing an entry to be greater than 1, but doing so is uncommon, since it is usually convenient to represent each element in the matrix by a single bit.

An adjacency matrix requires $|V|^2$ bits. If they are packed into computer words of length l bits, then each row of the matrix requires $\lceil |V|/l \rceil$ words. If each row starts at a new word, the number of words required to store the adjacency matrix is $|V| \cdot \lceil |V|/l \rceil$. For an undirected graph, the adjacency matrix is symmetric and it is sufficient to store only the upper triangle. A savings of almost 50% in storage results, but computation time may increase slightly because any reference to a_{ij} must be replaced by "**if** $i > j$ **then** a_{ji} **else** a_{ij}."

It is not difficult to see that in most graph algorithms every edge of the given graph G will need to be examined at least once; otherwise the one edge of G that we neglect to examine could possibly change the nature of G. Thus if a graph is represented by its adjacency matrix, most algorithms will require computation time at least proportional to $|V|^2$ (but see Exercise 1 for an exception).

Weight Matrix.   A graph in which a number w_{ij} is associated with every edge (i, j) in the graph is called a *weighted graph* and the number w_{ij} is called the *weight* of edge (i, j). In communication or transportation networks these weights represent some physical quantity, such as the cost, distance, efficiency, capacity, or reliability of the corresponding edge. A simple weighted graph can be represented by its weight matrix $W = [w_{ij}]$, where w_{ij} is the weight of the edge from vertex i to j. The weights of nonexistent edges are generally set to ∞ or 0, depending on the application (e.g., see the sections of Chapter 4 that deal with the traveling

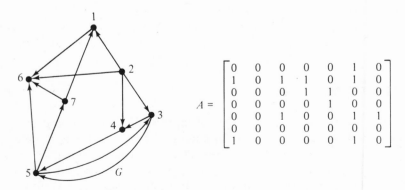

Figure 8.3   A directed graph and its adjacency matrix.

salesman problem). When the weight of a nonexistent edge is 0, weight matrices are a simple generalization of adjacency matrices.

List of Edges.   If a graph is *sparse* $[|E| = o(|V|^2)]$, it may be more efficient to list the edges present in the graph as pairs of vertices. This representation can be implemented with two arrays $g = (g_1, g_2, \ldots, g_{|E|})$ and $h = (h_1, h_2, \ldots, h_{|E|})$. Each entry is a vertex label, the ith edge in the graph going from vertex g_i to vertex h_i. For example, the digraph in Figure 8.3 would be represented as

$$g = (1, 2, 2, 2, 2, 3, 3, 4, 5, 5, 5, 7, 7)$$
$$h = (6, 1, 3, 4, 6, 4, 5, 5, 3, 6, 7, 1, 6).$$

Clearly, self-loops and parallel edges are easily represented.

Because it takes at least $\lceil \lg n \rceil$ bits to identify one among n vertices, we need at least $2|E|\lceil \lg |V| \rceil$ bits of storage for the edge list representation of a graph. If we use a word of l bits to identify one node (assuming $|V| \le 2^l$), this representation requires $2|E|l$ bits of storage.

With the number of bits required for storing a graph in the form of its adjacency matrix proportional to $|V|^2$, the question of whether the edge list is more economical (asymptotically) depends on how $|E|$ grows with $|V|$. If $|E| \sim c|V|^2$, the edge list is less economical for large $|V|$. If $|E| = o(|V|^2)$, the edge list may be more economical for large $|V|$. Thus the choice of representation often depends on whether the graphs to be encountered are sparse, having $|E| = o(|V|^2)$, or *dense*, having $|E| \sim c|V|^2$. For graphs of practical size, however, the word length l may play a deciding role.

Adjacency Structures.   A vertex y in a directed graph is called a *successor* of another vertex x if there is an edge directed from x to y; vertex x is then called a *predecessor* of y. In the case of an undirected graph, two vertices are *neighbors* of each other if there is an edge between them. A graph can be described by its *adjacency structure*, the list of all successors (neighbors) of each vertex; for each vertex v, Adj (v) is a list of all successors (neighbors) of vertex v. In most graph algorithms the relative order in Adj (v) of the vertices adjacent to a vertex v is unimportant, and in such cases it is convenient to consider Adj (v) as the *multiset* (or *set* if the graph is simple) of vertices adjacent to v. An adjacency structure of the digraph in Figure 8.3 is as follows.

v	Adj (v)
1:	6
2:	1, 3, 4, 6
3:	4, 5
4:	5
5:	3, 6, 7
6:	
7:	1, 6

If one computer word is used to store the label of a vertex, then an adjacency structure for a directed graph requires $|V| + |E|$ words. If the graph is undirected, $|V| + 2|E|$ words are needed, since each edge appears twice.

Adjacency structures can be conveniently implemented by an array of $|V|$ linearly linked lists, where each list contains the successors of a vertex. The data field contains the label of the successor vertex, and the pointer field indicates the next successor (see Section 2.2.2). Storing the adjacency lists as a linked list is desirable for algorithms in which vertices are added to and deleted from the graph.

In addition to these commonly used representations of graphs, there are other, less frequently used representations. One of them is discussed in Exercise 2.

In many graph problems the choice of representation is critical to the efficiency of the algorithm. On the other hand, converting from one representation into another is relatively simple and can be accomplished in $O(|V|^2)$ operations (see Exercise 3). Therefore if the solution of a graph problem inherently requires a number of operations at least proportional to $|V|^2$, then its time complexity is independent of the form in which the graph is presented, since it can be transformed to another form in $O(|V|^2)$ operations.

8.2   CONNECTIVITY AND DISTANCE

Two vertices x and y in a graph are said to be *adjacent* if there is an edge joining them. Similarly, two edges are said to be adjacent if they share (i.e., are incident upon) a common vertex. A *simple path*, or *path* for short, is a sequence of adjacent edges $(v_1, v_2), (v_2, v_3), \ldots, (v_{k-2}, v_{k-1}), (v_{k-1}, v_k)$, sometimes written $(v_1, v_2, \ldots, v_k)$, in which all the vertices $v_1, v_2, \ldots, v_k$ are distinct except possibly $v_1 = v_k$. In a digraph this path is said to be *directed from v_1 to v_k*; in an undirected graph this path is said to be *between* v_1 and v_k. The number of edges in a path, in this case, $k - 1$, is called the *length* of the path. In Figure 8.1 sequence (v_6, v_4), $(v_4, v_1), (v_1, v_2) = (v_6, v_4, v_1, v_2)$ is a path of length 3 between v_6 and v_2. In the digraph in Figure 8.3 the sequence $(3, 4), (4, 5), (5, 7), (7, 1) = (3, 4, 5, 7, 1)$ is a path of length 4 from vertex 3 to 1. The *distance* from vertex a to vertex b is defined as the length of the shortest path (i.e., a path of smallest length) from a to b. A *cycle* or *circuit* is a path with the first and the last vertices the same. In Figure 8.1 $(v_3, v_6, v_4, v_1, v_3)$ is a cycle of length 4. In Figure 8.3 $(3, 4, 5, 3)$ is a cycle of length 3. A graph that contains no cycle is called *acyclic*.

A *subgraph* of a graph $G = (V, E)$ is a graph whose vertices and edges are in G. The subgraph of G *induced* by $S \subseteq V$ is the subgraph of G that results when the vertices in $V - S$ and all edges incident on them are removed from G.

An undirected graph G is *connected* if there is at least one path between every pair of vertices v_i and v_j in G. A directed graph G is connected if the undirected graph obtained by ignoring the edge directions in G is connected. A directed graph is said to be *strongly connected* if for every pair of vertices v_i and v_j

there exists at least one directed path from v_i to v_j and at least one from v_j to v_i. A graph consisting of a single isolated vertex is (trivially) connected. A maximal[†] connected subgraph of a graph G is called a *connected component* or simply a *component* of G. A *disconnected* graph consists of two or more components. A maximal strongly connected subgraph is called a *strongly connected component*.

The most basic questions about a graph concern connectivity, paths, and distances. We may want to find out whether a graph is connected; if it is connected, we may want to find the shortest distance between a specified pair of vertices or to identify a shortest path between them. If the graph is disconnected, we may want to identify all its components. Algorithms for answering these and other related questions are developed in this section.

8.2.1 *Spanning Trees*

A connected, undirected acyclic graph is called a *tree*, and a set of trees is called a *forest*. We have already seen *rooted* trees and forests of rooted trees in the preceding chapters, but the (unrooted) trees and forests discussed in this chapter are graphs of a very special kind that play an important role in many applications.

In a connected undirected graph G there is at least one path between every pair of vertices in G; the absence of a cycle in G implies that there is at most one such path between any pair of vertices in G. Thus if G is a tree, there is exactly one path between every pair of vertices in G. The argument is easily reversed, and so an undirected graph G is a tree if and only if there is exactly one path between every pair of vertices in G. Since $n - 1$ edges are the fewest possible to connect n vertices (why?) and a tree with n vertices contains exactly $n - 1$ edges (Exercise 4), trees can be thought of as graphs that are minimally connected. Removing any edge from a tree renders it disconnected by destroying the only path between at least one pair of vertices.

Of particular interest are *spanning trees* of a graph G, trees that are subgraphs of G and contain every vertex of G. If G is not connected, a set consisting of a spanning tree for each component is called a *spanning forest* of G. To construct a spanning tree (forest) of a given undirected graph G, we examine edges of G one at a time and retain those that do not form a cycle with the edges already selected.

In a weighted graph $G = (V, E)$ it is often of interest to determine a spanning tree (forest) of minimum total edge weight—that is, such that the sum of the weights of all edges is minimum. Such a tree is called a *minimum spanning tree*. A minimum spanning tree can be found by applying the process described above with edges examined in increasing order of their weights. In other words, at each stage we choose the smallest unused edge that does not form a cycle with the edges

---

[†]We make a distinction between the two adjectives "maximal" and "maximum." A set S is a *maximal set* satisfying a certain property P if there is no set properly containing S that satisfies property P. Set S is *maximum* if there exists no set of greater cost, according to some cost function (such as cardinality), that satisfies property P. The corresponding distinction is also made between *minimal* and *minimum*.

already chosen and continue the process until $|V| - 1$ edges have been selected to form a spanning tree T. This process is known as the *greedy algorithm*.

It is easy to see that the greedy algorithm will terminate. The minimality of the resulting tree T is not so obvious, but it can be proved as follows. Suppose that T were not a minimum spanning tree. Let the greedy algorithm add edges to T in the order $e_1, e_2, \ldots, e_n$ so that these edges are in the order of nondecreasing weights. Let $T_{\min}$ be a minimum spanning tree that contains edges $e_1, e_2, \ldots, e_{i-1}$ for the largest possible i. Clearly $i \geq 1$, and since T is not a minimum spanning tree, $i \leq n$. Adding e_i to $T_{\min}$ causes a cycle that must include an edge $x \neq e_j$, $1 \leq j \leq i$. Since x and $e_1, e_2, \ldots, e_{i-1}$ are in $T_{\min}$, which is acyclic, and since the greedy algorithm adds the edge of least cost that does not cause a cycle, we know that the weight of x is at least the weight of e_i (otherwise x would have been added to $e_1, e_2, \ldots, e_{i-1}$ instead of e_i). If the weight of x is more than the weight of e_i, then $T'_{\min} = (T_{\min} - \{x\}) \cup \{e_i\}$ is a spanning tree of less total weight than $T_{\min}$, a contradiction. Thus the weight of x and the weight of e_i must be equal, making $T'_{\min}$ a minimum spanning tree and contradicting the assumption that i was as large as possible. Therefore T must be a minimum spanning tree.

The greedy algorithm can be executed in two passes. First the edges are sorted by weight and then a spanning tree is generated, selecting the smallest available edges. The time required in sorting $|E|$ edges is $O(|E| \log |E|)$ (see Chapter 7). Using the efficient set algorithms from Section 2.4 to find and merge the partially formed trees each time an edge is added, the tree can be "grown" in $o(|E| \log |E|)$ operations. Thus the choice of a sorting algorithm will essentially determine the efficiency of the algorithm.

Another method for obtaining a minimum spanning tree, which requires no sorting of edges nor checking for a circuit at each step, is the *nearest-neighbor algorithm*. We start at some arbitrary vertex a in the given graph. Let (a, b) be the edge of least weight incident on a; edge (a, b) is included in the tree. Next, from among all edges incident on either a or b we select the edge with the least weight and include this edge in the partially formed tree. This brings a third vertex into the tree, say c. We repeat the process, looking for a smallest edge connecting a, b, or c, to some other vertex in the graph. The process is continued until every vertex in G is included in the tree—that is, until the tree becomes a spanning tree. The proof that this procedure yields a minimum tree is similar to the proof for the greedy algorithm and is left as Exercise 5.

The worst case for this algorithm occurs when G is the *complete graph* (i.e., there is an edge joining every vertex to every other vertex); in this case, we must make the maximum number of comparisons in order to find the nearest neighbor at each stage. To pick the first edge, we compare the weights of all $|V| - 1$ edges incident on vertex a and select the smallest one; this step requires $|V| - 2$ comparisons. To pick the second edge from among a possible $2(|V| - 2)$ edges (incident on a or b), we perform as many as $2(|V| - 2) - 1$ comparisons to find the smallest. Continuing in this vein, we find that, in order to select the ith edge, we need $i(|V| - i) - 1$ comparisons, and thus a total of

$$\sum_{i=1}^{|V|-1} [i(|V| - i) - 1] = \tfrac{1}{6}|V|^3 + O(|V|^2)$$

comparisons would be required to construct a minimum spanning tree.

We can be more clever, however, and reduce the number of comparisons to approximately $|V|^2$. Suppose that the algorithm maintains, for each vertex v not yet in the tree, a record of the vertex in the tree to which v is closest. This information allows us to determine the ith vertex to be added in only $|V| - i$ comparisons for $i \geq 2$. The first vertex of the tree is chosen arbitrarily and hence requires no comparisons at all. Furthermore, after adding the ith vertex to the tree, the information about the closest vertices can be updated in $|V| - i$ comparisons by comparing, for each of the $|V| - i$ vertices v not in the tree, its distance from the vertex just added and its distance from what was previously the closest vertex to it in the tree. (Compare this process with the closest insertion algorithm for the traveling salesman problem, given in Section 4.3.) Thus, for $i \geq 2$, adding the ith vertex to the tree requires $2(|V| - i)$ comparisons for a total of

$$\sum_{i=2}^{|V|} 2(|V| - i) = (|V| - 1)(|V| - 2)$$

comparisons.

The record of the current closest vertex in the tree for each vertex not yet in the tree can be maintained by means of two arrays, *dist* (v) for the distances (weights) and *near* (v) for the vertex names. Assuming that the graph $G = (V, E)$ is given as a weight matrix $W = [w_{ij}]$ and that the weights of nonexistent edges are ∞, Algorithm 8.1 uses this technique to construct a minimum spanning tree $T = (V_T, E_T)$ of total weight W_T. In the ith iteration of the **while** loop the set $V - V_T$ has $|V| - i$ elements, and so the first step in the loop requires $|V| - i - 1$ comparisons. The next three steps in the loop require only a fixed number of operations, and the final step, the **for** loop, requires $|V| - i - 1$ comparisons "*dist* $(x) > w_{vx}$." Thus the total number of operations is $O(|V|^2)$.

Notice that if the graph G is disconnected, the algorithm produces a minimum spanning forest—that is, a set of minimum spanning trees, one for each connected component of G. These spanning trees are linked together by edges of infinite weight. So in this case $W_T = \infty$ at the end of the algorithm.

A different approach to finding a minimum spanning tree is outlined in Exercise 6. Exercise 7 shows how it leads to an algorithm requiring $O(|E| \log \log |V|)$ operations.

Generating All Spanning Trees.   It is occasionally of interest to examine all the spanning trees of a graph. This problem arises, for example, in the analysis of electrical networks when the networks are interpreted as graphs. Since the

choose v_0 arbitrarily

for $v \in V$ **do** $\begin{cases} dist\ (v) \leftarrow w_{v,v_0} \\ near\ (v) \leftarrow v_0 \end{cases}$

$V_T \leftarrow \{v_0\}$

$E_T \leftarrow \varnothing$

$W_T \leftarrow 0$

while $V_T \neq V$ **do** $\begin{cases} v \leftarrow \text{vertex in } V - V_T \text{ with the smallest value of } dist\ (v) \\[4pt] V_T \leftarrow V_T \cup \{v\} \\[4pt] E_T \leftarrow E_T \cup \{(v, near\ (v))\} \\[4pt] W_T \leftarrow W_T + dist\ (v) \\[4pt] \textbf{for } x \in V - V_T \textbf{ do if } dist\ (x) > w_{vx} \textbf{ then } \begin{cases} dist\ (x) \leftarrow w_{vx} \\ near\ (x) \leftarrow v \end{cases} \end{cases}$

Algorithm 8.1   Minimum spanning tree of a graph.

number of spanning trees grows exponentially with the number of vertices (the complete graph on n vertices has n^{n-2} spanning trees; see Exercise 10), the problem becomes infeasible even for moderate-sized graphs unless care is taken to do the generation efficiently.

The most successful algorithm for generating all spanning trees of a graph is similar to the branch and bound technique illustrated in Figure 4.17 for solving the traveling salesman problem: The set of all spanning trees of G is divided into two classes, those that contain a specific edge (u, v) and those that do not contain (u, v). The spanning trees that contain (u, v) are the ones consisting of (u, v) and a spanning tree of the graph $G_{u,v}$ obtained from G by merging vertices u and v into a "supervertex" (and removing any self-loops that might result). The other spanning trees are the spanning trees of the graph $G - (u, v)$, obtained by deleting edge (u, v) from G.

Notice that the two graphs $G_{u,v}$ and $G - (u, v)$ are smaller than G. Thus successive applications (by, say, recursion) of this basic step reduce the graphs until either all n vertices are merged together into one supervertex or the graphs become disconnected and have no spanning trees. In general, as the algorithm progresses, parallel edges can be created. These edges are retained, as illustrated in the example shown in Figure 8.4. Properly implemented, this algorithm requires $O(|V| + |E| + |E| \cdot \ell)$ operations, where ℓ is the number of spanning trees of G (Exercise 16).

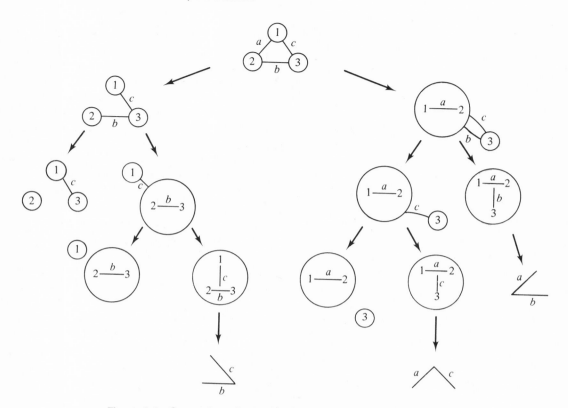

Figure 8.4 Generating all spanning trees of the complete graph of three nodes. Large circles denote "super-vertices."

8.2.2 *Depth-First Search*

One of the most natural ways to explore all the vertices and edges of a graph systematically comes from traversing the graph with a backtrack-style search, similar to Algorithm 4.1, which does a depth-first exploration of a tree. *Depth-first search* on an undirected graph $G = (V, E)$ explores the graph as follows. When we are "visiting" a vertex $v \in V$, we follow one of the edges (v, w) incident on v. If the vertex w has been previously visited, we return to v and choose another edge. If the vertex w has not been previously visited, we visit it and apply the process recursively to w. If all the edges incident on v have been thus examined, we go back along the edge (u, v) that led to v and continue exploring the edges incident on u. We are finished when we try to back up from the vertex at which the exploration began. As we shall see here and in later sections, this technique is extremely useful in determining various properties of both undirected and directed graphs.

Figure 8.5 illustrates how depth-first search examines an undirected graph G represented as an adjacency structure. We start with an arbitrarily chosen vertex a. From a we traverse the first edge that we encounter, which is (a, b). Since b is a vertex never visited before, we stay at b and traverse the first untraversed edge encountered at b, which is (b, c). Now at vertex c, the first untraversed edge that we find is (c, a). We traverse (c, a) and find that a has been previously visited. So we return to c, marking the edge (c, a) in some way (as a dashed line in Figure 8.5) to distinguish it from edges like (b, c), which led to new vertices. Back at vertex c, we look for another untraversed edge and traverse the first one that we encounter, say (c, d). Once again, since d is a new vertex, we stay at d and look for an untraversed edge, in this case (d, b). Since b has been previously visited, we mark edge (d, b) with a dashed line and return to d. Next, since there is no untraversed edge left at d, we back up to c. At c we find an untraversed edge (c, e), which we traverse. Since e is a new vertex, we remain at e and look for an untraversed edge at e, which is (e, a). And since a is an old vertex, we mark (e, a) with a dashed line. Now e is completely examined, and so we return to c. At c we find no untraversed edges left; thus we move back to b and finally, back to a. All the edges incident on a have been traversed, which means that we are finished.

As shown in Figure 8.5, the depth-first search converts the original undirected graph G into a digraph $\vec{G}$ by inducing a direction on each edge (the direction of traversal). The numbers shown next to the edges of $\vec{G}$ in Figure 8.5 indicate the order in which the edges were scanned. The edges of $\vec{G}$ that led to new vertices during the traversal, shown as solid lines in Figure 8.5, form a *directed spanning tree* (forest, if the graph is not connected)—that is, a tree T that spans $\vec{G}$ in which there is a distinguished vertex r, called the *root*, such that there is a directed path from r to every vertex in T but no directed path to r from any vertex in T. If there is a path from vertex x to vertex y in a directed spanning tree, we say that x is an *ancestor* of y and, conversely, that y is a *descendant* of x. If the path is not empty (i.e., $x \neq y$), we say that x is a *proper ancestor* of y and y is a *proper descendant* of x.

The directed spanning tree that results from depth-first search on a simple, connected, undirected graph will be referred to as a *DFS-tree*. The edges of $\vec{G}$ not

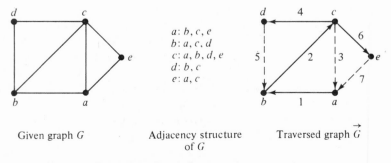

Given graph G Adjacency structure   Traversed graph $\vec{G}$
                            of G

Figure 8.5 Depth-first search on a graph.

in the DFS-tree are called the *back edges* because they lead back to vertices that have been previously visited. The back edges in Figure 8.5 are shown with dashed lines. Notice that a back edge must point from a descendant to an ancestor (why?).

To implement depth-first search, we need to distinguish between vertices that have not yet been visited and vertices that have been visited. This can be accomplished by successively numbering the vertices 1 to $|V|$ as we visit them. Initially, we set $num\ (v) = 0$ for all $v \in V$ to indicate that no vertex has been visited; and when we visit a vertex v for the first time, we set $num\ (v)$ to a nonzero value. Such a depth-first search is given as Algorithm 8.2. The recursive procedure $DFS\ (v,\ u)$ performs depth-first search on the graph $G = (V,\ E)$ containing v and constructs a directed spanning tree $(V,\ T)$ for the graph; u is the father of v in the tree. The graph is represented as an adjacency structure, Adj (v) being the set of vertices adjacent to v. The elements of the set T are the edges of the tree, and the elements of the set B are the back edges. If the graph G is not connected, $(V,\ T)$ will be a forest.

for $x \in V$ **do** *num* $(x) \leftarrow 0$

$i \leftarrow 0$

$T \leftarrow B \leftarrow \varnothing$

for $x \in V$ **do if** *num* $(x) = 0$ **then** *DFS* $(x, 0)$

procedure *DFS* (v, u)

   $i \leftarrow i + 1$

   num $(v) \leftarrow i$

   for $w \in$ Adj (v) **do if** *num* $(w) = 0$ **then** $\begin{cases} [\![(v, w) \text{ is a tree edge}]\!] \\ T \leftarrow T \cup \{(v, w)\} \\ DFS\ (w, v) \end{cases}$

                  else if *num* $(w) < $ *num* (v) **and** $w \neq u$

                          then $\begin{cases} [\![(v, w) \text{ is a back edge}]\!] \\ B \leftarrow B \cup \{(v, w)\} \end{cases}$

   return

Algorithm 8.2   Depth-first search on a simple undirected graph.

Notice that if *num* $(w) = 0$, then edge (v, w) is a tree edge, whereas if *num* $(w) \neq 0$, the test to determine if (v, w) is a back edge is "*num* $(w) < $ *num* (v) and $w \neq u$." If *num* $(w) < $ *num* (v), then the vertex w was visited before v. Thus (v, w) is a back edge, provided that it is not the edge in T that was followed from the father of v to v—that is, provided that $w \neq u$.

Since a call to *DFS* is made exactly once for each newly visited vertex, *DFS* is called $|V|$ times. For each call, the amount of work done is proportional to the number of edges incident on that vertex. Therefore the time required for a depth-first search on an undirected graph is $O(|V| + |E|)$.

Connected Components.   The simplest use of the depth-first search technique on undirected graphs is to determine the connected components of an undirected graph $G = (V, E)$. Algorithm 8.3 assigns a unique component number, *compnum* (v), to every vertex w belonging to the component in which the vertex v occurs.

> **for** $v \in V$ **do** *compnum* $(v) \leftarrow 0$
>
> $c \leftarrow 0$
>
> **for** $v \in V$ **do if** *compnum* $(v) = 0$ **then** $\begin{cases} c \leftarrow c + 1 \\ COMP\ (v) \end{cases}$
>
> **procedure** $COMP\ (x)$
>
> > *compnum* $(x) \leftarrow c$
> >
> > **for** $w \in$ Adj (x) **do if** *compnum* $(w) = 0$ **then** $COMP\ (w)$
> >
> > **return**

Algorithm 8.3   Determining the connected components of a graph.

It is easy to see that this algorithm, like Algorithm 8.2, requires $O(|V| + |E|)$ operations.

Topological Sorting.   The simplest use of the depth-first search technique on digraphs is to determine a labeling of the vertices of an acyclic digraph $G = (V, E)$ with integers $1, 2, \ldots, |V|$, such that if there is a directed edge from vertex i to vertex j, then $i < j$; such a labeling is called a *topological sort* of the vertices of G. For example, the vertices of the digraph in Figure 8.6(a) are topologically sorted but those of Figure 8.6(b) are not. Topological sorting can be viewed as the process of finding a linear order in which a given partial order can be embedded. It is not difficult to show that it is possible to topologically sort the vertices of a digraph if and only if it is acyclic (Exercise 17). Topological sorting is useful in the analysis of activity networks where a large, complex project is represented as a digraph in which the vertices correspond to the goals in the project and the edges correspond to activities. The topological sort gives an order in which the goals can be achieved.

Topological sorting begins by finding a vertex of $G = (V, E)$ with no outgoing edge (such a vertex must exist if G is acyclic) and assigning this vertex the highest number—namely, $|V|$. This vertex is deleted from G, along with all

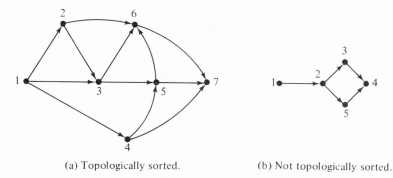

(a) Topologically sorted.   (b) Not topologically sorted.

Figure 8.6   Acyclic digraphs.

incoming edges. Since the remaining digraph is also acyclic, we can repeat the process and assign the next highest number, namely $|V| - 1$, to a vertex with no outgoing edges, and so on. To keep the algorithm $O(|V| + |E|)$, we must avoid searching the modified digraph for a vertex with no outgoing edges.

We do so by performing a single depth-first search on the given acyclic digraph G. In addition to the usual array *num*, we will need another array, *label*, of size $|V|$ for recording the topologically sorted vertex labels. That is, if there is an edge (u, v) in G, then *label* $(u) <$ *label* (v). The complete search and labeling procedure *TOPSORT* is given in Algorithm 8.4.

The time required by Algorithm 8.4 is $O(|V| + |E|)$, since every edge is traversed once and the procedure *TOPSORT* is called once for each vertex.

8.2.3   Biconnectivity

Sometimes it is not enough to know that a graph is connected; we may need to know how "well connected" a connected graph is. A connected graph, for example, may contain a vertex whose removal, along with its incident edges, disconnects the remaining vertices. Such a vertex is called an *articulation point* or a *cut-vertex*. A graph that contains an articulation point is called *separable*. Vertices *b*, *f*, and *i* in Figure 8.7(a), for example, are articulation points, and they are the only ones in the graph. A graph with no articulation points is called *biconnected* or *nonseparable*. A maximal biconnected subgraph of a graph is called a *biconnected component* or a *block*. Identification of the articulation points and biconnected components of a given graph is important in the study of the vulnerability of communication and transportation networks. It is also important in determining other properties, like planarity, of a graph G, since it is often advantageous to separate G into its biconnected components and examine each one individually (see Section 8.6).

A vertex v in an undirected connected graph is an articulation point if and only if there exist two other vertices x and y such that every path between x and y passes through v; in this case and only in this case does the deletion of v from G destroy all paths between x and y (i.e., disconnect G). This observation allows us

$$
\textbf{for } x \in V \textbf{ do } \begin{cases} num\,(x) \leftarrow 0 \\ label\,(x) \leftarrow 0 \end{cases}
$$

$j \leftarrow |V| + 1$

$i \leftarrow 0$

$$
\textbf{for } x \in V \textbf{ do if } num\,(x) = 0 \textbf{ then } \begin{cases} [\![x \text{ has no labeled ancestor}]\!] \\ TOPSORT\,(x) \end{cases}
$$

procedure *TOPSORT* (*v*)

$i \leftarrow i + 1$

$num\,(v) \leftarrow i$

$$
\textbf{for } w \in \text{Adj}\,(v) \textbf{ do} \begin{cases} [\![\text{process all descendants of } w]\!] \\ \textbf{if } num\,(w) = 0 \textbf{ then } TOPSORT\,(w) \\ \qquad \textbf{else if } label\,(w) = 0 \textbf{ then } [\![G \text{ has a cycle}]\!] \end{cases}
$$

$[\![v$ is labeled with a value less than
 the value assigned to any descendant$]\!]$

$j \leftarrow j - 1$

$label\,(v) \leftarrow j$

return

Algorithm 8.4   Topological sort of an acyclic digraph $G = (V, E)$.

to use depth-first search to find the articulation points and biconnected compo-
nents of a graph in $O(|V| + |E|)$ operations.

The central idea can be understood by studying the example in Figure 8.8,
which shows schematically a connected graph consisting of biconnected components G_i, $1 \le i \le 9$, and articulation points v_j, $1 \le j \le 5$. If we start the depth-
first search at, say, the vertex *s* in G_9, we might, perhaps, leave G_9 to go into G_4 by

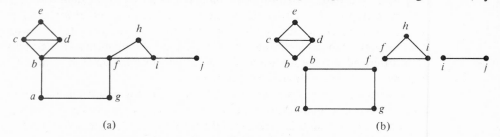

(a)   (b)

Figure 8.7 A separable graph (a) and its biconnected components
(b). The articulation points are *b*, *f*, and *i*.

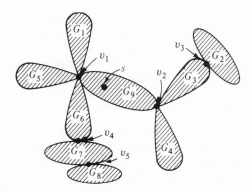

Figure 8.8 A schematic drawing of a graph with nine biconnected components and five articulation points.

passing through v_2. But by the depth-first nature of the search, all the edges in G_4 must be traversed before we back up to v_2; thus G_4 consists of exactly the edges traversed between visits to v_2. Matters are actually a little more complicated for the other biconnected components, since, for example, if we leave G_4 and go into G_3 and from there into G_2 through v_3, we would find ourselves in G_2, having traversed edges from G_9, G_3, and G_2. Fortunately, however, if we store the edges in a stack, by the time we pass through v_3 back into G_3 all the edges of G_2 will be on top of the stack. When they are removed, the edges on top of the stack will be from G_3, and we will once again be traversing G_3. Thus if we can recognize the articulation points, we can determine the biconnected components by applying depth-first search and storing the edges on a stack as they are traversed; the edges on top of the stack as we back up through an articulation point form a biconnected component.

In order to recognize an articulation point we need to compute, during the depth-first search, a new function *lowpt* (v) for every vertex v in the graph. We define *lowpt* (v) as the smallest value of *num* (x), where x is a vertex of the graph that can be reached from v by following a sequence of zero or more tree edges followed by at most one back edge. The function *lowpt* (v) is useful because of the following theorem.

Theorem 8.1

Let $G = (V, E)$ be a connected graph with a DFS-tree T and with back edges B. Then $a \in V$ is an articulation point if and only if there exist vertices $v, w \in V$ such that $(a, v) \in T$, w is not a descendant of v in T and *lowpt* $(v) \geq$ *num* (a).

Proof: Suppose that such vertices v and w exist. Since $(a, v) \in T$ and *lowpt* $(v) \geq$ *num* (a), any path starting at v that does not go through a must remain in the subtree with root v. Since w is not a descendant of v in T, such a path cannot contain w. Thus the only paths from v to w contain a and so a is an articulation point.

Conversely, suppose that a is an articulation point. If a is the root of T, then at least two edges of T start at a; otherwise there would be a path in G between every pair of vertices in $V - \{a\}$ that did not contain a. Let (a, v) and (a, w) be two of these edges; clearly, v and w satisfy the theorem. If a is not the root of T, then it has an ancestor w. One of the biconnected components containing a has all its nodes as descendants of a in T; in fact, they are all (except a) descendants of a vertex v; where (a, v) is an edge in T (why?). Clearly, v and w satisfy the theorem.

This theorem tells how to recognize articulation points if we have the values of *num* and *lowpt*: if we find a vertex v such that $(a, v) \in T$ and *lowpt* $(v) \geq$ *num* (a), then a is either an articulation point or the root of T. This result follows from Theorem 8.1 by observing that a suitable w can be chosen among the ancestors of a if a is not the root. Furthermore, computing the *lowpt* values during depth-first search is simple because

$$lowpt\,(v) = \min\,(\{num\,(v)\} \cup \{lowpt\,(x)|(v, x) \in T\} \cup \{num\,(x)|(v, x) \in B\}).$$

Thus Algorithm 8.5 determines the biconnected components of a graph $G = (V, E)$. Since the algorithm is a depth-first search with a constant amount of extra work done as each edge is traversed, the time required is clearly $O(|V| + |E|)$. A proof that the algorithm works correctly is left as Exercise 20.

8.2.4 Strong Connectivity

In digraphs, the properties involving paths, cycles, and connectivity become more complicated than in undirected graphs because of the orientations of the edges. The (weakly) connected components of a digraph are easily obtained by ignoring the edge directions and using Algorithm 8.3 to find the connected components of the resulting undirected graph. However, we are usually more interested in finding the *strongly connected components*—that is, the maximal strongly connected subgraphs. A digraph and its strongly connected components are illustrated in Figure 8.9.

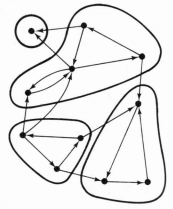

Figure 8.9 A digraph with its strongly connected components outlined in boldface.

$i \leftarrow 0$

$S \leftarrow$ empty stack

for $x \in V$ **do** $num\,(x) \leftarrow 0$

for $x \in V$ **do if** $num\,(x) = 0$ **then** $BICON\,(x, 0)$

procedure $BICON\,(v, u)$

  $i \leftarrow i + 1$

  $num\,(v) \leftarrow i$

  $lowpt\,(v) \leftarrow i$

  for $w \in \text{Adj}\,(v)$ **do if** $num\,(w) = 0$ **then**

$\left\{ \begin{array}{l} [\![(v, w) \text{ is a tree edge}]\!] \\ S \Leftarrow (v, w) \\ BICON\,(w, v) \\ lowpt\,(v) \leftarrow \min\,(lowpt\,(v), lowpt\,(w)) \\ \textbf{if } lowpt\,(w) \geq num\,(v) \textbf{ then} \left\{ \begin{array}{l} \text{At this point } v \text{ is either} \\ \text{the root of the tree or it} \\ \text{is an articulation point.} \\ \text{Form a new biconnected} \\ \text{component consisting of all} \\ \text{the edges on the stack above} \\ \text{and including } (v, w). \text{ Remove} \\ \text{these edges from the stack.} \end{array} \right. \end{array} \right.$

  else if $num\,(w) < num\,(v)$ **and** $w \neq u$ **then** $\left\{ \begin{array}{l} [\![(v, w) \text{ is a back edge}]\!] \\ S \Leftarrow (v, w) \\ lowpt\,(v) \leftarrow \min\,(lowpt\,(v), num\,(w)) \end{array} \right.$

return

Algorithm 8.5   Determining the biconnected components of $G = (V, E)$.

Depth-first search can be applied to digraphs to determine the strongly connected components, but the resulting structure will be more complicated than a set of tree edges and a set of back edges. This complication occurs because the edges in a digraph are already oriented, and we are not free to traverse them in the direction of our choosing, as we did in the depth-first search on an undirected graph.

Consider what happens when we are constrained to traverse the edges of a digraph G along their orientations during a depth-first search on G. As before, we assign a serial number $num(x)$ to each vertex x the first time that we visit it. These numbers are permanently assigned to the vertices and indicate the order

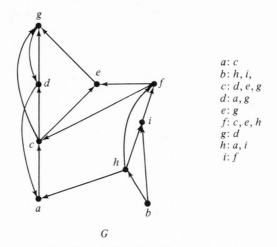

a: c
b: h, i,
c: d, e, g
d: a, g
e: g
f: c, e, h
g: d
h: a, i
i: f

(a) A digraph G and its adjacency structure.

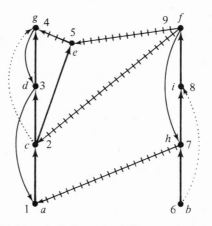

(b) Digraph G after a depth-first search on it.

Figure 8.10 A digraph (a) traversed by depth-first search. The result (b) consists of 2 trees (boldface), 3 back edges (lightface), 4 cross edges (crossed lines), and 2 forward edges (dotted lines).

in which the vertices were visited. When we encounter an edge (v, w) that has not yet been traversed, w may or may not have been previously visited. If w has not been previously visited, the edge (v, w) is marked as a tree edge, just as in an undirected case. If w has already been visited, then w may or may not be an ancestor of v, in contrast to the undirected case in which w was always an ancestor of v. If w is an ancestor of v, then clearly num $(w) < num$ (v) and (v, w) is a back edge. If w is not an ancestor of v and num $(w) > num$ (v), then w must be a descendant of v and the edge (v, w) is called a *forward edge*; it represents an alternate path from v to w. On the other hand, if num $(w) < num$ (v) and w is neither an ancestor nor a descendant of v, then the edge (v, w) is called a *cross edge*.

Figure 8.10 shows a depth-first search on a digraph and the resulting partitioning of the edges into four subsets. The digraph is assumed to be presented as the adjacency structure shown, and the depth-first search was started from an arbitrarily chosen vertex a. The numbers on vertices of the traversed digraph represent num values—that is, the order in which the vertices were visited. Boldface lines denote tree edges, lightface lines denote back edges, crossed lines denote cross edges, and dotted lines denote forward edges.

It is evident that the tree (or forest) generated by a depth-first search is not unique and that the classification of the edges depends on the arbitrary order in which the vertices and edges of the original digraph are given in the adjacency structure and on the choice of the starting vertex. Nevertheless, the partition of the edges that results from a depth-first search on a digraph can be used to determine its strongly connected components. Notice that the forward edges can be ignored, since they do not affect strong connectivity. Notice further that both back edges and cross edges starting at v can go only to vertices x for which num $(v) > num$ (x).

Let v and w be in the same strongly connected component of G and assume, without loss of generality, that num $(v) < num$ (w). By definition, there is a path p in G from v to w. The path p must contain no cross edges from one tree of the DFS-forest to another, since num $(v) < num$ (w) and cross edges go only to lower numbered vertices; that is,

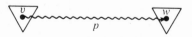

is impossible. Since there is a subtree which with its cross edges and back edges contains p, let r be the root of the minimal such subtree. Because of the minimality of the subtree, if p does not pass through r, we have

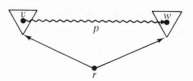

which is impossible. Thus *p* passes through *r*, implying that *r* is in the same strongly connected component as *v* and *w*. We conclude that if *S* is a strongly connected component of *G*, then the vertices of *S* define a tree which is a subgraph of the spanning forest. The converse of this statement is not true: not every subtree corresponds to a strongly connected component.

Identifying the roots of the subtrees corresponding to strongly connected components will allow us to recognize the strongly connected components themselves, just as recognizing articulation points allowed us to identify biconnected components. To recognize these roots, we define *lowlink* (*v*) to be the number of the smallest numbered vertex *in the same strongly connected component as v* that can be reached by following zero or more tree edges followed by at most one back edge or one cross edge. These *lowlink* values give us the information that we need to find the roots of the strongly connected components, for *v* is such a root if and only if *lowlink* (*v*) = *num* (*v*) (Exercise 21). Fortunately, these *lowlink* values can easily be computed during a depth-first search. The resulting procedure for determining the strongly connected components of a digraph $G = (V, E)$ is given in Algorithm 8.6.

When a boolean array is kept to determine whether a vertex *w* is on the stack (*w* is on the stack if and only if it is in the same strongly connected component as an ancestor of *v*), Algorithm 8.6 is obviously an elaboration of depth-first search during which a few extra operations are required as each edge of the graph is traversed. Each vertex is placed on the stack and removed from the stack exactly once. Thus Algorithm 8.6 requires time $O(|V| + |E|)$.

All we need to do to prove Algorithm 8.6 correct is to show that the *lowlink* values are calculated correctly and we can do this by induction (Exercise 22).

8.2.5   Transitive Closure

Given a graph $G = (V, E)$, it is occasionally useful to know which vertices are connected to which other vertices. If *G* is an undirected graph, the problem reduces to identifying its connected components and is solved by Algorithm 8.3. If *G* is a digraph, the problem is not so easily solved. In this section we discuss one method of solution and its generalization; other approaches are considered in Exercises 23, 24, and 25.

Suppose that $G = (V, E)$ is a digraph represented as a $|V| \times |V|$ adjacency matrix $A = [a_{ij}]$, where, for $i, j \in V$, $a_{ij} = 1$ if there is an edge $(i, j) \in E$ and $a_{ij} = 0$ if not. We would like to compute the *connectivity matrix* $A^* = [a_{ij}^*]$ defined by $a_{ij}^* = 1$ if there is a path in *G* from *i* to *j* and $a_{ij}^* = 0$ if not. If we view *E*, the edges of. *G*, as a binary relation on *V*, the vertices of *G*, then A^* is the adjacency matrix for the graph $G^* = (V, E^*)$ in which E^* is the *transitive closure* of the binary relation *E*. Figure 8.11 shows a digraph and its transitive closure G^*.

```
S ← empty stack
i ← 0
for x ∈ V do num (x) ← 0
for x ∈ V do if num (x) = 0 then STRONG (x)
procedure STRONG (v)
    i ← i + 1
    num (v) ← i
    lowlink (v) ← i
    S ⇐ v
    for w ∈ Adj (v) do if num (w) = 0 then  ⎧ [[(v, w) is a tree edge]]
                                            ⎨ STRONG (w)
                                            ⎩ lowlink (v) ← min (lowlink (v), lowlink (w))
        else if num (w) < num (v) then
            [[(v, w) is a back edge or a cross edge]]
            if w is on S then  ⎧ [[At this point w is in
                               ⎪    the same strongly
                               ⎪    connected component as
                               ⎨    v, since w ∈ Adj (v) and w
                               ⎪    on S means that
                               ⎪    there is a path from
                               ⎪    w to v.]]
                               ⎩ lowlink (v) ← min (lowlink (v), num (w))

    if lowlink (v) = num (v) then
        [[v is the root of a strongly connected component]]
        while x, the top vertex on S, satisfies num (x) ≥ num (v) do  ⎧ add x to the current strongly
                                                                      ⎨ connected component and
                                                                      ⎩ delete x from S

    return
```

Algorithm 8.6   Determining the strongly connected components of a digraph.

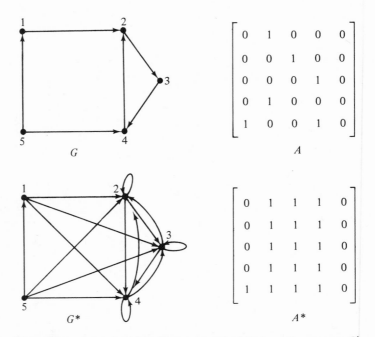

Figure 8.11 A graph G, its adjacency matrix A, its transitive closure G^*, and its adjacency matrix A^*.

We can compute A^* from A by defining a sequence of matrices $A^{(0)} = [a_{ij}^{(0)}]$, $A^{(1)} = [a_{ij}^{(1)}], \ldots, A^{(|V|)} = [a_{ij}^{(|V|)}]$ as follows.

$$\begin{aligned} a_{ij}^{(0)} &= a_{ij}, \\ a_{ij}^{(l)} &= a_{ij}^{(l-1)} \vee (a_{il}^{(l-1)} \wedge a_{lj}^{(l-1)}). \end{aligned} \tag{8.1}$$

We claim that $A^* = A^{(|V|)}$. To understand why, notice that it is easy to prove by induction that $a_{ij}^{(l)} = 1$ if and only if there is a path from $i \in V$ to $j \in V$ with the intermediate vertices on the path chosen only from $\{1, 2, \ldots, l\} \subseteq V$. For $l = 0$, this is obvious. If it is true for $l - 1$, then $a_{ij}^{(l)} = a_{ij}^{(l-1)} \vee (a_{il}^{(l-1)} \wedge a_{lj}^{(l-1)})$ is 1 if and only if either $a_{ij}^{(l-1)} = 1$ (there is a path from i to j using only vertices in $\{1, 2, \ldots, l - 1\}$) or both $a_{il}^{(l-1)}$ and $a_{lj}^{(l-1)}$ are 1 (there are paths from i to l and l to j using only vertices in $\{1, 2, \ldots, l - 1\}$). Thus it is true for l.

Although equation (8.1) is convenient for understanding the calculation of A^*, it is inconvenient for actually performing the calculation, since we would like to transform A to A^* *in place*; that is, we would like to transform A to A^* by using only a fixed amount of extra storage. We can do this by viewing equation (8.1) in a different way.

Notice that if we have $A^{(l-1)}$, we can transform it to $A^{(l)}$ as follows. If $a_{il}^{(l-1)} = 0$, we have simply $a_{ij}^{(l)} = a_{ij}^{(l-1)}$; and if $a_{il}^{(l-1)} = 1$, we have

$a_{ij}^{(l)} = a_{ij}^{(l-1)} \vee a_{lj}^{(l-1)}$. Letting $a_{i,*}$ denote the ith row of a matrix A, we have

$$
a_{i,*}^{(l)} = \begin{cases}
a_{i,*}^{(l-1)} & \text{if } a_{il}^{(l-1)} = 0, \\
a_{i,*}^{(l-1)} \vee a_{l,*}^{(l-1)} & \text{if } a_{il}^{(l-1)} = 1.
\end{cases}
\tag{8.2}
$$

Furthermore, for $i \neq l$, the value of $a_{i,*}^{(l-1)}$ enters into the computation only for $a_{i,*}^{(l)}$, and so its values can be changed without affecting the computation of $a_{k,*}^{(l)}$ for $k \neq i$. $a_{l,*}^{(l-1)}$ is used to compute the other rows, but, fortunately, equation (8.2) yields $a_{l,*}^{(l)} = a_{l,*}^{(l-1)}$.

Algorithm 8.7 uses equation (8.2) to implement the transformation of A to A^*. It obviously requires $O(|V|^3)$ operations, but notice that the basic operation in the innermost loop is the "or" of two rows of A. On most computers this can be done in a single operation if a row of A fits entirely into a computer word. Even if each row requires several words, there is still a considerable savings.

> **for** $l = 1$ **to** $|V|$ **do**
>
>   **for** $i = 1$ **to** $|V|$ **do if** $a_{il} = 1$ **then** $a_{i,*} \leftarrow a_{i,*} \vee a_{l,*}$
>
> *Algorithm 8.7*   Transitive closure.

8.2.6 Shortest Paths

In this section we will consider algorithms for answering questions about the distances between nodes in a simple, nonnegatively weighted digraph. Such digraphs are comprehensive of many types, for if a given digraph is not simple, it can be made so by discarding all self-loops and replacing every set of parallel edges by the shortest (least-weight) edge among them. If the graph is not directed, we can simply consider the directed graph that arises by replacing every undirected edge (i, j) by a pair of directed edges (i, j) and (j, i), each of the same weight as the original undirected edge. If the graph is not weighted, we can consider every edge as having unit weight. We will consider negative weights later.

The questions of interest are

1. What is the length of the shortest path between two specified vertices? What is the path?

2. What are the lengths of the shortest paths between a specified vertex and all other vertices in the graph? What are the paths?

3. What are the lengths of the shortest paths between all pairs of vertices? What are the paths?

We will assume that these questions are to be answered for a simple, weighted digraph in which we need not have $w_{ij} = w_{ji}$ and in which the triangle inequality

$w_{ik} + w_{kj} \leq w_{ij}$ need not hold. Nonexistent edges will be considered as having infinite weight. The sum of the weights of the edges in a path will be called the weight or length of that path.

The first problem, finding the length of the shortest path between two vertices s (start) and f (finish) and finding the path itself, can be solved by using a method similar to that of Algorithm 8.1, the nearest-neighbor method for finding a minimum spanning tree of a graph. We begin at the vertex s and fan out in a *breadth-first* manner (see Exercise 28), as contrasted to the depth-first approach used in Sections 8.2.2, 8.2.3, and 8.2.4, labeling vertices with their distance from s. When the vertex f is labeled with its distance from s, we are finished.

Before a vertex is labeled permanently, it will have temporary labels assigned to it; they represent the distance from s when the paths considered so far do not include all possible paths. Of course, the algorithm stops only when the label assigned to f can no longer change. Thus at any time during this algorithm some vertices will have their final labels assigned and other vertices will not.

To begin, the vertex s is assigned a final label of zero (it is distance zero from itself), and a temporary label of ∞ is assigned to the other $|V| - 1$ vertices. At each iteration one vertex with a temporary label is assigned a final label, and the search branches farther out. On each iteration the labels are modified as follows.

1. Every vertex j that does not have its final label is given a new temporary label of the smaller of its temporary label and w_{ij} + the final label of i, where i is the vertex that was given its final label on the previous iteration.

2. The smallest of all the temporary labels is found and made the final label of its vertex. In the case of a tie, any of the candidates can be chosen.

The iterations continue until f is assigned a final label.

The first vertex to be permanently labeled is s, which is at distance zero from itself. The second vertex to get a final label is the vertex closest to s. The third vertex to get a final label is the second closest vertex of s and so on. It is easy to see that the final label of each vertex is the shortest distance of that vertex from s; this can easily be proved by induction (Exercise 29).

To illustrate the procedure, let us find the distance from vertex b to g in the digraph shown in Figure 8.12. We use an array *dist* to hold the labels of the vertices as we go through the solution. The final labels will be shown enclosed in a box, and the most recently assigned final label is enclosed in a boldface box. The labeling progresses as shown in Figure 8.13.

The path itself can be computed as the algorithm progresses by keeping track of *pre* (v): when the label on v becomes final, *pre* (v) is set to the vertex with a final label that is closest to v; that is, *pre* (v) is the vertex preceding the vertex v along the shortest path. Hence the vertices of the shortest path are

$$s, \ldots, pre\ (pre\ (pre\ (f))), pre\ (pre\ (f)), pre\ (f), f.$$

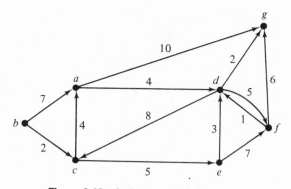

Figure 8.12 A simple weighted digraph.

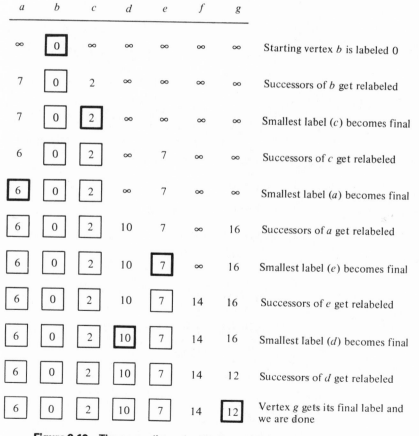

a	b	c	d	e	f	g	
∞	**0**	∞	∞	∞	∞	∞	Starting vertex b is labeled 0
7	0	2	∞	∞	∞	∞	Successors of b get relabeled
7	0	**2**	∞	∞	∞	∞	Smallest label (c) becomes final
6	0	2	∞	7	∞	∞	Successors of c get relabeled
6	0	2	∞	7	∞	∞	Smallest label (a) becomes final
6	0	2	10	7	∞	16	Successors of a get relabeled
6	0	2	10	**7**	∞	16	Smallest label (e) becomes final
6	0	2	10	7	14	16	Successors of e get relabeled
6	0	2	**10**	7	14	16	Smallest label (d) becomes final
6	0	2	10	7	14	12	Successors of d get relabeled
6	0	2	10	7	14	**12**	Vertex g gets its final label and we are done

Figure 8.13 The array *dist* as the shortest path from vertex *b* to vertex *g* is found.

Algorithm 8.8 finds the shortest path (and its length) from s to f in $G = (V, E)$, represented as a weight matrix $W = [w_{ij}]$. A boolean array *final* (v) is used to indicate which labels are final [*final* (v) = **true**] and which are temporary [*final* (v) = **false**]. The value of *last* is the latest vertex whose label has been made final. At the end of Algorithm 8.8, *label* (f) is the length of the shortest path between the vertices s and f; if *label* $(f) = \infty$, there is no such path and some of the *pre* (v) values will be undefined. Exercise 30 discusses the modification of Algorithm 8.8 to produce *all* the shortest paths from s to f.

$$\textbf{for } x \in V \textbf{ do} \begin{cases} \textit{label } (x) \leftarrow \infty \\ \\ \textit{final } (x) \leftarrow \textbf{false} \end{cases}$$

$\textit{label } (s) \leftarrow 0$

$\textit{final } (s) \leftarrow \textbf{true}$

$\textit{last} \leftarrow s$

$$\textbf{while not } \textit{final } (f) \textbf{ do} \begin{cases} \textbf{for } x \in V \textbf{ do if not } \textit{final } (x) \textbf{ and } \textit{label } (x) > \textit{label } (\textit{last}) + w_{\textit{last},x} \\ \qquad\qquad \textbf{then} \begin{cases} \textit{label } (x) \leftarrow \textit{label } (\textit{last}) + w_{\textit{last},x} \\ \\ \textit{pre } (x) \leftarrow \textit{last} \end{cases} \\ \\ \text{let } k \text{ be any vertex with minimum temporary label} \\ \text{[i.e., any vertex } k \text{ with } \textit{final } (k) = \textbf{false} \text{ and the} \\ \text{smallest such value of } \textit{label } (k)\text{]} \\ \\ \textit{final } (k) \leftarrow \textbf{true} \\ \\ \textit{last} \leftarrow k \end{cases}$$

Algorithm 8.8   Shortest path from s to f in a weighted digraph $G = (V, E)$ represented as a weight matrix $W = [w_{ij}]$.

Algorithm 8.8 executes the body of the **while** loop at most $|V| - 1$ times, and the number of operations required for each of these times is obviously $O(|V|)$. Thus the algorithm requires $O(|V|^2)$ operations.

If we want to find the lengths of the shortest paths from s to all the other vertices, we simply change the **while** loop to

$$\textbf{while } \bigvee_{x \in V} \textbf{ not } \textit{final } (x) \textbf{ do}.$$

Even here the algorithm requires only $O(|V|^2)$ operations. Exercise 31 discusses the modification needed to produce the paths themselves.

Negative Weights. So far we have assumed that all the edge weights are positive. If some of the weights are negative, Algorithm 8.8 will not work because once a vertex gets its final label, the label cannot be altered. It is possible to modify the algorithm for digraphs with negative weights, provided that the sum of the weights around every cycle in the digraph is positive. The problem is ill defined if a cycle of negative weight has a vertex on a directed path from s to f, for then there would be no shortest path from s to f.

The modification will require that on each iteration every vertex with a final label also gets a new temporary label if this temporary label turns out to be smaller than the final label. Such a modification increases the order of the computation time from $O(|V|^2)$ to $O(|V|2^{|V|})$, and there is a far more efficient method of finding shortest paths in the presence of negative weights. This algorithm, which is discussed next, has a time bound of $O(|V|^3)$.

Shortest Paths Between All Pairs of Vertices. To find shortest paths between all $|V|(|V| - 1)$ ordered pairs of vertices in a given digraph (or $|V|(|V| - 1)/2$ unordered pairs in the case of an undirected graph), we could modify Algorithm 8.8 to find all shortest paths starting at a vertex and use it $|V|$ times, once for each vertex. This would obviously yield an algorithm requiring $O(|V|^3)$ operations for graphs with nonnegative weights, but, as modified for negative weights, the algorithm requires $O(|V|^2 2^{|V|})$ operations. Instead we will present a more general method that requires $O(|V|^3)$ operations even for graphs with negative weights.

If $W = [w_{ij}]$ is the weight matrix of $G = (V, E)$, we want to compute $W^* = [w_{ij}^*]$, in which w_{ij}^* is the length of the shortest path from $i \in V$ to $j \in V$. In analogy to equation (8.1) in Section 8.2.5, we define a sequence of matrices $W^{(l)} = [w_{ij}^{(l)}]$ as follows:

$$
\begin{aligned}
w_{ij}^{(0)} &= w_{ij} \\
w_{ij}^{(l)} &= \min (w_{ij}^{(l-1)}, w_{il}^{(l-1)} + w_{lj}^{(l-1)}),
\end{aligned}
\tag{8.3}
$$

assuming that the weight of a nonexistent edge is ∞, that $x + \infty = \infty$, and that $\min (x, \infty) = x$ for all x. The value of $w_{ij}^{(l)}$ is the length of the shortest path from $i \in V$ to $j \in V$ with intermediate vertices on the path chosen only from $\{1, 2, \ldots, l\} \subseteq V$. Algorithm 8.9, based on equation (8.3), is very much like Algorithm 8.7 for computing the transitive closure of G, but the innermost loop, implicit in Algorithm 8.7 because of the use of $a_{i,*}$ to denote an entire row of A, has been expanded. Algorithm 8.9 obviously requires $O(|V|^3)$ operations.

for $l = 1$ **to** $|V|$ **do**
    for $i = 1$ **to** $|V|$ **do**
        if $w_{il} \neq \infty$ **then for** $j = 1$ **to** $|V|$ **do** $w_{ij} \leftarrow \min (w_{ij}, w_{il} + w_{lj})$
            Algorithm 8.9   All shortest paths.

Notice that negative weights have no effect on this algorithm but that if there is a cycle with negative length, the entire problem is meaningless. It is interesting to note that this algorithm, Algorithm 8.9, is the most efficient method known for computing a *single* shortest path in a dense digraph with negative weights.

Algorithm 8.9 does not actually produce the $|V|(|V| - 1)$ shortest paths; it only produces their lengths. The paths themselves can be constructed from the path matrix $P = [p_{ij}]$, in which p_{ij} is the second vertex along the shortest path from $i \in V$ to $j \in V$. The shortest path from i to j is then the sequence of edges $(v_1, v_2), (v_2, v_3), \ldots, (v_{q-1}, v_q)$, $v_1 = i$, $v_q = j$, in which $v_t = p_{v_{t-1},j}$ for $t \geq 1$.

The matrix P is easily calculated in Algorithm 8.9. Initially, we set

$$p_{ij} = \begin{cases} j & \text{if } w_{ij} \neq \infty, \\ 0 & \text{if } w_{ij} = \infty. \end{cases}$$

On the lth iteration, if $w_{il} + w_{lj} < w_{ij}$, then we set $p_{ij} \leftarrow p_{il}$. It is easy to prove that $p_{ij} = 0$ at the end of the algorithm if and only if $w_{ij}^* = \infty$—that is, there is no path from i to j.

8.3 CYCLES

In this section we examine algorithms to solve problems concerning the cycle structure of a graph. The simplest problem, of course, is to determine whether a given graph, directed or undirected, contains a cycle. Depth-first search suffices to answer the question for both directed and undirected graphs: a graph or digraph has a cycle if and only if the DFS-tree contains a back edge. A slightly harder problem is to determine a minimal set of cycles of an undirected graph from which all of the cycles in the graph can be constructed (a so-called *fundamental set of cycles*). It is still harder to determine all the cycles of a digraph.

8.3.1 *Fundamental Sets of Cycles*

Consider a spanning tree (V, T) of a connected undirected graph $G = (V, E)$. Any edge not in T, that is, any edge in $E - T$, will create exactly one cycle when added to T. Such a cycle is a member of the *fundamental set of cycles* of G with respect to T. Since every spanning tree of a graph G contains $|V| - 1$ edges, there are $|E| - |V| + 1$ cycles in the fundamental set of cycles with respect to any spanning tree of G.

The usefulness of a fundamental set of cycles stems from the fact that it completely determines the cycle structure of a graph: every cycle in the graph can be expressed as a combination of cycles in a fundamental set of cycles. Let $F = \{S_1, S_2, \ldots, S_{|E|-|V|+1}\}$ be a fundamental set of cycles, each cycle S_i being a subset of edges of G; that is, $S_i \subseteq E$. Then any cycle of G can be written as $((S_{i_1} \oplus S_{i_2}) \oplus \cdots) \oplus S_{i_r}$, where $\oplus$ denotes the operation of *symmetric difference*, $A \oplus B = \{x \mid x \in A \cup B, x \notin A \cap B\}$. For example, Figure 8.14 shows a graph and the fundamental set of cycles that arises from the boldface spanning tree of the

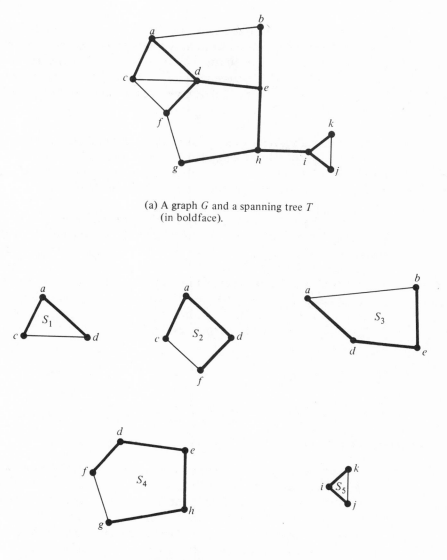

(a) A graph G and a spanning tree T
(in boldface).

(b) The 5 cycles in the fundamental
set of cycles with respect to T.

Figure 8.14   An example of a fundamental set of cycles.

graph; here the cycle (a, b, e, h, g, f, c, a) is $S_2 \oplus S_3 \oplus S_4$, while the cycle (c, d, f, c) is $S_1 \oplus S_2$. Notice that not every such sum is a cycle; for instance, $S_1 \oplus S_5$ consists of two disjoint cycles.

Depth-first search is a natural approach to use in generating a fundamental set of cycles; it produces a spanning tree (the DFS-tree), and each back edge forms a cycle with respect to that tree. When the algorithm arrives at a back edge (v, w), the cycle consists of the tree edges from w to v and the back edge (v, w). To keep track of the tree edges, we augment the depth-first search with a stack on which are stored all the vertices on the path from the root to the vertex currently being visited. When a back edge is encountered, the cycle formed consists of that back edge and the edges connecting the vertices on the top of the stack. Algorithm 8.10 produces the fundamental set of cycles $\{S_1, S_2, \ldots, S_{|E|-|V|+1}\}$ of a graph $G = (V, E)$ using this method.

Superficially, Algorithm 8.10 appears to be linear in $|E|$, but it is *not*. It is true that each edge (v, w) is examined only twice in the **for** loop, but that edge is placed into many of the cycles S_j and is thus "examined" many times. The number of operations required by Algorithm 8.10 is actually $O(|V| + |E| + l)$, where l is the sum of the lengths of all cycles generated.

This l depends on the graph as well as on the order of the vertices in the adjacency lists. Since each cycle can have length at most $|V| - 1$, it is clear that $l \leq (|V| - 1)(|E| - |V| + 1) \leq \frac{1}{2}(|V| - 1)^2(|V| - 2)$. In the complete graph of $|V|$ vertices with $|E| = \frac{1}{2}|V|(|V| - 1)$ edges ordered in the adjacency structure so that the first edge out of vertex i goes to vertex $i + 1, 1 \leq i < |V|$, the depth-first search generates $\frac{1}{2}(|V| - 1)(|V| - 2)$ cycles with lengths as follows: one of length $|V| - 1$, two of length $|V| - 2$, three of length $|V| - 3, \ldots, |V| - 2$ of length 2. The sum of the lengths of these cycles is

$$l = \sum_{i=1}^{|V|-2} i(|V| - i) = \frac{1}{6}|V|^3 + O(|V|^2).$$

Algorithm 8.10 thus requires $O(|V|^3)$ operations.

8.3.2   Generating All Cycles

In such applications as studying the flow through computer programs or breaking feedback paths in control systems, it is necessary to list all the cycles of a graph. One approach, examining the subsets of a set of fundamental cycles, is inefficient, since many of the subsets will lead to degenerate cycles; moreover, there is no analogy for digraphs. We will consider only the problem of generating the cycles of a digraph; any algorithm for generating all the cycles of a digraph can also be used for undirected graphs by converting the undirected graph into a digraph by replacing each edge with two oppositely directed edges between the same pair of vertices. (What happens to the cycles of the graph?)

As in the case of spanning trees, the number of cycles of a digraph can be exponential in the number of vertices. In fact, the complete digraph of n vertices and $n(n - 1)$ edges contains $\binom{n}{i}(i - 1)!$ cycles of length i for a total of

$$\sum_{i=2}^{n} \binom{n}{i}(i - 1)! > (n - 1)!$$

for $x \in V$ **do** $num\,(x) \leftarrow 0$

$i \leftarrow j \leftarrow k \leftarrow 0$

for $x \in V$ **do if** $num\,(x) = 0$ **then** $FNDCYC\,(x, 0)$

procedure $FNDCYC\,(v, u)$

    $i \leftarrow i+1$

    $num\,(v) \leftarrow i$

for $w \in \text{Adj}\,(v)$ **do if** $num\,(w) = 0$ **then**
$\begin{cases} k \leftarrow k + 1 \\ stack_k \leftarrow w \\ FNDCYC\,(w, v) \\ k \leftarrow k - 1 \end{cases}$

else if $num\,(w) < num\,(v)$ and $w \neq u$ **then**
$\begin{cases} j \leftarrow j + 1 \\ S_j \text{ is the cycle} \\ w, stack_k, stack_{k-1}, \ldots, stack_t, \\ \text{where } stack_t = w. \text{ Notice} \\ \text{that } stack_k = v. \end{cases}$

return

Algorithm 8.10   Fundamental cycles of a graph.

cycles. Thus we can hardly expect an algorithm to be polynomial in the size of the input. We can, however, try to develop an algorithm that generates the cycles in time $O((|V| + |E|)(c + 1))$, where c is the number of cycles in the digraph. Such an algorithm would be close to optimal, since the time taken just to print the cycles must be $\bar{\ell} \cdot c$, where $\bar{\ell}$ is the average cycle length, which may be expected to satisfy $\bar{\ell} = O(|V|)$.

The cycles in a digraph can be generated by a depth-first search in which edges are added to a path until a cycle is produced. If the backtracking is carelessly done, many more paths than necessary will be explored and the efficiency of the algorithm will suffer. Careful pruning of the search is necessary in order to ensure that every cycle is generated exactly once and that very little extra work is done.

In order to ensure that each cycle generated is new and not simply a permutation of a cycle generated earlier, we assume that the vertices are represented as integers from 1 to $|V|$, and we will consider each cycle to be rooted at its smallest vertex. We begin the search at a vertex s and build a (directed) path $(s, v_1, v_2, \ldots, v_k)$, such that $v_i > s, 1 \leq i \leq k$. A cycle is reported only when the next vertex $v_{k+1} = s$. After generating the cycle $(s, v_1, v_2, \ldots, v_k, s)$, we explore the next edge going out of v_k. If all edges going out of v_k have been explored, we back up to the previous vertex v_{k-1} and try extending paths from it and so on. This process is continued until we finally try to back up past the starting vertex s; all cycles containing vertex s have then been generated. This process is repeated for $s = 1, 2, \ldots, |V|$.

To prevent traversing cycles that originate at a vertex v_i during the search for cycles rooted at s, it is necessary to indicate that all the vertices on the current path (except s) are unavailable as extensions of that path. A vertex v will be marked as unavailable by setting *avail* (v) to **false** as soon as v is appended to the current path. The vertex v will remain unavailable at least until we have backed up past v to the previous vertex on the current path. If the current path up to vertex v did not lead to a cycle rooted at s, v will remain unavailable for awhile even after we back up past it. This prevents searching for cycles in parts of the graph on which such searches have been previously unsuccessful.

As the search progresses, the vertices on the current path are stored in a stack called *path*: a vertex v is added to the top of the stack before the search continues (by a recursive call) and is removed from the top of the stack in backing up (on return from the recursive call). It is marked unavailable as it is put on the stack and, if it leads to a cycle rooted at s, made available again after it is removed from the stack. If it did not lead to a cycle rooted at s, then it remains temporarily unavailable. A record is kept of the predecessors of all unavailable vertices that are not on the current path by maintaining sets $B(w)$ for each $w \in V$:

$$B(w) = \{v \in V | (v, w) \in E \text{ and } v \text{ is unavailable}$$
$$\text{and not on the current path}\}.$$

In other words, $B(w)$ is the set of unavailable predecessors of w.

Consider what happens when the current path is $(s, v_1, v_2, \ldots, v_k)$ and the next edge traversed is (v_k, v_{k+1}). There are three cases.

1. If $v_{k+1} = s$, then we have generated a new cycle $(s, v_1, v_2, \ldots, v_k, s)$. We output the cycle and then set $flag \leftarrow$ **true** to indicate that a cycle (rooted at s) going through v_k has been found.

2. If $v_{k+1} \neq s$ and v_{k+1} is available, we append v_{k+1} to the current path by making a recursive call to the cycle-generating procedure.

3. If v_{k+1} is unavailable, then we ignore it and look for another edge starting at v_k. If there are no unexplored edges starting at v_k, v_k is deleted from the path and we back up to v_{k-1}. In backing up to v_{k-1}, we make v_k available again *if* it led to a cycle (i.e., *flag* is **true**). Otherwise it is left unavailable and the vertices w that are successors of v_k have v_k added to their set $B(w)$.

This process is given as the recursive procedure *CYCLE* in Algorithm 8.11(a).

procedure *CYCLE* $(v, flag)$

> $flag \leftarrow$ **false**
>
> $path \Leftarrow v$
>
> $avail\,(v) \leftarrow$ **false**
>
> **for** $w \in \mathrm{Adj}\,(v)$ **do**
> > **if** $w = s$ **then** { output the cycle consisting of the vertices in *path* followed by s (*path* is not changed) $flag \leftarrow$ **true**
> >
> > **else if** *avail* (w) **then** { *CYCLE* (w, g) $flag \leftarrow flag \lor g$
>
> **if** *flag* **then** *UNMARK* (v)
>
> > **else for** $w \in \mathrm{Adj}\,(v)$ **do** $B(w) \leftarrow B(w) \cup \{v\}$
>
> ⟦delete v from top of *path*⟧
> $v \Leftarrow path$

return

Algorithm 8.11(a)   The recursive procedure to do the depth-first search for cycles. The procedure *UNMARK* is given in Algorithm 8.11(b).

In order to make a vertex u available again after it has been unavailable, we must change *avail* (u) to **true**. However, if u becomes available again, so do all its unavailable predecessors that are not on the current path (why?). Thus all the elements of $B(u)$ must be made available [the predecessors of u not in $B(u)$ are already available by the definition of $B(u)$], and $B(u)$ must be set to $\varnothing$. As these vertices are made available again, their unavailable predecessors that are not on the current path become available again and so on. This propagation of vertices becoming available again is given in Algorithm 8.11(b), the recursive procedure *UNMARK*.

> **procedure** *UNMARK* (u)
>
>   *avail* $(u) \leftarrow$ **true**
>
>   **for** $w \in B(u)$ **do** $\begin{cases} B(u) \leftarrow B(u) - \{w\} \\ \textbf{if not } avail\ (w) \textbf{ then } UNMARK\ (w) \end{cases}$
>
>   **return**

Algorithm 8.11(b)   The recursive procedure to make vertices available again as extensions to the current path in *CYCLE*.

To complete the algorithm, notice that every cycle in a digraph lies entirely within a strongly connected component. Since we can decompose a graph into such components by using Algorithm 8.6, we will apply the procedure *CYCLE* to

> *path* $\leftarrow$ empty stack
>
> $s \leftarrow 1$
>
> **while** $s < |V|$ **do** $\begin{cases} \text{Let } S = (V_S, E_S) \text{ be the nontrivial, strongly connected} \\ \text{component containing the smallest numbered vertex} \\ \text{in the subgraph of } G \text{ induced by the vertices} \\ \{s, s + 1, \ldots, |V|\}. \\[4pt] \textbf{if } V_S \neq \varnothing \textbf{ then} \begin{cases} Adj \leftarrow \text{adjacency structure for } S \\ s \leftarrow \text{smallest numbered vertex in } V_s \\ \textbf{for } v \in V_S \textbf{ do} \begin{cases} avail\ (v) \leftarrow \textbf{true} \\ B(v) \leftarrow \varnothing \end{cases} \\ CYCLE\ (s, f) \\ s \leftarrow s + 1 \end{cases} \\ \textbf{else } s \leftarrow |V| \end{cases}$

Algorithm 8.11(c)   The driver program to generate all cycles of a digraph $G = (V, E)$, $V = \{1, 2, \ldots, |V|\}$, using *CYCLE* and *UNMARK*.

one strongly connected component at a time. After calling $CYCLE\ (s, f)$, all cycles containing s will have been generated and s can be deleted from the graph. This, of course, changes the strongly connected components of the graph. Assuming that $G = (V, E)$, $V = \{1, 2, \ldots, |V|\}$, is represented by an adjacency structure Adj_G, the driver for $CYCLE$ is given as Algorithm 8.11(c).

We leave the proof that Algorithms 8.11(a), (b), and (c) work correctly to Exercise 36. From this exercise we can conclude that at most $O(|V| + |E|)$ time can elapse between successively generated cycles (Exercise 37), and thus the overall procedure requires $O((|V| + |E|)(c + 1))$ operations, where c is the number of cycles generated.

8.4   CLIQUES

A maximal complete subgraph of a graph G is called a *clique* of G; that is, a clique of G is a subset of the vertices of G such that there is an edge in G between every pair of vertices in the subset and, furthermore, such that the subset is not contained in any larger subset having the same property. For example, Figure 8.15 shows a graph and its cliques. The cliques of a graph represent the "natural" groupings of the vertices, and determining the cliques of a graph is useful in cluster analysis in such fields as information retrieval and sociology.

As with cycles in a digraph, the number of cliques in a graph can grow exponentially with the number of vertices. Consider the Moon-Moser graph M_n with $3n$ vertices $\{1, 2, \ldots, 3n\}$, in which the vertices are partitioned into triads $\{1, 2, 3\}, \{4, 5, 6\}, \ldots, \{3n - 2, 3n - 1, 3n\}$; M_n has no edge within any triad, but, aside from that, each vertex is connected to every other vertex. M_1, M_2, and

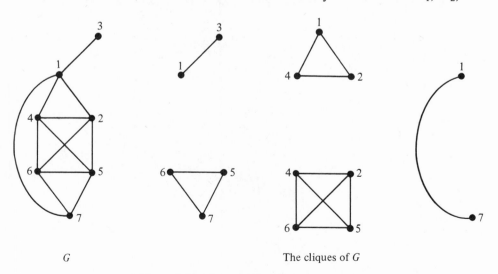

G The cliques of G

Figure 8.15   A graph G and all of its cliques.

M_3 are shown in Figure 8.16. It is easy to prove by induction that M_n has 3^n cliques, each consisting of n vertices. This is obviously true for M_1, in which the cliques are the vertices themselves. If M_{n-1} has 3^{n-1} cliques, each consisting of $n - 1$ vertices, then each of the three vertices added to form M_n forms a clique with each of the 3^{n-1} cliques of M_{n-1}; since they are obviously the only new cliques formed, M_n has $3 \cdot 3^{n-1} = 3^n$ cliques, each consisting of n vertices. M_n thus has a number of cliques that is exponential in the number of its vertices.

Because the number of cliques can be so large, we must be careful in generating them. Each clique must be generated exactly once so that no time will be wasted on repetitive work. The simple technique used in Section 8.3.2, considering a cycle to be rooted at its lowest numbered vertex, is too inefficient for cliques. Other techniques are needed; as a result, the algorithm for generating all

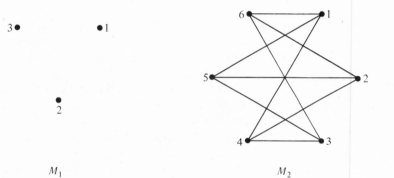

M_1 M_2

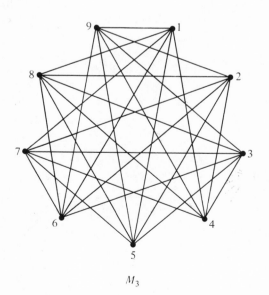

M_3

Figure 8.16   The first three Moon-Moser graphs.

the cliques of a graph is somewhat more intricate than any of the algorithms discussed so far in this chapter.

As a starting point, consider an unconstrained backtrack search for the cliques—that is, a search in which no attempt is made to prune the search tree. Each node in the search tree corresponds to a complete subgraph of the graph, and each edge corresponds to a vertex of the graph. A son of a given node C is obtained by adding to C a vertex $x \notin C$ that is adjacent to every vertex in C. The edge from C to the son $C \cup \{x\}$ corresponds to the vertex x. A sample graph G and the search tree T traversed by an unconstrained backtrack search are shown in Figure 8.17. Notice that each of the cliques is generated many times: $\{1, 2, 3\}$ and $\{3, 5, 6\}$ are each generated six times and $\{2, 4\}$ is generated twice; in general, a clique of size k is generated $k!$ times. All the edges of the tree in Figure 8.17 shown in lightface can be pruned by using Theorems 8.2 and 8.3.

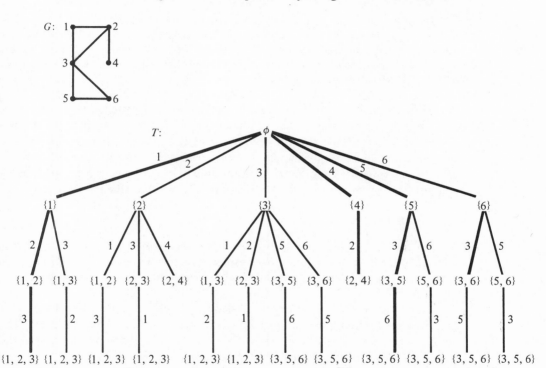

Figure 8.17 A graph G and the result of an unconstrained backtrack search for the cliques of G. Edges that can be pruned by applying Theorems 8.2 and 8.3 are shown in lightface.

Theorem 8.2

Let S be a node in the search tree T (i.e., S is a subset of the vertices of G that induces a complete subgraph of G) and let the first son of S in T to be explored be $S \cup \{x\}$ (i.e., x is adjacent to every vertex in S). Suppose that all the subtrees of the node $S \cup \{x\}$ in T have been explored, so that all cliques containing $S \cup \{x\}$ have been generated. Then among the sons $S \cup \{v\}$ of S, only those for which $v \notin$ Adj (x) need be explored. (See Figure 8.18.)

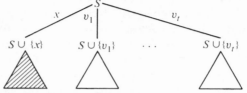

Figure 8.18   According to Theorem 8.2, the subtree rooted at $S \cup \{v_i\}$ need not be explored if the subtree rooted at $S \cup \{x\}$ (shaded) has been explored and $v_i \in$ Adj(x).

Proof: Let C be a clique generated in exploring a subtree with root $S \cup \{v_i\}$, where $v_i \in$ Adj (x). Obviously, $S \subseteq C$; and if C contains a vertex $v_j \notin$ Adj (x), then C will be found when the subtree rooted at $S \cup \{v_j\}$ is explored [note that $S \subseteq$ Adj (x), so that $v_j \notin S$]. If C contains no such vertex v_j, then it must have been found when the subtree rooted at $S \cup \{x\}$ was explored.

Notice that Theorem 8.2 cannot be applied iteratively to the tree as the sons of a node S are examined; that is, after using Theorem 8.2 to prune a subtree rooted at $S \cup \{v_i\}$ (see Figure 8.18) because $v_i \in$ Adj (x), we cannot prune a subtree rooted at $S \cup \{v_k\}$ when $v_k \in$ Adj (v_j) for some $v_j \notin$ Adj (x) (see Exercise 38). In other words, Theorem 8.2 applies only to the *first* son of S to be explored and not to the remaining sons of S. Fortunately, we have the trivial result:

Theorem 8.3

Let S be a node in the search tree T and let $\hat{S} \subset S$ be a proper ancestor of S in T. If all subtrees of the node $\hat{S} \cup \{x\}$ have been explored, so that all cliques containing $\hat{S} \cup \{x\}$ have been generated, then all unexplored subtrees with roots $S \cup \{x\}$ can be ignored.

Algorithm 8.12 is a backtrack procedure for generating all the cliques of a graph; it uses Theorems 8.2 and 8.3 to prune the tree to be searched. The recursive procedure CLIQUE has two parameters N and D. At a node S in the tree, $N \cup D$ is the set of vertices that are adjacent to every vertex in S. N consists of all the new vertices that can be added to S in looking for cliques; that is,

$$N = \{x \in V | (x, s) \in E \text{ for every } s \in S \subseteq V, \text{ and no subtree}$$
$$\hat{S} \cup \{x\} \text{ has been explored for any } \hat{S} \subseteq S\}.$$

Thus, to use Theorem 8.2, *CLIQUE* selects an arbitrary vertex $f \in N$, removes f from N, and explores the subtree $S \cup \{f\}$ by a call to *EXPLORE*; Theorem 8.2 tells us that we need only explore subtrees $S \cup \{x\}$, $x \in N$, when $x \notin$ Adj (f). subtrees of S are then explored by using *EXPLORE*.

$S \leftarrow \emptyset$

$CLIQUE\ (V, \emptyset)$

procedure $CLIQUE\ (N, D)$

    if $N \cup D = \emptyset$ **then** output S, which is a clique

$$
\textbf{else}
\begin{cases}
\textbf{if } N \neq \emptyset \textbf{ then}
\begin{cases}
[\![\text{explore first subtree}]\!] \\
f \leftarrow \text{vertex in } N \\
EXPLORE\ (f) \\
[\![\text{explore remaining subtrees, if any, not precluded by Theorem 8.2}]\!] \\
\textbf{while } N \cap (V - \text{Adj}\,(f)) \neq \emptyset \textbf{ do}
\begin{cases}
v \leftarrow \text{vertex in } N \cap (V - \text{Adj}\,(f)) \\
EXPLORE\ (v)
\end{cases}
\end{cases} \\
[\![\text{If } N = \emptyset \text{ and } D \neq \emptyset, \text{ then Theorem 8.3 tells us that no new clique will be found.}]\!]
\end{cases}
$$

return

procedure $EXPLORE\ (u)$

    $N \leftarrow N - \{u\}$

    $S \leftarrow S \cup \{u\}$

    $CLIQUE\ (N \cap \text{Adj}\,(u),\ D \cap \text{Adj}\,(u))$

    $S \leftarrow S - \{u\}$

    $D \leftarrow D \cup \{u\}$

    return

Algorithm 8.12   Generating the cliques of a graph.

The second parameter, D, of *CLIQUE* is a set consisting of all the vertices that are adjacent to all the vertices of S but that need not be added to S by Theorem 8.3; that is,

$$D = \{x \in V | (x, s) \in E \text{ for every } s \in S \subseteq V, \text{ and a subtree}$$
$$\hat{S} \cup \{x\} \text{ has been explored for some } \hat{S} \subseteq S\}.$$

As before, the sets $\hat{S}$ are ancestors of S in the search tree, and therefore the sets $\hat{S} \cup \{x\}$ are brothers, uncles, great uncles, great-great uncles, and so on of S in the search tree.

Since $S \subseteq V$ is always a complete subgraph of G and $N \cup D$ is the set of all vertices adjacent to every vertex in S, it follows that S is a clique if and only if $N \cup D = \varnothing$. From Theorem 8.3 it follows that if $N = \varnothing$ and $D \neq \varnothing$, then all the cliques containing S having already been generated. If $N \neq \varnothing$, there may still be cliques containing S that have not yet been generated. Thus *CLIQUE* outputs S as a clique when $N \cup D = \varnothing$, does nothing when $N = \varnothing$ and $D \neq \varnothing$, and explores (or prunes) the subtrees $S \cup \{x\}$, $x \in N$ when $N \neq \varnothing$.

Algorithm 8.12 can be improved by a more careful choice of the vertex f chosen as the root of the first subtree to be explored. It is obviously to our advantage, according to Theorem 8.2, to pick f so that $N \cap \text{Adj}(f)$ is as large as possible; this choice causes the most subtrees to be pruned. In fact, we need not even choose an $f \in N$. We can pick $f \in N \cup D$ so that $N \cap \text{Adj}(f)$ is as large as possible; if $f \in N$, then we explore that first subtree $S \cup \{f\}$. If $f \in D$, we need not explore $S \cup \{f\}$ by Theorem 8.3; we simply begin exploring the remaining subtrees. Thus in Algorithm 8.12 we replace

⟦explore first subtree⟧

$f \leftarrow$ vertex in N

EXPLORE (f)

by

$f \leftarrow$ vertex in $N \cup D$, which maximizes $|N \cap \text{Adj}(f)|$

if $f \in N$ **then** $\begin{cases} \text{⟦explore first subtree⟧} \\ \\ EXPLORE\ (f) \end{cases}$

This modification gives what is, in practice, the most efficient algorithm known for generating all the cliques of a graph. There is no known analysis of its running time; but it is known that the running time cannot be expressed as a polynomial in $|V|$ or $|E|$, as the Moon-Moser graphs (Figure 8.16) show. Furthermore (and more importantly), the running time of the unmodified Algorithm 8.12 cannot be expressed as a polynomial in the number of cliques. Consider the graph in Figure 8.19. This graph has 2^n cliques, each consisting of kn vertices, and the search tree contains at least $(k + 1)^n$ nodes for Algorithm 8.12 (but not its modification, why?). Choosing $n = \lg k$ gives a graph with $2k \lg k$ vertices, k cliques, and $k^{\lg k}$ nodes in the search tree.

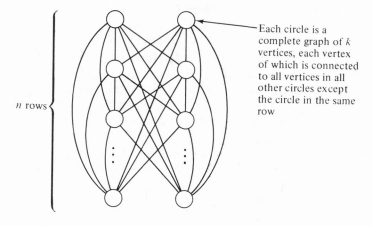

Each circle is a complete graph of k vertices, each vertex of which is connected to all vertices in all other circles except the circle in the same row

n rows

Figure 8.19   A case in which Algorithm 8.12 is inefficient.

Algorithm 8.12 and its modification can be proved correct by combining induction on the size of N and Theorems 8.2 and 8.3 (see Exercise 39).

8.5   ISOMORPHISM

Two graphs $G_A = (V_A, E_A)$ and $G_B = (V_B, E_B)$ are said to be *isomorphic*, written $G_A \cong G_B$, if there is a one-to-one, onto correspondence $f : V_A \to V_B$ such that $(v, w) \in E_A$ if and only if $(f(v), f(w)) \in E_B$—that is, there is a correspondence between the vertices of G_A and the vertices of G_B that preserves the adjacency relationships. For example, Figure 8.20 shows two isomorphic digraphs: vertices a, b, c, d, e, f in G_2 correspond to vertices 2, 3, 6, 1, 4, 5, respectively, in G_1. In general, there can be more than one correspondence between V_A and V_B, and, in fact, the graphs in Figure 8.20 have a second isomorphism, a, b, c, d, e, f corresponding, respectively, to 2, 3, 6, 1, 5, 4.

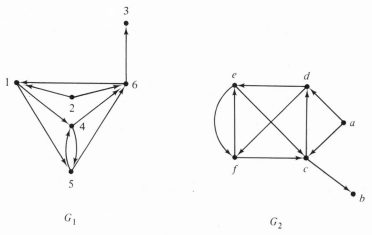

Figure 8.20   Isomorphic digraphs.

Isomorphic graphs differ only in the labeling of the vertices, and thus the problem of determining isomorphism occurs in a number of practical situations, such as information retrieval and identification of chemical compounds.

Notice that, as in the case of generating all cycles of a graph (Section 8.3.2), we can restrict our attention to digraphs. Any undirected graph can be transformed to a directed graph by replacing each edge with two oppositely directed edges. Two digraphs thus obtained are obviously isomorphic if and only if the original graphs are isomorphic.

The most straightforward method of checking for isomorphism of graphs is to use backtrack to examine all the $n!$ possible correspondences between the vertices of two n-vertex digraphs. When n is larger than about 13, this unconstrained backtrack is unsatisfactory because of the size of $n!$. However, with suitable pruning of the search tree, this method forms the basis of a reasonable digraph isomorphism algorithm

Define the *indegree* (*outdegree*) of a vertex in a digraph as the number of edges entering (leaving) the vertex. Obviously, a vertex v_1 in G_1 cannot be mapped by an isomorphism to the vertex v_2 in G_2 unless both indegree $(v_1) =$ indegree (v_2) and outdegree $(v_1) =$ outdegree (v_2). So in Figure 8.20 an isomorphism is limited to correspondences between the following subsets of vertices:

$$\{3\} \leftrightarrow \{b\}$$

$$\{2\} \leftrightarrow \{a\}$$

$$\{1, 4, 5\} \leftrightarrow \{d, e, f\}$$

$$\{6\} \leftrightarrow \{c\}$$

Thus only $3! = 6$ correspondences instead of $6! = 720$ need be checked, a considerable savings. It is this approach that we use in developing a digraph isomorphism algorithm.

Suppose that we are given two digraphs, $G_X = (V_X, E_X)$ and $G_Y = (V_Y, E_Y)$, to be tested to determine if $G_X \cong G_Y$. We assume that $V_X = V_Y = \{1, 2, \ldots, n\}$, for if $|V_X| \neq |V_Y|$, the digraphs cannot possibly be isomorphic. Let one of the digraphs, say G_X, be selected as the reference digraph. Let $G_X(k)$ denote the subgraph of G_X induced by the vertices $\{1, 2, \ldots, k\}$, $0 \leq k \leq n$. Clearly, $G_X(0)$ is the empty subgraph and $G_X(1)$ is the subgraph consisting of a single vertex 1 and no edges.

In determining whether $G_X \cong G_Y$, we use a backtrack technique. Obviously, $G_X(0)$ is isomorphic to the empty subgraph of G_Y. Suppose that at some point we have found a subgraph of G_Y consisting of vertices $S \subseteq V_Y$ that is isomorphic to $G_X(k)$. We try to extend the isomorphism to $G_X(k + 1)$ by choosing a vertex $v \in V_Y - S$ to correspond to $k + 1 \in V_X$. If we find such a v, we record the correspondence by setting $f_{k+1} \leftarrow v$ and try to extend the isomorphism to $G_X(k + 2)$. If there is no such v, we back up to $G_X(k - 1)$ and try to choose a different vertex to correspond to $k \in V_X$. This process continues until either we find an isomorphism between $G_X(n) = G_X$ and G_Y or we back up to $G_X(0)$ and conclude that $G_X \not\cong G_Y$.

The process is shown in Algorithm 8.13. The procedure *MATCH* used there returns the value **true** if k can be matched with v and **false** otherwise. The remainder of this section describes the preliminary work necessary to make the implementation of *MATCH* easy.

> *flag* ← **false**
>
> $k \leftarrow 0$
>
> *ISOMORPH* ($\varnothing$)
>
> **if** *flag* **then** $G_X \cong G_Y$, the correspondence is $i \leftrightarrow f_i$
>
> > **else** $G_X \not\cong G_Y$
>
> **procedure** *ISOMORPH* (*S*)
>
> > $k \leftarrow k + 1$
> >
> > **if** $S = V_Y$ **then** *flag* ← **true**
> >
> > **for** $v \in V_Y - S$ **while not** *flag* **do if** *MATCH* **then** $\begin{cases} f_k \leftarrow v \\ \textit{ISOMORPH } (S \cup \{v\}) \end{cases}$
> >
> > $k \leftarrow k - 1$
> >
> > **return**
>
> **procedure** *MATCH*
>
> > ⟦returns **true** if vertex $v \in V_Y - S$ can be matched to vertex $k \in V_X$. See Exercise 43.⟧
> >
> > **return**

Algorithm 8.13   Digraph isomorphism.

We begin by assuming that G_X and G_Y are given as adjacency structures Xadj and Yadj respectively. We compute the indegree and outdegree of the vertices, concatenating them into a single integer for convenience:

$$\text{degree} = 10^t \cdot \text{outdegree} + \text{indegree},$$

where $t = \lceil \log_{10}(n + 1) \rceil$. Let *Xdeg* (*i*) and *Ydeg* (*i*) denote the composite numbers for the *i*th vertices in G_X and G_Y, respectively. Notice that *Xdeg* (*i*) = *Ydeg* (*j*) if and only if vertex *i* in G_X and vertex *j* in G_Y have the same indegree and outdegree. Notice that the multisets $\{Xdeg\,(i)\,|\,1 \leq i \leq n\}$ and $\{Ydeg\,(i)\,|\,1 \leq i \leq n\}$ must be equal if $G_X \cong G_Y$.

In order to make our search tree as narrow as possible at the top (see Section 4.1.2), we want as few vertices as possible in G_Y to correspond to $G_X(1)$. For this reason, we renumber the vertices of G_X by performing a depth-first search on G_X, starting at a vertex $x \in V_X$ for which there are the fewest $v \in V_X$ such that $Xdeg\ (x) = Xdeg\ (v)$. From now on we will consider only this renumbered version of G_X, discarding the original labels on the vertices. The relabeled vertices have the additional desirable property that the vertex currently to be matched is usually adjacent to previously matched vertices; this fact limits the possible number of correspondences, thereby shortening the search.

As the edges are encountered during the depth-first search on G_X, we will arrange them in an array indicating the number of the vertex from which the particular edge originates and the vertex at which the edge terminates. In addition to the edges originally in G_X, this array will also contain "fictitious" edges going to the roots of the trees in the DFS-forest from a fictitious vertex 0. The edges are then sorted so that all edges belonging to the subgraph $G_X(k)$ appear before edges incident on vertex $i > k$.

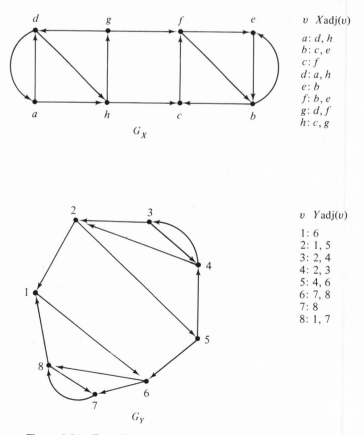

v	Xadj(v)
a:	d, h
b:	c, e
c:	f
d:	a, h
e:	b
f:	b, e
g:	d, f
h:	c, g

G_X

v	Yadj(v)
1:	6
2:	1, 5
3:	2, 4
4:	2, 3
5:	4, 6
6:	7, 8
7:	8
8:	1, 7

G_Y

Figure 8.21 Two digraphs and their adjacency structures.

Consider the example in Figure 8.21, which shows two digraphs G_X and G_Y and their adjacency structures. The outdegrees, indegrees, and composite degrees of the vertices are

G_X:

vertex	a	b	c	d	e	f	g	h
outdegree	2	2	1	2	1	2	2	2
indegree	1	2	2	2	2	2	1	2
Xdeg	21	22	12	22	12	22	21	22

G_Y:

vertex	1	2	3	4	5	6	7	8
outdegree	1	2	2	2	2	2	1	2
indegree	2	2	1	2	1	2	2	2
Ydeg	12	22	21	22	21	22	12	22

Since multisets consisting of the composite degrees are the same, the digraphs might be isomorphic.

We do a depth-first search on G_X to relabel the vertices; and because there are two vertices with Xdeg = 12, two with Xdeg = 21, and four vertices with Xdeg = 22, we begin the depth-first search at one of the vertices c, e, a, or g, for doing this will result in the fewest correspondences between $G_X(1)$ and subgraphs of G_Y. We arbitrarily choose c; the resulting DFS-tree is shown in Figure 8.22 with the old labels in parentheses. The figure shows tree edges in boldface and back edges, cross edges, and forward edges as broken lines. The two remaining edges connect the roots of the trees in the DFS-forest to the fictitious vertex labeled 0.

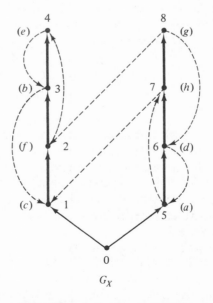

Figure 8.22 Digraph G_X after depth-first search.

As the edges are encountered during the depth-first search, they are added to the arrays *from* and *to*, each of size $|E_X| + t$, where t is the number of trees in the DFS-forest. The ith edge encountered goes from the vertex $from_i$ to the vertex to_i, $0 \leq i \leq n$. For the example in Figure 8.22, we have

i	1	2	3	4	5	6	7	8	9	10	11	12	13	14	15	16
$from_i$	0	1	2	3	3	4	2	0	5	6	6	7	7	8	8	5
to_i	1	2	3	1	4	3	4	5	6	5	7	1	8	6	2	7

These edges are then sorted into nondecreasing order according to $graph_i$, where $graph_i = \max(from_i, to_i)$ is the least k such that the edge $(from_i, to_i)$ is in $G_X(k)$. For the edges above, the reordered array is

i	1	2	3	4	5	6	7	8	9	10	11	12	13	14	15	16
$from_i$	0	1	2	3	3	4	2	0	5	6	6	5	7	7	8	8
to_i	1	2	3	1	4	3	4	5	6	5	7	7	1	8	6	2
$graph_i$	1	2	3	3	4	4	4	5	6	6	7	7	7	8	8	8

We are now ready to describe the procedure *MATCH* in Algorithm 8.13, which determines if the vertex $v \in V_Y$ can correspond to the vertex $k \in V_X$. Since we found an isomorphic copy of $G_X(k - 1)$ in G_Y, we know that, for all $i, j < k$, (i, j) is an edge of $G_X(k - 1)$ if and only if (f_i, f_j) is an edge of G_Y. The vertex v in G_Y can correspond to the vertex k in G_X, provided that (i, k) is an edge of $G_X(k)$ if and only if (f_i, v) is an edge of G_Y and, similarly, (k, i) is an edge of $G_X(k)$ if and only if (v, f_i) is an edge of G_Y. This is easy to check, since we have the arrays *from*, *to*, *graph*, and *f*. We leave the details of *MATCH* to Exercise 43.

The behavior of Algorithm 8.13 is obviously $O(n!)$ in the worst case, for the tree pruning may be fruitless and all (or almost all) the $n!$ possible correspondences may need to be tested.

★8.6   PLANARITY

A graph is said to be *planar* if there exists a mapping of its vertices and edges into the plane such that (a) each vertex v is mapped into a distinct point v' in the plane, and (b) each edge (v, w) is mapped onto a simple curve having endpoints (v', w'), so that (c) the mappings of edges meet only at common endpoints. The problem of determining whether a graph can be drawn on a plane without any edges crossing is of great practical interest (e.g., in designing an integrated circuit or a printed-circuit board, it is necessary to find out whether the resulting network can be embedded in a plane).

Determining the planarity of a graph differs from the other problems discussed in this chapter in that drawing points and lines in the plane involves continuous, rather than discrete, quantities. The interplay between the discrete and continuous aspects of planarity has intrigued mathematicians and has led to various characterizations of planar graphs (e.g., Exercise 45). These characteristics are mathematically elegant, but they do not lead to efficient algorithms for determining planarity. The most successful approach to determining planarity is simply to break a graph into subgraphs and then attempt an embedding in the plane by adding these subgraphs one by one while keeping the embedding planar.

Before proceeding, let us make a few simple but useful observations. Since a digraph is planar if and only if the undirected graph obtained by ignoring the direction of edges is planar, we need consider only undirected graphs. Since an undirected graph is planar if and only if all its connected components are planar, we need consider only connected, undirected graphs. Furthermore, it is easy to see that an undirected graph is planar if and only if all its biconnected components are planar. Thus if an undirected graph is separable, we can decompose it into its biconnected components and consider them individually. Finally, since parallel edges and self-loops can always be added or deleted without affecting planarity of a graph, we need consider only simple graphs. So in testing planarity, we will assume the graph to be undirected, simple, and biconnected.

It is not difficult to draw the complete graph of four vertices in the plane without crossing edges; consequently, a nonplanar graph must have at least five vertices. Similarly, it can be shown that a nonplanar graph has at least nine edges (see Exercise 45). In general, we can use Euler's formula relating the number of regions, vertices, and edges of a planar graph,

$$\text{regions} + \text{vertices} = \text{edges} + 2,$$

to derive the fact that a graph with $n > 2$ vertices and more than $3n - 6$ edges can never be planar (Exercise 44).

We will assume, therefore, that our planarity algorithm begins by using Algorithm 8.5 to decompose a graph into its biconnected components and by determining which such components satisfy $|V| \geqq 5$ and $9 \leqq |E| \leqq 3|V| - 6$. The planarity algorithm must then consider only these biconnected graphs with $O(|V|)$ edges. The algorithm for determining planarity in this case is quite complex, and the remainder of this section is a step-by-step development.

Our basic strategy is first to find a cycle C in G, to embed C on the plane as a simple closed curve, to decompose the remaining graph $G - C$ into edge-disjoint paths, and then to attempt to embed each of these paths either entirely inside C or entirely outside C. If we succeed in embedding the entire graph G, the graph is planar; otherwise it is nonplanar. The difficulty of this method is that in embedding the paths we can choose either the inside of C or the outside of C, and we must make sure that an early wrong choice does not preclude embedding a path later on, thereby forcing us to assert incorrectly that a planar graph is nonplanar. For example, in Figure 8.23(a) suppose that the starting cycle was (a, b, c, d, e, a) and that (e, b) is embedded outside whereas (b, f, e) and (f, c) are embedded inside as shown. Now we find that the last edge (a, d) can be added neither inside nor outside without a crossing. It would be wrong to conclude that G is nonplanar because a different embedding, Figure 8.23(b), shows G to be planar. Thus it is necessary to be able to generate paths systematically, to choose

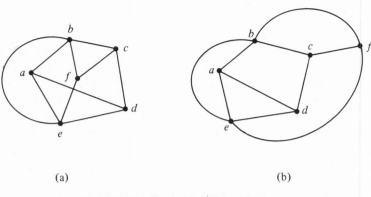

(a)   (b)

Figure 8.23   Two embeddings of a graph.

appropriate areas for embedding them, and, perhaps, to rearrange already embedded paths to accommodate new ones.

Path Generation.   In order to generate the paths in a certain desired order, we will reorder the vertices and the adjacency lists. The precise purpose of this reordering will be clear later.

First we perform a depth-first search on the given graph G (which is simple and biconnected), converting it into a digraph $\vec{G}$ that consists of $|V| - 1$ tree edges and $|E| - |V| + 1$ back edges, as discussed in Section 8.2.2. From now on we will ignore the original labels of the vertices and identify the vertices by their *num* values. This depth-first search and vertex relabeling are important because of the simple structure of paths in $\vec{G}$. As in Section 8.2.3, let *lowpt* (v) be the lowest numbered vertex reachable from vertex v or from any of its descendants (in the DFS-tree) by means of at most one back edge. When it is not possible to reach below v by means of a single back edge, v itself is *lowpt* (v). Similarly, let *nexlopt* (v) be the next lowest vertex below v, excluding *lowpt* (v), that can be reached in this manner. If there is no such vertex, *nexlopt* (v) is made equal to v. More precisely, if S_v is the set of all vertices lying on any path from v consisting of zero or more tree edges and ending with at most one back edge, then

$$lowpt\ (v) = \min\ (S_v)$$

and

$$nexlopt\ (v) = \min\ (\{v\} \cup (S_v - \{lowpt\ (v)\}))$$

Figure 8.24 shows a graph G converted into $\vec{G}$, with its vertices renumbered. The same digraph $\vec{G}$ is shown in Figure 8.25, in which the old vertex labels have been dropped and their *num* values are used as vertex names. The *lowpt* and *nexlopt* values are shown as an ordered pair (*lowpt* (v), *nexlopt* (v)) next to each vertex v. Notice that since G is biconnected, Theorem 8.1 guarantees that $v \geq nexlopt\ (v) > lowpt\ (v)$ except when v is the root—that is, when $v = 1$. When v is the root, we have *lowpt* $(v) = nexlopt\ (v) = v = 1$.

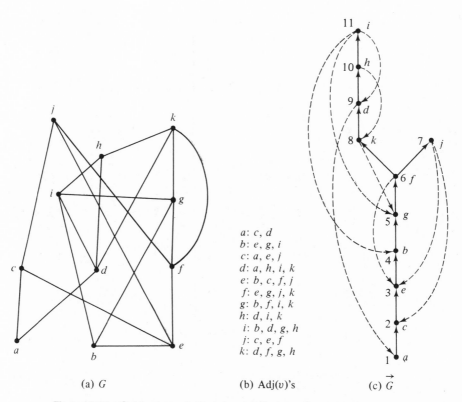

a: c, d
b: e, g, i
c: a, e, j
d: a, h, i, k
e: b, c, f, j
f: e, g, j, k
g: b, f, i, k
h: d, i, k
i: b, d, g, h
j: c, e, f
k: d, f, g, h

(a) *G*   (b) Adj(*v*)'s   (c) $\vec{G}$

Figure 8.24 Original graph *G* given as adjacency lists, converted into digraph $\vec{G}$ through depth-first search. Vertex numbers in $\vec{G}$ are *num* values of the vertices.

The next step is to reorder the adjacency lists of $\vec{G}$ so that during a depth-first search on $\vec{G}$ the paths in $\vec{G}$ will be generated in a certain desirable order. For this purpose, we compute an integer-valued function ϕ for each edge (v, w) in $\vec{G}$:

$$\phi[(v, w)] = \begin{cases} 2w & \text{if } (v, w) \text{ is a back edge.} \\ 2\, lowpt\,(w) & \text{if } (v, w) \text{ is a tree edge and } nexlopt\,(w) \geq v. \\ 2\, lowpt\,(w) + 1 & \text{if } (v, w) \text{ is a tree edge and } nexlopt\,(w) < v. \end{cases}$$

Then for each vertex v we sort all edges (v, w) into nondecreasing order according to the ϕ values and use this order in the adjacency lists.† The new adjacency lists

---

† The sorting can be done in $O(|V| + |E|)$ operations by a distribution sort of the type described in Section 7.1.4.

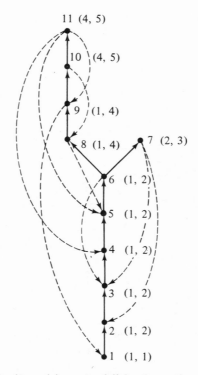

Figure 8.25 The (*lowpt*(v), *nexlopt*(v)) for the vertices of the graph in Figure 8.23(b).

for $\vec{G}$ will be called Padj (v)'s, the *properly ordered adjacency lists*. Figure 8.26 shows the ϕ values of the edges of $\vec{G}$ of Figure 8.25, and Figure 8.27 shows the Padj (v)'s. Notice that in the reordered adjacency lists a back edge going to a lower vertex always precedes a back edge going to a higher vertex; and, roughly speaking, tree edges (v, w) appear in nondecreasing order of their abilities to lead to a vertex below v through a single back edge.

Having obtained the properly ordered adjacency lists representing digraph $\vec{G}$, we now apply a depth-first search to decompose $\vec{G}$ into one cycle C and a number of edge-disjoint paths p_i. Starting at vertex 1, we continue the path with tree edges until we encounter a vertex z such that the first edge in list Padj (z) is a back edge (which, as we will see, goes to vertex 1). This back edge, together with the path formed so far (from 1 to z), forms the first path p_0, the cycle C. Next, we start from z and begin a path with the next edge out of z. Each time that we traverse a tree edge we continue building the current path: when we traverse a back edge, it becomes the last edge of the current path. Thus each path consists of a sequence of zero or more tree edges followed by a single back edge. A new path is started from the initial vertex of the last back edge; if this vertex has no more unexplored edges, we back up to the previous vertex on the last path. The process is continued until $\vec{G}$ has no more untraversed edges. Algorithm 8.14 does this

Edge (v, w)	$\phi[(v, w)]$
Tree edges $\Big\{$	
(1, 2)	2
(2, 3)	2
(3, 4)	3
(4, 5)	3
(5, 6)	3
(6, 7)	5
(6, 8)	3
(8, 9)	3
(9, 10)	9
(10, 11)	9
Back edges $\Big\{$	
(9, 1)	2
(7, 2)	4
(6, 3)	6
(7, 3)	6
(11, 4)	8
(8, 5)	10
(11, 5)	10
(10, 8)	16
(11, 9)	18

Figure 8.26 ϕ values for the edges of digraph $\bar{G}$ in Figure 8.24.

cycle/path decomposition. Figure 8.27 shows how this algorithm decomposes the digraph of Figure 8.25 (represented by the properly ordered adjacency lists shown in Figure 8.26) into a cycle $C = p_0$ and paths $p_1, \ldots, p_8$.

$i \leftarrow 0$

$p_i \leftarrow \varnothing$

$PATH\,(1)$

procedure $PATH\,(v)$

$\qquad$ **for** $w \in \text{Padj}\,(v)^\dagger$ **do** $\begin{cases} p_i \leftarrow p_i \cup \{(v, w)\} \\[1em] \textbf{if } v < w \textbf{ then} \begin{cases} [\![(v, w) \text{ is a tree edge}]\!] \\[0.5em] PATH\,(w) \end{cases} \\[2em] \textbf{else} \begin{cases} [\![(v, w) \text{ is a back edge}]\!] \\[0.5em] i \leftarrow i + 1 \\[0.5em] p_i \leftarrow \varnothing \end{cases} \end{cases}$

$\qquad$ **return**

Algorithm 8.14 Decomposing a digraph represented by properly ordered adjacency lists into a cycle p_0 and paths $p_1, p_2, \ldots$.

† By use of Padj (v) instead of Adj (v), we mean to imply that the edges are considered in increasing order by ϕ value.

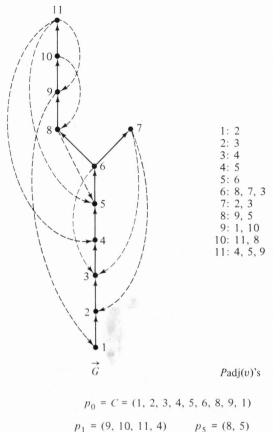

$$p_0 = C = (1, 2, 3, 4, 5, 6, 8, 9, 1)$$

$p_1 = (9, 10, 11, 4)$ $p_5 = (8, 5)$
$p_2 = (11, 5)$ $p_6 = (6, 7, 2)$
$p_3 = (11, 9)$ $p_7 = (7, 3)$
$p_4 = (10, 8)$ $p_8 = (6, 3)$

Figure 8.27   Digraph $\vec{G}$, its properly ordered adjacency lists Padj(v)'s,
and its decomposition into a cycle p_0 and paths $p_1, \ldots, p_8$.

Path Properties.   For a given graph, the decomposition into a cycle $C = p_0$ and a
sequence of edge-disjoint paths $p_1, p_2, \ldots, p_m$ is not generally unique, but the
number of such paths is always $m = |E| - |V|$, since each path p_i contains exactly
one back edge. The cycle/path decomposition also has a number of other
invariant properties that are important to the planarity algorithm.

Let the cycle generated be $C = p_0 = (v_1, v_2, \ldots, v_k, v_1)$. We know that
$v_1 = 1$ and that $v_1 < v_2 < \cdots < v_k$ (Exercise 49). It is also clear from the
depth-first method of path generation that every path p_i has only its two end
vertices in common with the union of all previously generated paths—that is, with
subgraph $p_0 \cup p_1 \cup p_2 \cup \cdots \cup p_{i-1}$ (why?). Later we will see that all we actually need to know are the two end vertices of each of the paths $p_1, p_2, \ldots, p_m$ and
not the paths themselves.

In the procedure *PATH*, each path ends with a back edge. From among all the back edges available at a particular vertex, Algorithm 8.14 always selects the unused back edge that leads to the lowest vertex. This follows from the fact that Padj (v)'s are each in nondecreasing order of ϕ values, and $\phi[(v, w)]$ of a back edge (v, w) is $2w$. It also explains why path $(11, 5)$ is generated before $(11, 9)$ in Figure 8.26.

The following properties of the paths generated are due to the special ordering of Padj (v)'s.

Property 1

Let p_i, $i \geq 1$, be a generated path going from s_i (its start vertex) to f_i (its finish vertex). If we consider all the back edges that have not been used in any path when the first edge in path p_i is traversed, then f_i is the lowest vertex reachable from s_i by a sequence of tree edges and any one of these back edges. Furthermore, if v is an intermediate vertex in path p_i (i.e., if $v \in p_i$, $v \neq s_i$, $v \neq f_i$), then f_i is the lowest vertex in the DFS-tree reachable from v or any of its descendants by a sequence of tree edges and *any* back edge in $\vec{G}$.

Proof: This property follows from the fact that all edges from v and its descendants are untraversed when the first edge in p_i is traversed, and from the reordering of the edges according to the ϕ values.

Property 2

Let p_i and p_j, $j > i \geq 1$, be two generated paths with start and finish vertices s_i, s_j, and f_i, f_j, respectively. If s_i is an ancestor (not necessarily proper) of s_j, then $f_i \leq f_j$.

Proof: This property is true because the back edge ending p_j was unused when p_i was being generated, and, from Property 1, p_i takes the back edge reaching farthest down in $\vec{G}$.

Property 3

Let p_i and p_j, $j > i \geq 1$, be two generated paths having the same pair of start and finish vertices s and f. Let x_i be the second vertex in path p_i and x_j be the second vertex in path p_j (the first vertex in each path being the common start vertex s). If edge (s, x_i) is not a back edge (i.e., $x_i \neq f$) and *nexlopt* $(x_i) < s$, then (s, x_j) is not a back edge and *nexlopt* $(x_j) < s$.

Proof: From Property 1, $f = lowpt(x_i)$. Since (s, x_i) is a tree edge $(x_i \neq f)$ and *nexlopt* $(x_i) < s$, its ϕ value is, by definition,

$$\phi[(s, x_i)] = 2\,lowpt\,(x_i) + 1 = 2f + 1.$$

Since p_j is generated after p_i, vertex x_j must appear later than x_i in the properly ordered adjacency list Padj (s). Hence we must have

$$\phi[(s, x_j)] \geq \phi[(s, x_i)] = 2f + 1$$

Thus (s, x_j) cannot be a back edge and so $x_j \neq f$. Also, *nexlopt* $(x_j) < s$, for otherwise we would have

$$\phi[(s, x_j)] = 2 \ lowpt \ (x_j) = 2f,$$

a contradiction.

The usefulness of these properties will become clear later on. Notice that it is Property 3 that requires us to compute the *nexlopt* values. These values allow us to break a tie between two tree edges (u, v) and (u, w), starting from a common vertex u, when there are identical *lowpt* values for v and w.

Segments.   When the edges of the cycle $C = p_0$ are removed from $\vec{G}$, the edges in the remaining digraph $\vec{G} - C$ will fall into one or more connected pieces. Each of these connected pieces will consist of one or more segments, defined as follows. A *segment* (of $\vec{G}$ with respect to cycle C) is either (a) a single back edge (v_i, w) not on C but where $v_i, w \in C$, or (b) a subgraph consisting of a tree edge (v_i, w), $v_i \in C$, $w \notin C$, and the directed subtree rooted at w, together with all the back edges from this subtree. The vertex v_i in C at which a segment originates is called the *base vertex* of the segment. Figure 8.28 shows the initial circuit C and the four segments of the digraph of Figure 8.27.

Each segment is a connected subgraph in $G - C$, but not every connected component of $G - C$ is a segment, since two or more segments may have a common base vertex, as do segments S_3 and S_4 in Figure 8.28. The procedure

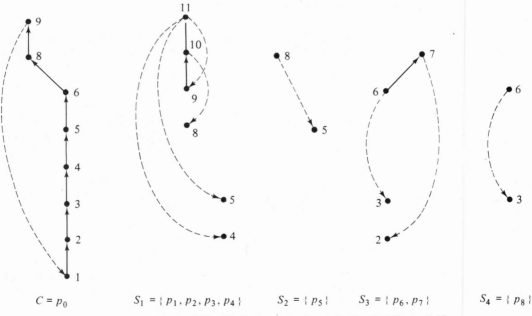

$$C = p_0 \qquad S_1 = \{ p_1, p_2, p_3, p_4 \} \qquad S_2 = \{ p_5 \} \qquad S_3 = \{ p_6, p_7 \} \qquad S_4 = \{ p_8 \}$$

Figure 8.28   Cycle C and segments in $G - C$ of Figure 8.27.

PATH generates segments in decreasing order of their base vertices, and all paths in one segment are generated before paths in the next segment. Clearly, all paths belonging to a segment must be embedded together—either all inside C or all outside C. This is the reason for grouping paths into segments.

We will see later that, while embedding a segment, we may apply the algorithm recursively and generate segments within a segment (with respect to another cycle). Thus the concept of a segment is crucial for understanding the planarity algorithm.

Embedding.   The cycle C consisting of edges $(1, v_2), (v_2, v_3), \ldots, (v_{k-1}, v_k)$, $(v_k, 1)$ is embedded in the plane as a simple closed curve and then the segments are embedded. C divides the plane into an inside and an outside. A segment whose first edge is (v_i, w) is said to be embedded *inside C* if, while going clockwise around the base point v_i in the plane, the order of the edges encountered is (v_{i-1}, v_i), (v_i, w), (v_i, v_{i+1}). The segment is said to be embedded *outside C* if this order is (v_{i-1}, v_i), (v_i, v_{i+1}), (v_i, w).[†] (See Figure 8.29.) A back edge (x, v_j) belonging to a segment embedded on the inside is said to be entering C at vertex v_j from

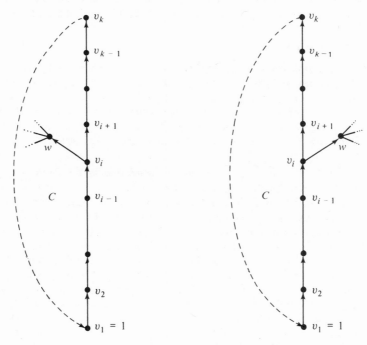

A segment embedded inside C.   A segment embedded outside C.

Figure 8.29 Segments embedded inside and outside C.

[†]The subscripts $i - 1$ and $i + 1$ are taken modulo k: if $v_i = v_k$, then $v_{i+1} = v_1$; and if $v_i = v_1$, then $v_{i-1} = v_k$.

inside. Clearly, the order of edges as we go clockwise around v_j will be (v_{j-1}, v_j), (x, v_j), (v_j, v_{j+1}). Similarly, for a back edge (x, v_j) entering C at v_j from outside, this order will be (v_{j-1}, v_j), (v_j, v_{j+1}), (x, v_j).

After embedding C, we attempt to embed the segments of $\bar{G} - C$, one at a time, in the order in which they are generated by the procedure *PATH*. To embed a segment S, we consider the first path p in S generated by *PATH* and choose a side of C, say the inside, on which to embed p. By examining the previously embedded paths, we can determine whether p can be embedded on the inside of C. If p can be so embedded, we embed it; otherwise all those segments previously embedded on the inside of C that are blocking the embedding of p are moved to the outside of C. Bringing these segments outside may force some other segments from outside to inside and so forth. If p still cannot be embedded inside C, even after this rearrangement of segments, then G is nonplanar. If p can be embedded inside C, we embed it and then try to embed the rest of the segment S by applying the embedding algorithm recursively. If successful, we continue with the next segment.

In order to do this efficiently, we need an easily tested criterion for determining whether the first path of a segment can be embedded on a specific side of C at a given stage in the embedding. The following theorem provides such a criterion.

Theorem 8.4

Let p be the first path in the current segment S and suppose that p goes from v_i (the base vertex of S) to v_j (another vertex in C). If all segments generated before S have already been embedded, then p can be embedded on the inside of C if there is no previously embedded back edge (x, v_t) entering C from inside between vertices v_i and v_j—that is, no back edge (x, v_t) embedded inside satisfies $v_j < v_t < v_i$. Furthermore, if there is such a back edge (x, v_t), then S cannot be embedded on the inside of C.

Proof: Since the segments are generated in the decreasing order of their base vertices, none of the edges embedded so far can be leaving any of the vertices below (smaller than) v_i in C. Thus if no back edge enters C from inside between v_i and v_j, nothing prevents us from embedding p on the inside of C by placing p sufficiently close to C.

Suppose that there is an embedded back edge (x, v_t) entering C from the inside such that $v_j < v_t < v_i$. This back edge must belong to a previously generated segment, say S', which is embedded inside C. Let the base vertex of this segment S' be v_q. [Segment S' might be a single back edge (x, v_t), in which case $v_q = x$.] Since S' was generated earlier than S, then $v_q \geq v_i$. If $v_q > v_i$, clearly the sequence of edges in S' going from v_q to v_t will prevent the embedding of p on the inside of C [see Figure 8.30(a)].

The other possibility is $v_q = v_i$, that is, segments S' and S have a common base vertex. In this case, let us consider the first path generated in segment S', say p'. Path p' goes from v_q to v_r (both v_q and v_r being in C). From Property 2 of the paths it follows that $v_r \leq v_j$. Thus vertex v_r is distinct from v_t, and there are at least two paths in segment S', one entering C at v_t and the other at v_r. Therefore the first path p' in S' cannot consist solely of a back edge.

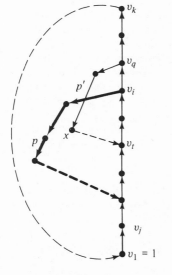

(a) $v_q > v_i$

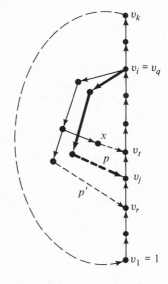

(b) $v_q = v_i$ and $v_r < v_j$

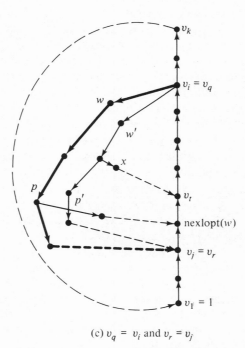

(c) $v_q = v_i$ and $v_r = v_j$

Figure 8.30 Embedding path p going from v_i to v_j (in boldface).

Next, if $v_r < v_j$ [see Figure 8.30(b)], it is clear that path p cannot be embedded inside C. This leaves only one possibility—namely, paths p and p' have both their end vertices in common. That is, $v_i = v_q$ and $v_j = v_r$. Property 3 of the paths will handle this case. Let w and w' be the second vertices on paths p and p', respectively. Since the first edge in p' is not a back edge, we know that $w' \neq v_r$. Furthermore, there are at least two back edges leading from S' and entering C below $v_q (= v_i)$—namely, at v_r and v_t—so that $nexlopt(w') < v_q$. Thus from Property 3 we know that $w \neq v_i$ (i.e., path p is more than just a back edge) and $v_j < nexlopt(w) < v_i$. In other words, there are at least two back edges in segment S (leading from w or its descendants), one going to v_j and the other going to vertex $nexlopt(w)$, which lies above v_j but below v_i. The sequence of edges from v_i to $nexlopt(w)$ in S, together with path p, blocks the embedding of segment S on the inside of C [see Figure 8.30(c)].

As a result of this theorem, it is clear that the names of the start and finish vertices of the paths are sufficient to test planarity, since the embedability of a path is determined by its end vertices.

Recursion.   So far we have considered embedding only the first path p in a segment S. We must also determine whether the rest of the segment, that is, $S - p$, can be added to the planar embedding. To do this, we observe that the subgraph $S - p$ can be added to the planar embedding (which now consists of the cycle C, all segments generated before S, and path p) if and only if subgraph $\hat{G} = S \cup C$ is planar (Exercise 50). Thus we are faced with determining the planarity of the subgraph $\hat{G}$. We can determine it by applying the embedability criterion recursively to $\hat{G}$. For this recursion, the path p together with the tree edges in the path from f (the finish vertex of p) to s (the start vertex of p) will serve as the initial cycle $\hat{C}$ of $\hat{G}$. Removing cycle $\hat{C}$ from $\hat{G}$ may further partition the remaining digraph $\hat{G} - \hat{C}$ into segments (of $\hat{G}$ with respect to $\hat{C}$), which, in turn, are handled by recursion. This process is continued until all paths in segment S are embedded in the plane or some path (the first path of a segment of a segment of a segment . . .) cannot be embedded.

As a simple example, consider segment S_1 in Figure 8.28. Its first path $p = (9, 10, 11, 4)$ with $s = 9$ and $f = 4$. We form graph $\hat{G} = S_1 \cup C$, which consists of 14 edges, and examine the planarity of $\hat{G}$. The initial cycle of $\hat{G}$ is $\hat{C} = (4, 5, 6, 8, 9, 10, 11, 4)$. The remaining digraph $\hat{G} - \hat{C}$ consists of seven edges, out of which four are part of C. Each of the remaining three edges $(11, 5)$, $(11, 9)$, and $(10, 8)$ is a back edge and constitutes a segment (in $\hat{G}$ with respect to $\hat{C}$) by itself. Another level of recursion is not necessary.

Data Structures.   When embedding a segment that starts at a vertex v_i (in C), we need to know which vertices in C below v_i (smaller than v_i) have back edges entering from either the inside or the outside. With this information we can use Theorem 8.4 to decide if the segment can be embedded. We will maintain two push-down stacks IST and OST for keeping track of vertices with back edges entering from the inside and outside respectively. We will see that these two stacks have the property that the vertices in them are in increasing order (highest vertex on the top).

When the first path in a segment S is generated, going from vertex s to vertex f (both on C), vertex f must be added on top of stack IST,[†] since we always try to embed a new segment on the inside. Similarly, during the recursive application of the algorithm, finish vertices of other paths in S must be added to the stacks. Observe that we need to add the finish vertex f of the current path to a stack if and only if f is greater than the finish vertex of the earliest path, say p, that passes through the start vertex of the current path. Otherwise if f is equal to the finish vertex of p, then f has already been added to a stack and is accounted for. Vertex f, of course, can never be smaller than the finish vertex of p.

As we back down the tree edge (v_{i-1}, v_i) during the generation (and embedding) of segments, we know that we have generated and successfully embedded all segments starting at (or above) vertex v_i. Since segments are generated in decreasing order of their base vertices (in C), no segment yet to be explored starts at a vertex greater than v_{i-1}. Thus by Theorem 8.4 no back edge entering vertex v_{i-1} or above can interfere with the embedding of the next segment. And so we can pop up the stacks IST and OST to remove all occurrences of vertex v_{i-1} as we back down the edge (v_{i-1}, v_i).

As a segment is moved from inside to outside, or vice versa, the corresponding entries in the stacks IST and OST must be shifted. This step may result in moving the same entries back and forth many times, and it is more efficient to move these entries in groups rather than individually. For this purpose, let us define a *bundle* as a maximal set of entries in IST and OST that correspond to back edges such that the embedding of one of the back edges determines the embedding of all others.

It is clear from the definition that the bundles change as the contents of the stacks change. However, the bundles always form a partition of the entries in the two stacks. Furthermore, from the order in which the paths are explored and from the fact that the entries to a stack for a new segment are always added at the top of the stack, all entries in each stack belonging to a specific bundle are adjacent.

Consider the embedding, on the inside of C, of a new segment S having the first path p from s to f. When all entries corresponding to back edges in S are added to IST, a new bundle B is formed that contains these entries for S as well as all those entries for back edges that interfere with S—that is, all vertices v_l satisfying $f < v_l < s$. In other words, every bundle that contains any entry v_l must be combined with the new entries to form the new bundle B. The vertices v in all other bundles must satisfy $v < f$. Thus every vertex z in either IST or OST

[†]Except when $f = 1$; then we know that embedding p will not interfere with the embedding of any subsequent path.

must satisfy two conditions.

1. If $f < z < s$, then z is in bundle B,

2. If $z > s$ or $z < f$, then z is not in bundle B.

In general, some members of the latest bundle B will be in IST and others will be in OST, but they will be on the top in both stacks. Keeping track of bundles is useful because embedding of any back edge belonging to a bundle B completely determines the embedding of all back edges in B, and embedding of a back edge from one bundle has no effect on the embedding of a back edge not in that bundle.

The information regarding bundles can be conveniently kept in a third stack called BST. Each entry on BST is an ordered pair (x, y), with x being the bottommost entry in IST belonging to that bundle and y being the bottommost entry in OST belonging to the same bundle. $x = 0$ implies that the particular bundle has no entries in IST, and $y = 0$ implies that the bundle has no entries in OST. Figure 8.31 shows the contents of the three stacks IST, OST and BST at various times during the application of the embedding procedure to the decomposed digraph in Figure 8.28. As we will see shortly, the entries (x, y) kept in BST are actually pointers to the entries in IST and OST rather than the entries themselves.

Complete Planarity Algorithm.   We are ready to give a detailed description of the planarity algorithm in its entirety. Using the recursive path-finding procedure described earlier, we generate one path at a time and attempt to embed it before generating the next path. Just after the first path p from vertex s to vertex f (both in the cycle C) in the first segment S is embedded on the inside of C, as described above, the top entry on stack IST is f. We now try to embed the rest of S recursively. To distinguish the stack entries added during the second level of recursion, we put an end-of-stack marker on top of stack OST. For IST, the top entry f itself serves as an end-of-stack marker because no back edge coming out of S can lead to a vertex smaller than f.

Suppose that the recursion is successfully completed on segment S and that we have just come back to the top level of recursion. Embedding S would, in general, add entries to the stacks IST and OST (above f and the end-of-stack marker, respectively). If there is a new bundle containing back edges entering on both sides of the cycle $\hat{C}$, formed by the path p and the portion of C from f to s, then we cannot add the rest of the cycle C to the planar embedding of $S \cup \hat{C}$. (See Figure 8.32.) In this case, the subgraph $S \cup C$ is nonplanar and therefore G is nonplanar. Conversely, if each new bundle has back edges entering the cycle $\hat{C}$ inside or outside but not on both, then outside entries can and must be moved inside so that the remaining portion of cycle C can be added on the outside of $\hat{C}$ (C must be outside of $\hat{C}$, since we embed S inside C).

In implementing the two stacks IST and OST, it is most convenient to keep them as linked lists, using two arrays called *stack* and *next*. *stack* (i) gives an

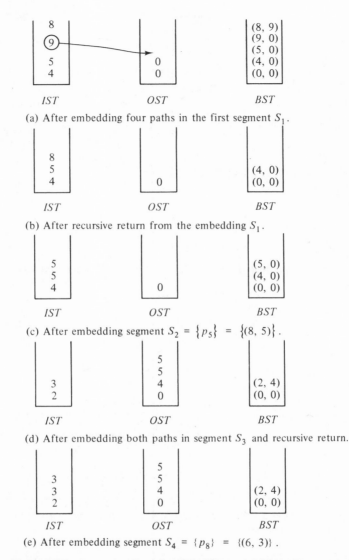

(a) After embedding four paths in the first segment S_1.

(b) After recursive return from the embedding S_1.

(c) After embedding segment $S_2 = \{p_5\} = \{(8, 5)\}$.

(d) After embedding both paths in segment S_3 and recursive return.

(e) After embedding segment $S_4 = \{p_8\} = \{(6, 3)\}$.

Figure 8.31 Contents of stacks *IST, OST,* and *BST* for Figure 8.27 at various stages.

actual stack entry, and *next* (*i*) points to the next (lower) entry on the same stack. Element *next* (0) points to the top entry on *IST* and *next* (− 1) points to the top entry on *OST*. The example in Figure 8.33 illustrates how the two stacks are implemented with the two arrays. The advantage of this representation is that, as we switch a bundle of entries from one stack to another, we need not actually move the entries; we need only switch the pointers at the beginning and end of the bundle. Since there may be much movement of the bundles back and forth, this

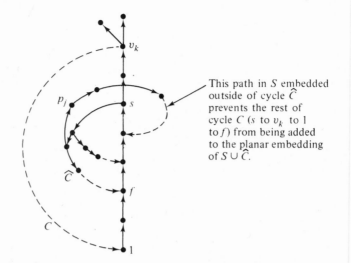

This path in S embedded outside of cycle $\widehat{C}$ prevents the rest of cycle C (s to v_k to 1 to f) from being added to the planar embedding of $S \cup \widehat{C}$.

Figure 8.32   Embedding of the first segment S together with cycle C.

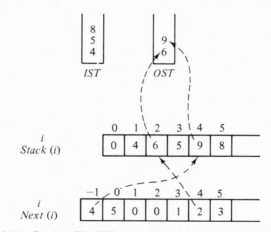

Figure 8.33   Stacks *IST, OST,* and their representation as two linear arrays *stack* and *next*.

representation is important. Observe that entries in the array *stack* need never be altered once they are initially made; only the entries in the array *next* are changed. With this implementation, the bundle stack *BST* contains pairs of points (x, y) pointing to the *stack* array.

In addition to the arrays *stack* and *next*, the bundle stack *BST*, and the Padj (v)'s, we use several other variables in the planarity algorithm. The integer variable p denotes the serial number of the current path. The array f of size $|E| - |V| + 1$ contains entries $f(t)$ denoting the finish vertex on the path numbered t. Another array, *path*, of size $|V|$, is such that *path* (v) denotes the number of the first path passing through vertex v. The integer variable s is used to keep track of the starting vertex of the current path; and when that path is completely embedded, *s* is reset to zero. Finally, the variable *avail* is used as the index of the first available location in the arrays *stack* and *next*.

An outline of the complete planarity procedure, which combines both the path-finding and the embedding procedure, is given in Algorithm 8.15. The various details are outlined below and given in Algorithms 8.16(a) through (f).

Step A.   After having explored all edges going out of vertex *w*, we back down to vertex *v*. While backing down the tree edge (v, w), we must delete all occurrences of vertices that are equal to or greater than *v*, since, according to Theorem 8.4, no back edge entering the cycle *C* at or above vertex *v* can interfere with the embedding of a segment starting at or below *v*. This deletion process is given in Algorithm 8.16(a).

Step B.   This step is executed only when we have embedded the entire segment whose first edge is (v, w), and we are backing down this edge. In this case, *path* (v), the first path passing through *v* (*path* $(v) = 1$ if $v \in C$), will be different from the first path passing through *w*. Observe that the base vertex of any segment is always contained in a path generated before the first path in the segment. As noted earlier, when an entire segment *S* (with respect to cycle *C*) has been successfully embedded, we must ensure that the portion of *C* not included in $\hat{C}$ can be added to the planar embedding of $S \cup \hat{C}$ (see Figure 8.32). To do this, we must move all bundles above the end-of-stack marker from to *and IST*. If any bundle has entries on both stacks *IST* and *OST*, we cannot make this shift and therefore *G* is nonplanar. The shifting of bundles from outside to inside is accomplished as shown in Algorithm 8.16(b).

Step C.   This step is executed repeatedly as long as the *IST* and *OST* are nonempty and their top entries interfere with embedding of the current path *p*, which ends in the back edge (v, w). If the top entries in both *IST* and *OST* are greater than *w*, then *p* cannot be embedded and *G* is nonplanar. If, however, only the top entry on *OST* is greater than *w*, that is, if *stack* $(next \, (-1)) > w$, then we can embed the path inside without any problem. If only the top entry on *IST* is greater than *w*, that is, if *stack* $(next \, (0)) > w$, then we move the entire bundle from *IST* to *OST* by interchanging two pairs of pointers. This procedure is given in detail in Algorithm 8.16(c).

Step D.   Vertex *s* is the start vertex of the current path *p*, and *path*(s) is the earliest path that contains vertex *s*. Unless *p* is the initial cycle, we have *path* $(s) < p$, and so either $f(path \, (s))$ has already been added to stack *IST* or it need not be added at all. Therefore the finish vertex $f(p) = w$ of the current path is to be added to the *IST* only if $w > f \, (path(s))$. Since *avail* is the index of the first available space in *next* and *stack* arrays, this step is accomplished as shown in Algorithm 8.16(d).

$next(-1) \leftarrow next(0) \leftarrow stack(0) \leftarrow s \leftarrow path(1) \leftarrow 0$

$avail \leftarrow 1$

$p \leftarrow -1$

$BST \leftarrow$ empty stack

$PLANAR(1)$

procedure $PLANAR(v)$

if $s = 0$ **then** $\begin{cases} \llbracket \text{new path starting at } v \rrbracket \\ s \leftarrow v \\ p \leftarrow p + 1 \end{cases}$

$\begin{cases} \llbracket \text{tree edge} \rrbracket \\ path(w) \leftarrow p \\ PLANAR(w) \\ \text{A: Delete stack entries corresponding to vertices no smaller than } v; \text{ see Algorithm 8.16(a).} \end{cases}$

if $v < w$ **then**

$\llbracket \text{all the segment with first edge } (v, w) \text{ has been embedded} \rrbracket$

if $path(w) \neq path(v)$ **then** $\begin{cases} \text{B: Move new bundles from outside to inside and combine the} \\ \text{new bundles into a single bundle; see Algorithm 8.16(b).} \end{cases}$

for $w \in \text{Padj}(v)$ **do**

$\llbracket \text{back edge—current path, } p \text{ is complete. Move bundles, if necessary, to embed it on the inside.} \rrbracket$

else $\begin{cases} f(p) \leftarrow w \\ in \leftarrow 0 \\ out \leftarrow -1 \\ \textbf{while } [next(in) \neq 0 \textbf{ and } stack(next)(in)) > w] \\ \quad \textbf{or } [next(out) \neq 0 \textbf{ and } stack(next(out)) > w] \textbf{ do} \begin{cases} \text{C: Move bundles so that } p \text{ can be embedded} \\ \text{on the inside; see Algorithm 8.16(c).} \end{cases} \\ \llbracket w > \text{finish vertex of the earliest path containing } s \rrbracket \\ \textbf{if } w > f(path(s)) \textbf{ then } \begin{cases} \text{D: Add } w \text{ to } IST; \text{ see Algorithm 8.16(d).} \end{cases} \\ \text{E: Add new bundle to } BST, \text{ if necessary, corresponding to combined old bundles; see Algorithm 8.16(e).} \\ \text{F: Add an end-of-stack marker to } OST \text{ if the current path is more than just a back edge;} \\ \quad \text{see Algorithm 8.16(f).} \\ s \leftarrow 0 \end{cases}$

Algorithm 8.15 Outline of the planarity algorithm. The details of the statements labeled A through F are given in Algorithms 8.16(a) through (f).

⟦delete vertices from *BST*⟧

$(x, y) \Leftarrow BST$

while $[(stack\ (x) \geqq v)\ \textbf{or}\ (x = 0)]$

$\qquad\qquad$ **and** $\quad [(stack\ (y) \geqq v)\ \textbf{or}\ (y = 0)]\ \textbf{do}\ (x, y) \Leftarrow BST$

if $stack\ (x) \geqq v$ **then** $x \leftarrow 0$

if $stack\ (y) \geqq v$ **then** $y \leftarrow 0$

⟦ delete entries from *IST* and *OST* ⟧

while $next\ (-1) \neq 0$ **and** $stack\ (next\ (-1)) \geqq v$ **do** $\begin{cases} ⟦\text{pop } OST⟧ \\ next\ (-1) \leftarrow next\ (next\ (-1)) \end{cases}$

while $next\ (0) \neq 0$ **and** $stack\ (next\ (0)) \geqq v$ **do** $\begin{cases} ⟦\text{pop } IST⟧ \\ next\ (0) \leftarrow next\ (next\ (0)) \end{cases}$

$BST \Leftarrow (x, y)$

Algorithm 8.16(a)   Deletion of stack entries.

⟦move new bundles from outside to inside⟧

$in \leftarrow 0$

$(x, y) \Leftarrow BST$

while $stack\ (x) > f(path\ (w))$ **or** $\left\{ \begin{array}{l} \textbf{if}\ stack\ (x) > f(path\ (w)) \\[4pt] \quad \textbf{then if}\ stack\ (y) > f(path)\ (w)\ \textbf{then}\ ⟦G\ \text{is nonplanar}⟧ \\ \qquad\qquad\qquad\qquad\qquad\qquad\qquad \textbf{else}\ in \leftarrow x \\[4pt] \quad \textbf{else} \begin{cases} ⟦\text{interchange}⟧ \\ temp \leftarrow next\ (in) \\ next\ (in) \leftarrow next\ (-1) \\ next\ (-1) \leftarrow next\ (y) \\ next\ (y) \leftarrow temp \\ in \leftarrow y \end{cases} \\[4pt] (x, y) \Leftarrow BST \end{array} \right.$

$\qquad\quad [stack\ (y) > f(path\ (w))]$

$\qquad$ **and** $stack\ (next\ (-1)) \neq 0]$ **do**

⟦combine topmost bundle on *BST* with the new bundles just deleted (those created by embedding the segment whose first edge was (v, w))⟧

if $x = 0$ **and** $(in \neq 0$ **or** $y \neq 0)$ **then** $BST \Leftarrow (in, y)$

$\qquad\qquad\qquad\qquad\qquad\qquad$ **else** $BST \Leftarrow (x, y)$

⟦delete end-of-stack marker from *OST*⟧

$next\ (-1) \leftarrow next\ (next\ (-1))$

Algorithm 8.16(b)   *Combining and moving bundles.*

$$(x, y) \Leftarrow BST$$

if $x \neq 0$ **and** $y \neq 0$ **then if** $stack\ ((next\ (in)) > w$ **then if** $stack\ (next\ (out)) > (w)$

$$\textbf{then}\,[\![\,G \text{ is nonplanar}\,]\!]$$

$$\textbf{else} \begin{cases} next\ (out) \leftrightarrow next\ (in) \\ next\ (x) \leftrightarrow next\ (y) \\ in \leftarrow y \\ out \leftarrow x \end{cases}$$

$$\textbf{else} \begin{cases} in \leftarrow x \\ out \leftarrow y \end{cases}$$

else if $x \neq 0$ **then** $\begin{cases} temp \leftarrow next\ (x) \\ next\ (x) \leftarrow next\ (out) \\ next\ (out) \leftarrow next\ (in) \\ next\ (in) \leftarrow temp \\ out \leftarrow x \end{cases}$

else if $y \neq 0$ **then** $out \leftarrow y$

Algorithm 8.16(c)   Moving bundles.

$$stack\ (avail) \leftarrow w$$

$$next\ (avail) \leftarrow next\ (0)$$

$$next\ (0) \leftarrow avail$$

if $in = 0$ **then** $in \leftarrow avail$

$$avail \leftarrow avail + 1$$

Algorithm 8.16(d)   Add w to *IST*.

Step E.   This step updates the bundle stack to reflect the embedding of the current path [ending with back edge (v, w)]. The updating is required only when w has been added to stack *IST*, when bundles have been shifted in step C, or when the entire segment containing the current path consists of more than a single back edge. Observe that the segment consists of a single back edge if and only if $s = v$. Thus this step is as shown in Algorithm 8.16(e).

if $out = -1$ **then** $out \leftarrow 0$

if $in \neq 0$ **or** $out \neq 0$ **or** $v \neq s$ **then** $BST \Leftarrow (in, out)$

Algorithm 8.16(e)   Update *BST*.

Step F.   If the segment containing the current path is more than a single back edge, we must attempt to embed this segment by going into the next level of recursion. For this purpose, we put a zero on top of *OST* as the end-of-stack marker. Since the current segment is a single back edge (v, w) if and only if $v = s$, this is done as shown in Algorithm 8.16(f).

$$\textbf{if } v \neq s \textbf{ then do} \begin{cases} stack\ (avail) \leftarrow 0 \\ next\ (avail) \leftarrow next\ (-1) \\ next\ (-1) \leftarrow avail \\ avail \leftarrow avail + 1 \end{cases}$$

Algorithm 8.16(f)   Put an end-of-stack marker on *OST.*

Conclusion.   It is not difficult to see that the path-finding part of Algorithm 8.15 requires $O(|V| + |E|)$ operations. The information used by the embedding part of the algorithm consists of the endpoints of the $|E| - |V|$ paths generated (i.e., the paths other than the cycle p_0). The embedding part of the algorithm consists solely of a sequence of stack manipulations, and adding/deleting an element requires constant time. The total number of entries on *IST*, *OST*, and *BST* is $O(|V| + |E|)$; therefore the entire algorithm requires only $O(|V| + |E|)$ operations. However, since $|E| \leq 3|V| - 6$ (otherwise the graph must be nonplanar), the algorithm requires $O(|V|)$ operations. Similarly, the storage requirements are $O(|V|)$.

We have only shown how to test a graph for planarity, not actually how to embed it in the plane if it is planar. A planar embedding can be constructed with a little additional work (Exercise 51).

8.7 REMARKS AND REFERENCES

The first published paper on graph theory is Euler's 1736 solution of the Königsberg bridge problem, but graph theory did not become an organized branch of mathematics until the 1930s. A history of the development of graph theory and a description of its numerous applications are given in

DEO, N., *Graph Theory with Applications to Engineering and Computer Science*, Prentice-Hall, Englewood Cliffs, N.J., 1974.

Three classic texts on graph theory are

BERGE, C., *The Theory of Graphs and Its Applications*, Wiley, New York, 1962

HARARY, F., *Graph Theory*, Addison-Wesley, Reading, Mass., 1969

and

ORE, O., *Theory of Graphs* (Cambridge Colloquium Publications, Vol. 38), American Mathematical Society, Providence, R.I., 1962.

A very elementary exposition can be found in

ORE, O., *Graphs and Their Uses*, Random House and the L. W. Singer Company, New York, 1963.

Although a great deal of work was done in graph theory in the 1930s and 1940s, attempts to find practical algorithms for manipulating large graphs began only in the 1950s. This effort developed as part of operations research and was a direct consequence of the growing popularity and availability of computers. Some of the first problems considered were minimum spanning tree construction, shortest path determination, traveling salesman tour determination, assignment problems, and network flow problems.

The greedy algorithm for generating a minimum spanning tree is due to

KRUSKAL, J. B., "On the Shortest Spanning Subtree of a Graph and the Traveling Salesman Problem," *Proc. Amer. Math. Soc.*, **7** (1956), 48–50.

The nearest-neighbor minimum spanning tree algorithm (Algorithm 8.1) was first published in

PRIM, R. C., "Shortest Connection Networks and Some Generalizations," *Bell Syst. Tech. J.*, **36** (1957), 1389–1401

and independently in

DIJKSTRA, E., "Two Problems in Connexion with Graphs," *Num. Math.*, **1**, (1959), 269–271.

Kruskal's algorithm requires $O(|E| \log |E|)$ operations and the Dijkstra-Prim algorithm requires $O(|V|^2)$ operations. Recently algorithms requiring $O(|E| \log \log |V|)$ operations have been discovered. The first such algorithm was given in

YAO, A. C. C., "An $O(|E| \log \log |V|)$ Algorithm for Finding Minimum Spanning Trees," *Info. Proc. Let.*, **4** (1975), 21–23

(Exercise 7); alternative methods are discussed in

CHERITON, D., and R. E. TARJAN, "Finding Minimum Spanning Trees," *SIAM J. Comput.*, to appear.

The basic idea used to generate all the spanning trees of a graph was described in

MINTY, G. J., "A Simple Algorithm for Listing All the Trees of a Graph," *IEEE Trans. Circuit Theory*, **12** (1965), 120.

A relatively inefficient recursive program based on the same idea was published in

MCILROY, M. D., "Algorithm 354: Generator of Spanning Trees," *Comm. ACM*, **12** (1969), 511.

More efficient algorithms are described in

READ, R. C., and R. E. TARJAN, "Bounds on Backtrack Algorithms for Listing Cycles, Paths, and Spanning Trees," *Networks*, **5** (1975), 237–252 (see Exercise 16) and in

GABOW, H. N., "Two Algorithms for Generating Weighted Spanning Trees in Order," *SIAM J. Comput.*, to appear.

Gabow's algorithms generate the spanning trees in increasing order of cost.

The technique of depth-first search (backtrack) has been used in the graph algorithms for a long time, but the extent to which it reveals the properties of a graph has only recently been discovered. The pioneering paper in this area is

TARJAN, R. E., "Depth-First Search and Linear Graph Algorithms," *SIAM J. Comput.*, **1** (1972), 146–160.

This paper presents the basic depth-first search procedure (Algorithm 8.2) and the algorithms for finding biconnected components (Algorithm 8.5) and strongly connected components (Algorithm 8.6). Some of these algorithms also appeared in

HOPCROFT, J. E., and R. E. TARJAN, "Algorithm 447: Efficient Algorithms for Graph Manipulation," *Comm. ACM*, **16** (1973), 372–378.

Additional applications of depth-first search, including Algorithm 8.4 for topologically sorting an acyclic digraph, are described in

TARJAN, R. E., "Finding Dominators in Directed Graphs," *SIAM J. Comput.*, **3** (1974), 62–89.

Still other applications are discussed in

TARJAN, R. E., "Testing Flow Graph Reducibility," *J. Comput. Sys. Sci.*, **9** (1974), 335–365.

A k-connected graph is one in which any k vertices can be removed without disconnecting the graph. Triconnectivity ($k = 3$) can be determined in $O(|V| + |E|)$ operations by depth-first search. See

HOPCROFT, J. E., and R. E. TARJAN, "Dividing a Graph into Triconnected Components," *SIAM J. Comput.*, **2** (1973), 135–158.

For $3 < k < \sqrt{|V|}$, the best algorithm known requires $O(k|V|^3)$ operations; it can be found in

EVEN, S., "An Algorithm for Determining Whether the Connectivity of a Graph is at Least k," *SIAM J. Comput.*, **4** (1975), 393–396.

The idea behind the transitive closure algorithm (Algorithm 8.7) was first published in

KLEENE, S. C., "Representation of Events in Nerve Nets and Finite Automata," in *Automata Studies*, C. E. Shannon and J. McCarthy (Eds.), Princeton University Press, Princeton, N.J., 1956.

as a technique for proving a theorem about finite automata. The algorithm itself was discovered independently later and appeared in

WARSHALL, S., "A Theorem on Boolean Matrices," *J. ACM.*, **9** (1962), 11–12.

When the adjacency matrix is packed so that the innermost loop can be implemented by only one or two computer instructions this algorithm is generally the most efficient known. For empirical data supporting this claim, see

SYSŁO, M. M., and J. DZIKIEWICZ, "Computational Experience with Some Transitive Closure Algorithms," *Computing*, **15** (1975), 33–39.

A modified version of the algorithm, suitable for very large, sparse digraphs, particularly in a paging environment, can be found in

WARREN, H. S., "A Modification of Warshall's Algorithm for the Transitive Closure of Binary Relations," *Comm. ACM*, **18** (1975), 218–220.

The question of how fast the transitive closure of a graph can be computed has been studied in connection with $o(n^3)$ algorithms for multiplying $n \times n$ matrices. A transitive closure algorithm requiring $O(|V|^{\lg 7} \log |V|)$ operations (Exercise 25) and some further references are given in

MUNRO, J. I., "Efficient Determination of the Transitive Closure of a Directed Graph," *Info. Proc. Let.*, **1** (1971), 56–58.

This result appears to be of theoretical interest only, since it becomes faster than Algorithm 8.7 only for extremely large $|V|$.

Under the assumption that each of the $2^{|V|^2}$ boolean matrices of size $|V| \times |V|$ are equally probable, it can be shown that the expected number of operations for a variant of Algorithm 8.7 is $O(|V|^2 \log |V|)$. See

O'NEIL, P. E., and E. J. O'NEIL, "A Fast Expected Time Algorithm for Boolean Matrix Multiplication and Transitive Closure," *Information and Control*, **22** (1973), 132–138.

A precursor of the shortest path algorithm of Section 8.2.6 (Algorithm 8.8) was discovered by E. F. Moore in 1956 and first published in

MOORE, E. F., "The Shortest Path Through a Maze," pp. 285–292 in *Proceedings of the International Symposium on the Theory of Switching*, Part II, Harvard University Press, Cambridge, Mass., 1959.

The improved version given here was discovered a few years later by E. Dijkstra and published in his paper cited above. It was erroneously assumed that this algorithm could be modified to work in $O(|V|^3)$ operations for graphs with negative edge weights. This was disproved in

JOHNSON, D. B., "A Note on Dijkstra's Shortest Path Algorithm," *J. ACM*, **20** (1973), 385–388,

in which the bound of $O(|V|2^{|V|})$ is established.

Algorithm 8.9 for finding shortest paths between all pairs of vertices in a graph is from

FLOYD, R. W., "Algorithm 97: Shortest Path," *Comm. ACM*, **5** (1962), 345.

This algorithm is also the most efficient known algorithm for finding a single shortest path in a graph with negative edge weights. Another algorithm for finding all shortest paths in a graph, just as efficient as Algorithm 8.9, is described in

> TABOURIER, Y., "All Shortest Distances in a Graph: An Improvement to Dantzig's Inductive Algorithm," *Discrete Math.*, **4** (1973), 83–87.

Surveys of shortest path problems and results are given in

> JOHNSON, D. B., "Algorithms for Shortest Paths," *Technical Report* 73–169, Department of Computer Science, Cornell University, Ithaca, N.Y., 1973

and in

> YEN, J. Y., *Shortest Path Network Problems* (*Mathematical Systems in Economics*, Heft 18), Verlag Anton Hain, Meisenheim am Glan, 1975.

An extensive bibliography of papers on the shortest path, minimum spanning tree, and related problems can be found in

> PIERCE, A. R., "Bibliography on Algorithms for Shortest Path, Shortest Spanning Tree, and Related Circuit Routing Problems (1956–1974)," *Networks*, **5** (1975), 129–149.

Algorithm 8.10, which finds a fundamental set of cycles of a graph, is due to

> PATON, K., "An Algorithm for Finding a Fundamental Set of Cycles of a Graph," *Comm. ACM*, **12** (1969), 514–518.

A slightly different version than the one presented in Section 8.3.1 can be found in

> GIBBS, N. W., "Algorithm 491: Basic Cycle Generation," *Comm. ACM*, **18** (1975), 275–276.

The algorithm given in Section 8.3.2 for generating all the cycles of a digraph is from

> JOHNSON, D. B., "Finding All the Elementary Circuits of a Directed Graph," *SIAM J. Comput.*, **4** (1975), 77–84.

Many other algorithms have been proposed for generating all the cycles of a digraph. A survey of these algorithms and their relative merits can be found in

> MATETI, P., and N. DEO, "On Algorithms for Enumerating All circuits of a Graph," *SIAM J. Comput.*, **5** (1976), 90–99.

An account of the use of cliques in cluster analysis and in information storage and retrieval is given in

> AUGUSTSON, G. J., and J. MINKER, "Analysis of Some Graph-Theoretic Cluster Techniques," *J. ACM*, **17** (1970), 571–588.

A correction to this paper was published in

> MULLIGAN, G. D., and D. G. CORNEIL, "Correction to Bierstone's Algorithm for Generating Cliques," *J. ACM*, **19** (1972), 244–247.

The Moon–Moser graphs, exemplified in Figure 8.16, have the most cliques possible for a graph with a given number of vertices. More exactly, a graph with n vertices can have at most $f(n)$ cliques, where

$$f(n) = \begin{cases} 3^{n/3} & n \equiv 0 \pmod 3 \\ 4 \cdot 3^{(n-4)/3} & n \equiv 1 \pmod 3 \\ 2 \cdot 3^{(n-2)/3} & n \equiv 2 \pmod 3 \end{cases}$$

[the construction in Figure 8.16 is for $n \equiv 0 \pmod 3$; what are the corresponding constructions for the other two cases?]. This result was first obtained in 1960 by Raymond Miller and David Muller but was never published. It was rediscovered independently and published in

MOON, J. W., and L. MOSER, "On Cliques in Graphs," *Israel J. Math.*, **3** (1965), 23–28.

This paper also contains a proof that the graphs achieving these bounds are unique.

Algorithm 8.12 for finding all the cliques in a graph is from

BRON, C., and J. KERBOSCH, "Algorithm 457: Finding All Cliques of an Undirected Graph," *Comm. ACM*, **16** (1973), 575–577.

The formulation given in Section 8.4 is due largely to Daniel Watanabe. The result that the performance of Algorithm 8.12 is not polynomial in the number of cliques is due to Robert Tarjan. The worst case performance of the modified Algorithm 8.12 is not known, but extensive testing has indicated that it is competitive with a clever implementation of the algorithm in

PAULL, M. C., and S. H. UNGER, "Minimizing the Number of States in Incompletely Specified Sequential Switching Functions," *IRE Trans. Elect. Comput.* **8** (1959), 356–367.

This algorithm can be implemented in time that is linear in the number of cliques (Exercise 40); such an implementation is given in

TSUKIYAMA, S., M. IDE, H. ARIYOSHI, and H. OZAKI, "A New Algorithm for Generating All Maximal Independent Sets," *SIAM J. Comput.*, to appear.

The main idea of the isomorphism procedure given in Section 8.5 is due, independently and simultaneously, to

SUSSENGUTH, E. H., JR., "A Graph-Theoretic Algorithm for Matching Chemical Structures," *J. Chem. Doc.*, **5** (1965), 36–43

and

UNGER, S. H., "GIT—A Heuristic Program for Testing Pairs of Directed Line Graphs for Isomorphism," *Comm. ACM*, **7** (1964), 26–34.

A completely different approach to the problem can be found in

BERTZISS, A. T., "A Backtrack Procedure for Isomorphism of Directed Graphs," *J. ACM*, **20** (1973), 365–377.

All known algorithms for digraph isomorphism require, in the worst case, a number of operations exponential in the number of vertices.

In certain special cases, however, it is possible to do much better. If the graphs are planar, isomorphism can be determined in $O(|V| \log |V|)$ operations; see

HOPCROFT, J. E., and R. E. TARJAN, "A $V \log V$ Algorithm for Isomorphism of Triconnected Planar Graphs," *J. Comput. Syst. Sci.*, **7** (1973), 323–331.

In fact, it has proven possible to test planar graphs for isomorphism in $O(|V|)$. For the details of this result, see

HOPCROFT, J. E., and J. K. WONG, "Linear Time Algorithm for Isomorphism on Planar Graphs," *Proc. Sixth Annual ACM Symposium on Theory of Computing* (1974), 172–184.

Much progress has been made in the past ten years in finding an efficient algorithm for planarity testing. The earliest characterization of a planar graph was given by Kuratowski (in 1930) in terms of two forbidden subgraphs (see Exercise 45). To use Kuratowski's characterization for planarity testing would require at least time proportional to $|V|^6$. Until 1970 the best algorithms for determining planarity were $O(|V|^3)$, or $O(|V|^2)$ if carefully implemented. In 1970 Hopcroft and Tarjan discovered an $O(|V| \log |V|)$ algorithm, which they eventually improved to $O(|V|)$ in

HOPCROFT, J. E., and R. E. TARJAN, "Efficient Planarity Testing," *J. ACM*, **21** (1974), 549–568.

The planarity algorithm of Section 8.6 is from this paper, which also describes how to construct a planar representation if the given graph is planar (see Exercise 51). Some corrections to this paper are given in

DEO, N., "Note on Hopcroft and Tarjan Planarity Algorithm," *J. ACM*, **23** (1976), 74–75.

Graph-theoretic techniques have become important in the solution of large, sparse systems of linear equations by Gaussian elimination. This application has suggested many new and interesting graph algorithms; see

TARJAN, R. E., "Graph Theory and Gaussian Elimination," in *Sparse Matrix Computations*, by J. R. Bunch and D. J. Rose (eds.), Academic Press, New York, 1976.

Epilogue. A number of problems connected with graphs and algorithms for their solution have been presented in this chapter, and it is instructive to classify these problems according to the order of magnitude of the number of operations they require in their worst cases.

1. Problems for which there are algorithms requiring proportional to $|V| + |E|$ operations.

2. Problems for which the known algorithms require $f(|V|, |E|)$ operations for some nonlinear but polynomially bounded function f.

3. Problems that do not appear to be inherently exponential in complexity but for which no algorithms are known that require less than an exponential number of operations.

4.   Problems that inherently require a number of operations that are exponential in the size of the input because they involve the generation of an exponential number of subgraphs.

Problems in the first class include finding a spanning tree, the connected components, the biconnected components, or the strongly connected components, topological sorting, and determining planarity. Since most nontrivial algorithms require examining all the vertices and edges of a graph, we cannot expect algorithms to be more efficient than $O(|V| + |E|)$ except in some special cases. See Exercise 1 for one of these exceptions and

RIVEST, R., and J. VUILLEMIN, "A Generalization and Proof of the Aanderaa–Rosenberg Conjecture," in *Theoretical Computer Sci.*, to appear.

for a discussion of the properties that any such exception must have.

The division of problems among the first three classes depends on the state of the art. Until 1971 planarity testing was in the second class, but the discovery of a new algorithm shifted it into the first class. Similarly, it is possible but unlikely that polynomial-time algorithms will be discovered for problems in the third class and that they will be shifted into the second class. Problems *currently* in this second class include finding a minimum spanning tree, shortest paths, or a fundamental sets of cycles.

The fourth class of problems includes generating all of the spanning trees of a graph, all of the cycles of a graph, or all of the cliques of a graph. There is no chance of new algorithms being discovered that would cause problems in this class to be shifted into one of the first two classes: they are inherently exponential in the size of the input (although perhaps linear in the size of the output).

The third class of problems, which includes the traveling salesman problem, determining isomorphism, finding the maximum clique, and many others, is an enigma. Although none of the problems in this class seems to be inherently exponential in its time requirements, there are no known polynomial-time algorithms. Further, many of the problems in this class are known to have the property that a polynomial-time algorithm for any one of them would yield polynomial-time algorithms for all problems in the class. The third class of problems is the subject of Chapter 9.

8.8 EXERCISES

1.   A *sink* in a simple digraph $G = (V, E)$ is a vertex with $|V| - 1$ incoming edges and no outgoing edges. Find an algorithm that, given the adjacency matrix of G, determines in $O(|V|)$ bit inspections whether G contains a sink.

2.   The *incidence matrix* of an undirected graph $G = (V, E)$ is the $|V| \times |E|$ bit matrix $A = [a_{ij}]$ defined by $a_{ij} = 1$ if and only if the ith vertex is incident of the jth edge; otherwise $a_{ij} = 0$. Can self-loops and parallel edges be represented? Can the idea be extended to digraphs? What are the advantages and disadvantages of incidence matrices?

3. Design algorithms to convert between all pairs of the following graph representations.
   (a) Adjacency matrix
   (b) List of edges
   (c) Adjacency structure
   (d) Incidence matrix
   How many operations are needed for the various conversions?

4. Prove that if an undirected graph $G = (V, E)$ consists of k connected components, then $|E| \geq |V| - k$. Prove that if $|E| > |V| - k$, then G contains a cycle.

5. Prove that the nearest-neighbor method, as used in Algorithm 8.1, does indeed produce a minimum spanning tree:
   (a) Prove that it produces a spanning tree.
   (b) Prove that the tree produced is minimum.

6. Design an algorithm that finds a minimum spanning tree in $O(|E| \log |V|)$ operations as follows. First the minimum weight edge incident on each vertex is found; these edges form part of the minimum spanning tree. The connected components formed by these edges are collapsed into "supervertices" (as in Figure 8.4). The process is repeated on the supervertices, and then on the resulting "super-supervertices," and so on, until only a single vertex remains. Show that the algorithm requires $O(|E| \log |V|)$ operations. (*Hint:* Use the efficient set algorithms of Section 2.4.)

★7. Improve the algorithm described in Exercise 6 so that it requires $O(|E| \log \log |V|)$ operations. (*Hint:* For every vertex v partition the edges incident on v into $\lceil \lg |V| \rceil$ subsets $S_1(v), S_2(v), \ldots, S_{\lceil \lg |V| \rceil}(v)$, which have the property that the weight of any edge in $S_i(v)$ is greater than or equal to the weight of any edge in $S_j(v)$ for $i < j$. This can be done in $O(|E| \log \log |V|)$ operations by repeatedly using the algorithm illustrated in Figure 7.11. How many edges now need to be examined at every stage of the algorithm outlined in Exercise 6?)

8. Define a *1-tree* of an undirected graph to be spanning tree with one extra edge added (this edge causes a circuit). Design, analyze, and prove to be correct an algorithm to find a minimum 1-tree of a weighted undirected graph.

9. Suppose we are given a symmetric $n \times n$ cost matrix C which represents an n-city traveling salesman problem satisfying the triangle inequality. We can consider C to be the weight matrix of a weighted undirected graph G, and use Algorithm 8.1 to find a minimum spanning tree for G in $O(n^2)$ operations.
   (a) Why is the cost of a minimum spanning tree a lower bound on the cost of an optimal tour?
   (b) How can a minimum spanning tree be used to obtain, in $O(n^2)$ operations, a traveling salesman tour whose cost is at most twice that of an optimal tour?

★10. Prove that there are n^{n-2} distinct trees of n vertices labeled $1, 2, \ldots, n$. [*Hint:* Consider the following procedure. Starting with a tree on the vertices $1, 2, \ldots, n$, remove the lowest-numbered vertex v_1 that has only one incident edge and that edge. Repeat this process, removing vertices $v_2, v_3, \ldots, v_{n-2}$ until the graph is reduced to $\underset{v_{n-1} \qquad\quad v_n}{\bullet\!\!-\!\!-\!\!-\!\!-\!\!-\!\!-\!\!\bullet}$. Let $w_1, w_2, \ldots, w_{n-2}$ be, respectively, the vertices to which $v_1, v_2, \ldots, v_{n-2}$ were attached at the time of their removal. Show that each of the n^{n-2} sequences $(w_1, w_2, \ldots, w_{n-2})$, $1 \leq w_i \leq n$, corresponds to a unique tree.]

11. Implement the algorithm outlined at the end of Section 8.2.1 for generating all the spanning trees of a graph. How does your algorithm behave on the graph

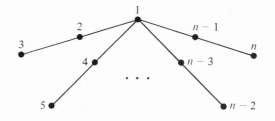

12. Devise a nonrecursive version of Algorithm 8.2.

13. Define *highpt* (v), in analogy with *lowpt* (v), as the number of the highest numbered vertex reachable from v by a sequence of zero or more tree edges followed by at most one back edge. Show how Algorithm 8.2 can be modified to compute the *highpt* values.

14. Let $\vec{G} = (V, \vec{E})$ be a DFS-forest resulting from a depth-first search on $G = (V, E)$ in which every vertex $v \in V$ is assigned a serial number *num* (v). Let *nd* (v) be the number of descendants of v (including itself). Prove that a vertex w is a descendant of v if and only if *num* $(v) \leq$ *num* $(w) <$ *num* $(v) +$ *nd* (v). Prove that *nd* $(v) = 1 + \sum_{(v,w) \in \vec{E}} nd \ (w)$. Show how Algorithm 8.2 can be modified to compute the *nd* values.

15. A *bridge* in a connected, undirected graph $G = (V, E)$ is an edge whose removal disconnects G. Design a depth-first search algorithm requiring $O(|V| + |E|)$ operations to determine the bridges of G. [*Hint*: Prove that an edge (v, w) of G is a bridge if and only if (v, w) is a tree edge and *lowpt* $(w) \geq$ *num* (w).]

16. Prove that the bridges of a graph $G = (V, E)$ must be in every spanning tree of G. Use this result to improve your algorithm in Exercise 11 so that it requires only $O(|V| + |E| + |E| \cdot \ell)$ operations, where ℓ is the number of spanning trees of G. How does your algorithm behave on the graph shown in Exercise 11?

17. Prove that a digraph can be topologically sorted if and only if it is acyclic.

18. Design, analyze, and prove correct an algorithm to find all possible ways of topologically sorting an acyclic digraph G.

19. Given an acyclic digraph $G = (V, E)$, two vertices v and w are said to be *unrelated* if there are sequences of vertices $v_1 = v, v_2, \ldots, v_m = w$ and $w_1 = w, w_2, \ldots, w_n = v$ such that there is no path in G from v_i to v_{i+1} for $1 \leq i < m$ or from w_j to w_{j+1} for $1 \leq j < n$.
    (a) Prove that "unrelatedness" is an equivalence relation on V. The equivalence classes give the finest partition of V on which the edges E induce a linear ordering (this is in contrast to the equivalence classes based on "strong-connectedness" which give the finest partition of V on which the edges E induce a partial ordering).
    (b) Prove that the labels assigned by the topological sort of Algorithm 8.4 to the vertices in an equivalence class of unrelated vertices form a set of consecutive integers.

20. Prove that Algorithm 8.5, for determining the biconnected components of an undirected graph, is correct. (*Hint*: Use induction on the number of edges in the graph, along with Theorem 8.1.)

21. Prove that a vertex v is the root of a strongly connected component if and only if *num* (v) = *lowlink* (v). [*Hint*: If v is the root of a strongly connected component, then what would *num* (v) > *lowlink* (v) imply? On the other hand, suppose that *num* (v) = *lowlink* (v) and $u \neq v$ is the root of the strongly connected component. Consider the first edge on any path from v to u that leads to a vertex w not in the subtree rooted at v; this edge is a back edge or a cross edge.]

22. Prove that Algorithm 8.6 correctly determines the *lowlink* values and hence correctly determines the strongly connected components of a graph. [*Hint*: Use induction as follows. Prove that, for all k, the vertices v such that v is a proper descendant of vertex k or *num* (v) < *num* (k), *lowlink* (v) is correctly computed. This induction hypothesis corresponds to considering vertices in the order that they are examined for the *last* time during the depth-first search process.]

23. Does the following algorithm correctly transform A into A^*?

    for l = 1 **to** $|V|$ **do**
    　　for i = 1 **to** $|V|$ **do**
    　　　　for j = 1 **to** $|V|$ **do** $a_{ij} \leftarrow a_{ij} \vee (a_{il} \wedge a_{lj})$

    Prove your answer. If the algorithm is correct, under what condition is it better than Algorithm 8.7?

24. Define the *logical product of $n \times n$ bit matrices* as

    $$C = A \circ B \quad \text{has entries} \quad c_{ij} = \bigvee_{k=1}^{n} (a_{ik} \wedge b_{kj}).$$

    Prove that logical product is an associative operation. Prove that

    $$A^* = A \vee A^2 \vee A^3 \vee \cdots \vee A^{n-1}$$

    where the exponentiation is with respect to logical product. How can this identity be used to compute A^* in a reasonable fashion?

25. Let

    $$A = \begin{pmatrix} a_{11} & a_{12} \\ a_{21} & a_{22} \end{pmatrix} \quad \text{and} \quad B = \begin{pmatrix} b_{11} & b_{12} \\ b_{21} & b_{22} \end{pmatrix}$$

    be 2×2 matrices and let

    $$M_1 = (a_{11} + a_{22})(b_{11} + b_{22})$$
    $$M_2 = (a_{21} + a_{22})b_{11}$$
    $$M_3 = a_{11}(b_{12} - b_{22})$$
    $$M_4 = a_{22}(- b_{11} + b_{21})$$
    $$M_5 = (a_{11} + a_{12})b_{22}$$
    $$M_6 = (- a_{11} + a_{21})(b_{11} + b_{12})$$
    $$M_7 = (a_{12} - a_{22})(b_{21} + b_{22}).$$

Verify that

$$AB = \begin{pmatrix} M_1 + M_4 - M_5 + M_7 & M_3 + M_5 \\ M_2 + M_4 & M_1 + M_3 - M_2 + M_6 \end{pmatrix}.$$

Show how this leads to an algorithm requiring $O(n^{\lg 7})$ operations, $\lg 7 \approx 2.81$, for multiplying two $n \times n$ matrices (instead of the usual n^3). Show how the resulting technique can be applied to Exercise 24 to obtain an algorithm for transitive closure that requires $O(|V|^{\lg 7} \log |V|)$ operations.

26. A digraph $G = (V, E)$ is said to be *transitive* if $(u, v), (v, w) \in E$ implies $(u, w) \in E$ for all $u, v, w \in V$. Find an algorithm that uses $O(|V|)$ operations to determine whether a transitive digraph is strongly connected.

27. A *transitive reduction* of a digraph $G = (V, E)$ is a graph $G^T = (V, E^T)$ having the property that there is a path from v to w in G^T if and only if there is a path from v to w in G and, furthermore, that there is no graph with fewer edges having that property. How is the problem of finding a transitive reduction of G related to the problem of computing G^* from G?

28. *Breadth-first search* explores the edges of a graph $G = (V, E)$ and produces a BFS-tree as follows. Start at an arbitrarily chosen vertex r (the root) and traverse all edges incident on r; let these edges be $(r, a_1), (r, a_2), \ldots, (r, a_k)$. Then explore the edges incident on $a_1, a_2, \ldots, a_k$. Let these edges be $(a_1, a_{1,1}), (a_1, a_{1,2}), \ldots, (a_1, a_{1,r_1}), (a_2, a_{2,1}), (a_2, a_{2,2}), \ldots, (a_2, a_{2,r_2}), \ldots, (a_k, a_{k,r_k})$. The edges incident on $a_{1,1}, a_{1,2}, \ldots, a_{k,r_k}$ are then explored and so on. This process continues until all the edges of the tree have been explored. (If G is not connected, the process can be repeated to produce a BFS-forest.) Devise an algorithm to explore an undirected graph according to this scheme. (*Hint*: Use a queue.) Use this type of graph traversal to determine all vertices at a fixed distance d from a vertex s in an undirected graph in which every edge has weight 1.

29. Prove that the value of the final label of a vertex in Algorithm 8.8 is the distance of that vertex from s, the starting vertex. (*Hint*: Use the induction on the number of vertices with final labels.)

30. Modify Algorithm 8.8 so that it finds *all* shortest paths from s to f. (*Hint*: Instead of the *pre* values, maintain a digraph P in which any path in P from s to any other vertex in P is a shortest path. Consider *last* to be a set of all vertices whose labels became final on the previous pass.)

31. Extend the idea in Exercise 30 so that all shortest paths from s to all other vertices are found.

32. Give an example of a digraph in which some of the edges have negative weights but where there is no cycle with a total negative weight, such that direct application of Algorithm 8.8 produces an incorrect result.

33. Show that if all the edges of $G = (V, E)$ have weight 1 (i.e., $W = [w_{ij}]$ is the adjacency matrix of G with each zero replaced by ∞), then Algorithm 8.8 can be modified to use only two bits per vertex.

34. Use the dynamic programming approach of section 4.1.7 to obtain an algorithm for finding the shortest path between two vertices of a weighted digraph. How does your algorithm compare to Algorithm 8.8 in the number of operations required?

★35. Design, analyze, and prove correct an efficient algorithm to generate all the cycles of an undirected graph, given a fundamental set of cycles.

36. In Algorithm 8.11(c), prove that during the execution of $CYCLE$ (s, f), for any vertex $x \neq s$, there is a call to $UNMARK$ (v) that causes $avail$ $(x) \leftarrow$ **true** if and only if

    (a) there is a path containing v from x to s on which only v and s are on the path stack,
    (b) there is no path from x to s on which only s is on the path stack.
    Use this result to prove that Algorithm 8.11 generates every cycle exactly once.

37. Prove that in Algorithm 8.11 the number of operations performed between a call to $CYCLE$ and either the return or the generation of a cycle is $O(|V| + |E|)$. (*Hint:* Use the result of Exercise 36.) Use this result to prove that the total number of operations required by Algorithm 8.11 to generate all the cycles is $O((|V| + |E|) (1 + c))$, where c is the number of cycles.

38. Give an example of a graph in which applying Theorem 8.2 iteratively causes a clique to be missed. (*Hint:* One such example has five vertices and seven edges.)

39. Prove that the modified Algorithm 8.12 generates each of the cliques of a graph exactly once. [*Hint:* Using Theorems 8.2 and 8.3, prove by induction on $|N|$ that the call $CLIQUE$ (N, D) generates each clique of G in the subgraph induced by $N \cup S$ exactly once.]

40. Let $G = (V, E)$ be an undirected graph with $V = \{1, 2, \ldots, n\}$, and let G_i, $1 \leq i \leq n$, be the subgraph of G induced by the subset of vertices $\{1, 2, \ldots, i\}$. Consider the following inductive method for finding the cliques of G: Assume that the set $\mathscr{C}_{i-1}$ of all cliques of G_{i-1} has been found. For each clique $C \in \mathscr{C}_{i-1}$ we have

    (i) If $C \subseteq$ Adj (i) then $C \cup \{i\} \in \mathscr{C}_i$.
    (ii) If $C \nsubseteq$ Adj (i) then $C \in \mathscr{C}_i$.
    (iii) If $C \nsubseteq$ Adj (i) then $[C \cap$ Adj $(i)] \cup \{i\}$ is a complete subgraph of G_i and may or may not be a clique.

    Thus each clique $C \in \mathscr{C}_{i-1}$ generates a distinct clique in $\mathscr{C}_i$ by either (i) or (ii). A second clique in $\mathscr{C}_i$ may be generated by (iii).
    (a) Prove that if $C \in \mathscr{C}_i$ then C can be generated from some $\hat{C} \in \mathscr{C}_{i-1}$ by either (i), (ii), or (iii).
    (b) Use this technique to obtain an algorithm to determine all the cliques of graph.

41. A clique C of a graph G is called a *maximum clique* if no clique in G contains more vertices than C. Devise a branch-and-bound algorithm to find a maximum clique of a graph.

42. A subset of vertices $S \subseteq V$ in a graph $G = (V, E)$ is called *independent* if no pair of vertices is adjacent in G. S is called a *maximal independent set* if S is not contained in any larger independent set. Design an algorithm to generate all maximal independent sets of a graph $G = (V, E)$.

43. Implement Algorithm 8.13, including the preliminary work and the details of the procedure *MATCH*.

★44. Use Euler's formula

$$\text{regions} + \text{vertices} = \text{edges} + 2$$

relating the number of regions, vertices, and edges of a planar graph to prove that a simple graph with $n > 2$ vertices and more than $3n - 6$ edges can never be planar.

★45. **Kuratowski's Theorem.** Two graphs are *homeomorphic* if both can be obtained from the same graph by a sequence of subdivisions of lines; for example, any two cycles are homeomorphic. Prove that a graph is nonlinear if and only if it contains a homeomorphic subgraph to

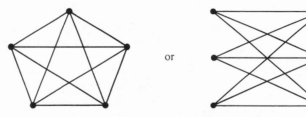

or

★46. The *complement* $\bar{G} = (V, \bar{E})$ of a graph $G = (V, E)$ is defined by $(u, v) \in \bar{E}$ if and only if $(u, v) \notin E$, for all $u, v \in V$.
(a) Prove that the complement of a planar graph $G = (V, E)$ with $|V| \geq 11$ is nonplanar.
(b) Use exhaustive search to find the smallest value of n such that the complement of every planar graph $G = (V, E)$ with $|V| \geq n$ is nonplanar.

★47. Prove that a graph can be embedded in the plane if and only if it can be embedded on the surface of a sphere.

★48. Prove that any simple planar graph (i.e. no loops or parallel edges) can be drawn in the plane with straight-line segments as edges.

★49. Show that, in the planarity testing algorithm, the renumbering of the vertices and the subsequent reordering of the adjacency lists of $\bar{G}$ cause the vertices in the cycle $C = p_0$ generated by Algorithm 8.14 to be in increasing order. In other words, show that if $C = (1, v_2, v_3, \ldots, v_k, v_{k+1})$, then $1 < v_2 < v_3 < \cdots < v_k$ and $v_{k+1} = 1$.

★50. In the planarity testing algorithm, suppose that cycle C, all segments before S (the current segment), and the first path p in S have been successfully embedded. Show that $S - p$ can be added to the embedding if and only if subgraph $S \cup C$ is a planar.

★51.   Modify Algorithm 8.15 so that it not only tests the planarity of a graph G but also finds a planar embedding if G is planar.

52.   An *Eulerian path* in an undirected, connected graph is a path that traverses every edge of the graph exactly once. Prove that a connected graph $G = (V, E)$ has an Eulerian path if and only if it has at most two vertices of odd degree (a vertex has odd degree if and only if there is an odd number of edges incident on it). Design an algorithm for finding an Eulerian path that requires $O(|V| + |E|)$ operations.

53.   An undirected graph $G = (V, E)$ is *bipartite* if V can be partitioned into two sets V_1 and V_2, so that every edge in E joins a vertex in V_1 to a vertex in V_2. Use depth-first search to determine whether a given graph is bipartite.

54.   Let $G = (V, E)$ be a rooted, acyclic digraph. A vertex $x \in V$ is called a *dominator* of another vertex $y \in V$ if every directed path from the root to y must pass through x. Every vertex is a dominator of itself and the root is a dominator of every vertex in G. The dominators of a vertex z can be linearly ordered (from the root to z itself) by their order of occurrence on a shortest path from the root to z. The relationships "a dominates b" can thus be represented by a tree, called the *dominator tree* of G. Devise an algorithm for constructing a dominator tree of a rooted, acyclic digraph G, assuming that the root of G is known.

55.   A *vertex basis* of a digraph $G = (V, E)$ is a minimal subset $B \subseteq V$ with the property that there is a path of length zero or more to every vertex in V from some vertex in B. Show that
   (a) a vertex that is not on a cycle and has nonzero indegree cannot be in any vertex basis.
   (b) an acyclic digraph has a unique vertex basis.
   (c) every vertex basis of a digraph contains the same number of vertices.
   Design an algorithm for finding a vertex basis of a digraph in $O(|V| + |E|)$ operations.

56.   A *matching* in an undirected graph G is a set of edges, no two of which have a vertex in common. A matching M in G is called a *maximum matching* if no matching in G contains more edges than M. Design an algorithm to find a maximum matching in a given connected undirected graph G when
   (a) G is a bipartite graph.
   (b) G is any graph.

57.   In an undirected graph $G = (V, E)$ a *cut-set* with respect to a vertex pair x, y is a minimal subset $S \subseteq E$ such that at least one edge of every path between x and y is in S. (Thus removing S from G disconnects x and y but no proper subset of S has this property.) Devise an algorithm to find a cut-set with respect to given vertices x and y.

58.   Let G be a nonnegatively weighted, connected digraph in which the weight of an edge represents its *capacity* (i.e., the amount of some commodity that can flow through the edge) and let the capacity of a cut-set be the sum of the weights of its edges. If there is exactly one vertex x with no incoming edges and exactly one vertex y with no outgoing edges, then G is called a *transport network*. A *flow* in

such a network is an assignment of nonnegative numbers f_{ij} to edges (i, j) such that $f_{ij} \leqq W_{ij}$ for each edge (i, j) and $\sum_i f_{ij} = \sum_k f_{jk}$ for each vertex $j \neq x, j \neq y$ (if there is no edge from i to j then $f_{ij} = 0$). The quantity $\sum_j f_{xj}$ is called the *volume* of the flow.

(a) **Max-flow-min-cut theorem.** Show that the maximum value of a flow is equal to the minimum of the capacities of all cut-sets with respect to x and y.

(b) Design an algorithm to determine a maximum flow (i.e., a flow whose value is maximum).

59. Let $G = (V, E)$ be an undirected, connected graph. Suppose that there is exactly one vertex x with no incoming edges and exactly one vertex y with no outgoing edges. Design an algorithm to find a cut-set with respect to x and y that has the smallest number of edges.

60. A *feedback vertex set* in a digraph G is a set of vertices S such that every cycle of G contains at least one vertex from S. Thus deleting S from G would destroy all the cycles in G. Design an algorithm for finding a *minimum feedback vertex set*—that is, a feedback vertex set with the fewest possible vertices.

61. Define a *minimum feedback edge set* in analogy to the definition of a minimum feedback vertex set given in the previous exercise. Design an algorithm for finding such a set of edges.

62. An undirected graph G is said to be k-*colorable* if all vertices of G can be colored using k different colors so that no two adjacent vertices have the same color. The *chromatic number* of G is the smallest integer k for which G is k-colorable. Show that

(a) the chromatic number of a tree is at most 2.

(b) the chromatic number of a bipartite graph with at least one edge is 2.

★(c) every planar graph is 5-colorable.

63. Design an algorithm for coloring the vertices of a given graph G using the minimum possible number of colors such that no adjacent vertices in G have the same color.

64. Design an $O(|V| + |E|)$-time algorithm to determine whether a graph is 2-colorable. (*Hint*: See Exercise 53.)

chapter 9

The Equivalence of Certain Combinatorial Problems

In this chapter we take a close look at the third class of problems described in the epilogue of Chapter 8: those problems that do not seem inherently difficult but for which only exponential-time algorithms are known. As a result of this examination, we will see the dramatic consequences the discovery of a polynomial-time algorithm for the traveling salesman problem would have. Thus it is not surprising that the clever tricks of Chapter 4 failed to yield a polynomial-time algorithm.

9.1 THE CLASSES $\mathscr{P}$ AND $\mathscr{NP}$

For the first eight chapters of this book we have ignored the details of what an algorithm is and what a time bound is, under the reasonable assumption that the reader would know from experience what was meant, even without formal definitions. But in order to prove results about the consequences of the existence of polynomial-time algorithms for various combinatorial problems, a more precise notion of a "polynomial-time algorithm" is needed.

We can view any algorithm in this book as a "black box" that, when presented with an input string, produces an output string. This situation is depicted in Figure 9.1. In particular, we will insist that the input and output strings be sequences of

Figure 9.1 An algorithm as a black box.

zeros and ones that encode the problem's inputs and the algorithm's outputs as binary numbers. The algorithm then can be viewed as a sequence of elementary bit operations (such as *or*, *and*, *not*, *print*, etc.), operating on a memory of bits that can be arbitrarily large.

These assumptions about algorithms contrast greatly with the high-level language presentations that we have been using for algorithms. However, these assumptions are equivalent in that the distinction between polynomial-time and exponential-time algorithms remains unchanged: since all of our data types contain finitely many bits, translating our high-level language algorithms down to the bit level increases the number of operations only polynomially because the operations we have used require time polynomial in the length of the operands. If an algorithm requires at most $T(l)$ high-level operations on an input of length l, then it requires at most $P(T(l))$ bit operations on such an input sequence. Here P is a polynomially bounded function that reflects the increase in the number of operations in going from high-level operations to bit operations. Thus the polynomial or exponential character of an algorithm's running time is invariant, since $P(T(l))$ is bounded by a polynomial in l if and only if $T(l)$ is bounded by a polynomial in l. Furthermore, the same argument shows that as long as the input and output sequences are encoded in a "reasonable" way,[†] the polynomial or exponential character of the length of the output sequence as a function of the length of the input sequence is also invariant.

To make this invariance under polynomial changes clear, consider the n-city traveling salesman problem. The input is an $n \times n$ distance matrix and the output is a permutation of the set $\{1, 2, \ldots, n\}$. If k is the largest intercity distance (all of which are assumed to be integers), then the input can be encoded in about $n^2 \lg k$ bits and the output can be encoded in about $n \lg n$ bits. The dynamic programming approach of Section 4.1.7 uses $O(n^2 2^n)$ arithmetical and logical operations on the entries in the distance matrix, and thus only $O(n^2 2^n \log k)$ bit operations. It is clear that changing the encoding to decimal, say, would not affect the polynomial or exponential character of the functions involved, but it might increase or decrease them polynomially.

We define a polynomial-time algorithm as one whose running time—that is, the number of elementary bit operations it performs—on an input string of length l is bounded above by some polynomial $P(l)$. $\mathscr{P}$ is the class of all problems that can be solved by such an algorithm. All problems discussed in this book with algorithms whose running times were shown to be bounded by a polynomial in the *number* of inputs are obviously in $\mathscr{P}$ and all problems with algorithms whose running times or number of outputs were shown to be *necessarily* exponential in the number of inputs are not in $\mathscr{P}$.

[†]There is no accepted, precise definition of "reasonable," but, for example, it would certainly be *unreasonable* to encode an integer n in more than $O(\log n)$ characters (Exercise 20). In general, it would be unreasonable to encode the input for a problem in more characters than the minimum number necessary by information theory.

To understand the consequences of $\mathcal{P}$ containing the traveling salesman problem or one of many other enigmatic problems, it is necessary to introduce the class $\mathcal{NP}$ of problems that can be solved in polynomial time by a *nondeterministic* algorithm. Informally, we define the *state* of an algorithm as the combination of the location of the instruction currently being executed and the values of all variables. All the algorithms considered so far have been *deterministic*; that is, for any given state there is at most one valid "next" state.[†] In other words, a deterministic algorithm can do only one thing at a time. A *nondeterministic algorithm* is one in which, for any given state, there may be more than one valid next state; in other words, nondeterministic algorithms can do more than one thing at a time. Nondeterministic algorithms are *not* in any sense probabilistic or random algorithms; they are algorithms that can be in many states simultaneously.

It is easiest to understand nondeterminism by considering an algorithm that does its calculations until it reaches a point at which a choice must be made among several alternatives. A deterministic algorithm would have to explore a single alternative and then return later to explore the remaining alternatives. A nondeterministic algorithm can explore all the alternatives simultaneously, essentially by "creating" a copy of itself for each alternative. All copies act independently without communicating with one another in any way. These copies can, of course, create further copies and so on. If a copy finds that it has made an incorrect (or fruitless) choice, it stops executing. If any copy finds a solution, it announces its success and *all* copies stop executing.

We will represent a nondeterministic algorithm by using the three primitives **choice**, **failure**, and **success**. **choice** (S) is a multiple-valued function whose values are the elements of a finite set S. **failure** causes that copy of the algorithm to stop executing; **success** causes all instances of the algorithm to stop executing and indicates a successful completion of the computation.

Obviously, no physical device is capable of unbounded nondeterministic behavior; nondeterministic algorithms are abstractions that allow us to ignore some of the usual considerations of backtrack programming. For example, Algorithm 9.1 is a nondeterministic version of Algorithm 4.1 (the generalized backtrack search). Comparing Algorithm 9.1 with Algorithm 4.1, we see a striking difference: Algorithm 9.1 contains no provision for backing up when a dead end is reached because the function **choice** causes all paths to be explored.

We define $\mathcal{NP}$ to be the class of all problems that can be solved by nondeterministic algorithms that run in polynomial time—that is, nondeterministic algorithms in which there is always a successful computational path that requires time polynomial in the length of the input string; obviously, $\mathcal{P} \subseteq \mathcal{NP}$. Since there may be exponentially many computational paths, it is plausible that the algorithms allowed are much more powerful than the deterministic

---

[†]Strictly speaking, this is not true. Many algorithms contain statements like "choose any value such that . . . ." Such algorithms work correctly for any valid choice, and as such are nondeterministic. However, these algorithms do not use nondeterminism in an essential way and are easily made deterministic by specifying the value to be chosen.

$k \leftarrow 1$

compute S_1

$$\textbf{while } S_k \neq \varnothing \textbf{ do} \begin{cases} a_k \leftarrow \textbf{choice } (S_k) \\[4pt] \textbf{if } (a_1, a_2, \ldots, a_k) \text{ is a solution } \textbf{then success} \\[8pt] \qquad\qquad\qquad\qquad \textbf{else} \begin{cases} k \leftarrow k + 1 \\[4pt] \text{compute } S_k \end{cases} \end{cases}$$

$[\![(a_1, a_2, \ldots, a_{k-1}) \text{ leads to no solution}]\!]$

failure

Algorithm 9.1   Nondeterministic generalized backtrack to determine whether a problem has a solution.

algorithms allowed for problems in $\mathscr{P}$. For example, a problem is in $\mathscr{NP}$ if it can be solved by a nondeterministic backtrack algorithm (Algorithm 9.1) in which

1.   for all k, the operations "compute S_k" and the tests "Is $S_k = \varnothing$?" and "Is $(a_1, a_2, \ldots, a_k)$ a solution?" can each be done in time polynomial in the length of the input string, and

2.   the length of each computational path is bounded by a polynomial in the length of the input string.

$S \leftarrow \{1, 2, \ldots, n\}$

$a_1 \leftarrow \textbf{choice } (S)$

$S \leftarrow S - \{a_1\}$

$k \leftarrow 1$

$cost \leftarrow 0$

$$\textbf{while } S \neq \varnothing \textbf{ do} \begin{cases} k \leftarrow k + 1 \\[4pt] a_k \leftarrow \textbf{choice } (S) \\[4pt] S \leftarrow S - \{a_k\} \\[4pt] cost \leftarrow cost + C_{a_{k-1}, a_k} \end{cases}$$

if $cost \leqq b$ **then success**

else failure

Algorithm 9.2   A polynomially bounded nondeterministic algorithm to determine whether an n-city traveling salesman problem (represented by the $n \times n$ distance matrix C) has a solution of cost at most b.

To display the power of polynomially bounded nondeterministic algorithms, and hence the extent of $\mathcal{NP}$, we will show that a variant of the traveling salesman problem is in $\mathcal{NP}$. Given an $n \times n$ matrix C of the intercity distances in an n-city traveling salesman problem, it is easy to see that Algorithm 9.1 can be modified to produce a polynomially bounded nondeterministic algorithm to determine, for any bound b, whether there is any tour of cost at most b; such an algorithm is given in Algorithm 9.2.

9.2 $\mathcal{NP}$-HARD AND $\mathcal{NP}$-COMPLETE PROBLEMS

Roughly speaking, a problem P is $\mathcal{NP}$-hard if a deterministic polynomial-time algorithm for its solution can be used to obtain a deterministic polynomial-time algorithm for every problem in $\mathcal{NP}$. In other words, P is $\mathcal{NP}$-hard if it is at least as hard as any problem in $\mathcal{NP}$. An $\mathcal{NP}$-hard problem in $\mathcal{NP}$ is called $\mathcal{NP}$-complete; such problems are at least as hard as any problem in $\mathcal{NP}$, but no harder.

The consequences of the existence of a deterministic polynomial-time algorithm for an $\mathcal{NP}$-hard or an $\mathcal{NP}$-complete problem are dramatic: it would mean that the classes $\mathcal{P}$ and $\mathcal{NP}$ are the same, so that any problem that can be solved by a backtrack algorithm in which each of the computation paths can be examined in polynomial time can be solved by a deterministic algorithm that runs in polynomial time. This would be *very* surprising, and thus it provides strong evidence against the existence of deterministic polynomial-time algorithms for any $\mathcal{NP}$-hard or $\mathcal{NP}$-complete problem.

In this section we examine various combinatorial problems, including the traveling salesman problem, and prove them to be $\mathcal{NP}$-hard or $\mathcal{NP}$-complete. To prove that a problem is $\mathcal{NP}$-hard, we must prove that, if we are given a deterministic polynomial-time algorithm for the problem, we can somehow use that algorithm to obtain deterministic polynomial-time algorithms for every problem in $\mathcal{NP}$. To establish that a problem is $\mathcal{NP}$-complete, we must further prove that the problem is in $\mathcal{NP}$.

The idea of using an algorithm for one problem in order to obtain an algorithm for a different problem is one of the most important ideas discussed here.

Definition

A problem P_1 is *transformable* to a problem P_2 if any instance of P_1 can be transformed in polynomial time into an instance of P_2, such that the solution of P_1 can be obtained in polynomial time from the solution to the instance of P_2.

This concept is illustrated schematically in Figure 9.2.

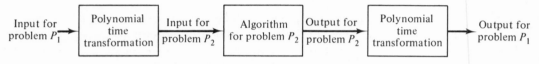

Figure 9.2   A schematic diagram showing transformability of P_1 to P_2.

Clearly, if P_1 is transformable to P_2 and $P_2 \in \mathscr{P}$, then $P_1 \in \mathscr{P}$. This allows us to define the notions of $\mathscr{NP}$-hard and $\mathscr{NP}$-complete precisely:

Definition

A problem is $\mathscr{NP}$-*hard* if every problem in $\mathscr{NP}$ is transformable to it, and a problem is $\mathscr{NP}$-*complete* if it is both $\mathscr{NP}$-hard and in $\mathscr{NP}$.

Therefore to prove that a problem is $\mathscr{NP}$-hard, it is only necessary to prove that some $\mathscr{NP}$-hard problem is transformable to it. The difficulty is to establish, as a starting point, that some particular problem is $\mathscr{NP}$-hard; we can then use that problem to prove that other problems are $\mathscr{NP}$-hard.

9.2.1 Satisfiability

The *satisfiability problem* from mathematical logic will be our starting point. Let $X_1, X_2, X_3, \ldots$ be boolean variables and let $\bar{X}_1, \bar{X}_2, \bar{X}_3, \ldots$ denote the complements of these variables, respectively, so that X_i is **true** if and only if $\bar{X}_i$ is **false**. The symbols $X_1, \bar{X}_1, X_2, \bar{X}_2, \ldots$ are called *literals*. Also, as usual, let $\vee$ and $\wedge$ denote the boolean operations of "or" and "and," respectively. A *clause* is the "or" of a set of literals; for example, $X_2 \vee \bar{X}_5 \vee \bar{X}_7 \vee X_{10}$ is a clause. A boolean expression in *conjunctive normal form* is the "and" of a set of clauses; for example,

$$(X_1 \vee X_2 \vee X_3) \wedge (X_1 \vee \bar{X}_2 \vee \bar{X}_4) \wedge (\bar{X}_3 \vee X_4) \wedge (\bar{X}_1)$$

is a boolean expression in conjunctive normal form. It is easy to see that *any* boolean expression can be expanded into conjunctive normal form by De Morgan's laws

$$\overline{A \vee B} = \bar{A} \wedge \bar{B}$$

$$\overline{A \wedge B} = \bar{A} \vee \bar{B}$$

and the identity

$$A \vee (B \wedge C) = (A \vee B) \wedge (A \vee C).$$

A boolean expression is called *satisfiable* if there is some assignment of the values **true** and **false** to the variables so that the expression has the value **true**. Thus the expression

$$(X_1 \vee X_2 \vee X_3) \wedge (X_1 \vee \bar{X}_2 \vee \bar{X}_4) \wedge (\bar{X}_3 \vee X_4) \wedge (\bar{X}_1)$$

is satisfiable because its value is **true** if $X_1 = X_3 = X_4 =$ **false** and $X_2 =$ **true**. On the other hand, the expression

$$(X_1 \lor X_2) \land (\bar{X}_1) \land (\bar{X}_2)$$

is not satisfiable because every assignment of the values **true** and **false** to the variables causes the expression to have the value **false**. The *satisfiability problem* is to determine whether a boolean expression in conjunctive normal form is satisfiable. We have the following basic result.

Theorem 9.1

The satisfiability problem is $\mathcal{NP}$-complete.

The proof of this theorem is beyond the scope of this book, but we can outline its major components. It is easy to show that satisfiability is in $\mathcal{NP}$; Algorithm 9.3 is a nondeterministic polynomial-time algorithm that determines whether the boolean expression $E(X_1, X_2, \ldots, X_n)$ is satisfiable. The proof that a deterministic polynomial-time algorithm for satisfiability implies a deterministic polynomial-time algorithm for every problem in $\mathcal{NP}$ is not difficult, but it involves a very detailed model of nondeterministic algorithms that would be out of place here. Essentially, the proof gives a construction that, for any polynomial-time nondeterministic algorithm whose output is simply **true** or **false** and for any input sequence, produces a single boolean expression in conjunctive normal form that is satisfiable if and only if the algorithm computes the answer **true** for that input sequence. Furthermore, the proof shows that the construction of this boolean expression from the formal description of the nondeterministic algorithm and its input sequence can be done in polynomial time. From this result it can be shown that there is a sequence of such boolean expressions corresponding to the bits of the output sequence from any algorithm in $\mathcal{NP}$ and any input sequence.

> **for** $i = 1$ **to** n **do** $X_i \leftarrow$ choice ({**true**, **false**})
>
> **if** $E(X_1, X_2, \ldots, X_n)$ **then success**
>
> **else failure**

Algorithm 9.3 A nondeterministic polynomially bounded algorithm for satisfiability.

Let us now return to the idea of a problem P_1 being transformable to another problem P_2 and illustrate that notion by proving that the full power of satisfiability is not necessary: even a very restricted type of satisfiability is $\mathcal{NP}$-complete. *k-satisfiability* is satisfiability in which the boolean expression in conjunctive normal form is allowed to have at most k literals per clause.

Corollary

3-satisfiability is $\mathcal{NP}$-complete.

Proof: It is immediate from Algorithm 9.3 that 3-satisfiability is in $\mathcal{NP}$. We must show that satisfiability is transformable to 3-satisfiability; that is, given a boolean expression $E(X_1, X_2, \ldots, X_n)$ in conjunctive normal form, we want to construct, in polynomial time, a boolean expression $\hat{E}(Y_1, Y_2, \ldots, Y_m)$ in conjunctive normal form, in which each clause has at most three literals, so that E is satisfiable if and only if $\hat{E}$ is satisfiable. Such an $\hat{E}$ can be constructed from E as follows. Replace each clause $\alpha_1 \vee \alpha_2 \vee \cdots \vee \alpha_k$ in E, where the α_i are literals and $k \geq 4$, by the expression

$$(\alpha_1 \vee \alpha_2 \vee Z) \wedge (\alpha_3 \vee \cdots \vee \alpha_k \vee \bar{Z}),$$

where Z is a new variable; repeat this process until no clause has more than three literals. The resulting expression $\hat{E}$ is satisfiable if and only if the original expression E is satisfiable [Exercise 7(a)], and, furthermore, $\hat{E}$ can be constructed in time polynomial in the length of E [Exercise 7(b)].

It is natural to ask now whether we can restrict satisfiability even further, say to only two literals per clause, and still have its membership in $\mathcal{P}$ open to question. The answer is no.

Theorem 9.2

2-satisfiability is in $\mathcal{P}$.

Proof: Exercise 8.

Problems and Their Complements.   Before considering various other $\mathcal{NP}$-complete problems, let us discuss an anomalous property of nondeterministic algorithms. Consider decision problems whose solution is a simple "yes" or "no" (e.g. "Is this boolean expression satisfiable?"), and define the complement of such a problem P to be the problem $\bar{P}$ in which the answers are reversed (e.g. "Is this boolean expression unsatisfiable?"). If there is a deterministic polynomial time algorithm for P, then by simply complementing the output we obtain a deterministic polynomial time algorithm for the complement $\bar{P}$; thus $P \in \mathcal{P}$ if and only if $\bar{P} \in \mathcal{P}$. There is no such result known for $\mathcal{NP}$. In other words, the existence of a nondeterministic polynomial time algorithm for a problem P is not known to guarantee the existence of such an algorithm for $\bar{P}$.

That a decision problem and its complement might not be of the same complexity violates intuition. The difficulty stems from the unusual nature of nondeterministic computation: an answer of "yes" *must* correspond to **success**; we cannot simply complement the output of a nondeterministic algorithm by interchanging instances of **success** and **failure** since ultimate success requires only one successful computational path, while ultimate failure requires that *all* computational paths lead to failure. Nondeterministic computation thus appears to be more suited to existence questions like "Does there exist an element x such that a property of x holds?" than to the complementary nonexistence questions like "Does a property hold for all x?"

There are examples of decision problems $P \in \mathcal{NP}$ not known to be in $\mathcal{P}$ for which $\bar{P} \in \mathcal{NP}$ (see Exercise 3), but these are exceptional. For most decision problems in $\mathcal{NP}$ that are not known to be in $\mathcal{P}$, it is unknown whether the complementary problem is in $\mathcal{NP}$. Algorithm 9.3, for example, clearly demonstrates that satisfiability is in $\mathcal{NP}$, but *unsatisfiability* (determining if all possible assignments of truth values cause a boolean expression to be false) is not known to be in $\mathcal{NP}$. In fact, it is easy to prove that for any $\mathcal{NP}$-complete problem P, $\bar{P}$ is in $\mathcal{NP}$ if and only if $\mathcal{NP}$ is closed under complementation, that is, if and only if the complement of every problem in $\mathcal{NP}$ is also in $\mathcal{NP}$.

The unresolved asymmetry between most decision problems and their complements is why the optimization problems discussed in Sections 9.2.2 and 9.2.3 are proved only to be $\mathcal{NP}$-hard, not $\mathcal{NP}$-complete. For example, the problem "Is there a traveling salesman tour of cost at most b?" is in $\mathcal{NP}$ by Algorithm 9.2; but it is not known if the complementary problem "Do all tours cost more than b?" is in $\mathcal{NP}$. Determining an optimal tour requires answering both questions, and so Theorem 9.6 in Section 9.2.3 states only that the problem of determining an optimal traveling salesman tour is $\mathcal{NP}$-hard.

9.2.2 Some 𝒩𝒫-complete Problems

We are now ready to examine some combinatorial problems similar to those considered in Chapter 8 (i.e., based on graphs), in light of the notion of $\mathcal{NP}$-completeness.

Recall from Chapter 8 that a complete subgraph of an undirected graph $G = (V, E)$ is a graph $G' = (V', E')$ for which $V' \subseteq V$, $E' \subseteq E$, and there is an edge in E' between every pair of vertices in V'.

Theorem 9.3

The problem of determining whether an undirected graph $G = (V, E)$ has a complete subgraph of size k is $\mathcal{NP}$-complete.

Proof: First we describe a simple nondeterministic polynomial-time algorithm for determining whether G has a complete subgraph of size k: The algorithm selects, nondeterministically, a subset of k vertices and tests it, in time $O(k^2)$, to determine whether it is complete. This proves that the problem is in $\mathcal{NP}$. We must now show that the problem is $\mathcal{NP}$-hard; we do so by showing that satisfiability is transformable to it.

Suppose that we are given a boolean expression $C_1 \wedge C_2 \wedge \cdots \wedge C_k$ in conjunctive normal form, in which each C_i is a clause. We will transform the question of its satisfiability to a question about the existence of a complete subgraph of size k in a graph that can be constructed from the expression in time polynomial in its length. Consider the undirected graph $G = (V, E)$ defined by

$$V = \{(\alpha, i)|\alpha \text{ is a literal in clause } C_i\}$$
$$E = \{((\alpha, i), (\beta, j))|i \neq j \text{ and } \alpha \neq \bar{\beta}\}.$$

Intuitively, G has a vertex for each occurrence of a literal in the expression and an edge connecting pairs of literals in different clauses, whose values can be chosen **true** simultaneously. A complete subgraph of size k (the number of clauses) in G corresponds to k literals, each in a different clause, whose values can all be chosen **true** simultaneously so that the literals, the clauses, and hence the expression itself are **true**. Thus the expression is satisfiable if G contains a complete subgraph of size k. Conversely, if the expression is satisfiable, then there is an assignment of the values **true** and **false** to the variables so that the expression is **true**. In other words, each of the k clauses contains a literal whose value is **true**, and the vertices corresponding to these literals form a complete subgraph of size k. Thus $G = (V, E)$ has a complete subgraph of size k if and only if the expression is satisfiable. Obviously, G can be constructed from the expression in polynomial time.

A clique (see Section 8.4) is a maximal complete subgraph of a graph. A complete subgraph C of a graph G is a *maximum clique* of G if no other complete subgraph of G contains more vertices than C.

Corollary

The problem of determining a maximum clique of an undirected graph $G = (V, E)$ is $\mathcal{NP}$-hard.

A set $R \subseteq V$ is a *vertex cover* for G if every edge of G is incident with some vertex in R. A problem that is quite similar to determining whether there is a complete subgraph of size k in an undirected graph is determining whether an undirected graph $G = (V, E)$ has a vertex cover of size l. For example, the set of vertices $\{2, 4, 5, 6, 7\}$ is a vertex cover for the graph in Figure 9.3(a). Each vertex cover of G corresponds to a complete subgraph in the complement $G' = (V, E')$ of G formed by taking the vertices of G and the set of all edges *not* in G (why?). Figure 9.3(b) shows the complement of the graph in Figure 9.3(a); the complete subgraph corresponding to the vertex cover $\{2, 4, 5, 6, 7\}$ is the subgraph induced by the set of vertices $\{1, 3\}$.

The duality between complete subgraphs of a graph and the vertex covers of the complement graph makes it trivial to prove the following results.

Theorem 9.4

The problem of determining whether an undirected graph $G = (V, E)$ has a vertex cover of size l is $\mathcal{NP}$-complete.

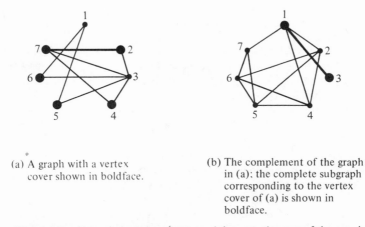

(a) A graph with a vertex
    cover shown in boldface.

(b) The complement of the graph
    in (a); the complete subgraph
    corresponding to the vertex
    cover of (a) is shown in
    boldface.

Figure 9.3   A graph, a vertex cover, and the complement of the graph
with the corresponding complete subgraph.

Corollary

The problem of determining a minimum vertex cover of an undirected graph $G = (V, E)$ is NP-hard.

A *Hamiltonian cycle* in a graph $G = (V, E)$ is a cycle in G containing each of the vertices in V. If G is directed, then the Hamiltonian cycle is directed; if G is undirected, then the Hamiltonian cycle is undirected. Figure 9.4(a) and (b) shows directed and undirected graphs, respectively, in which Hamiltonian cycles are in boldface. Notice that a graph need not have a Hamiltonian cycle: the graph in Figure 9.4(c) does not.

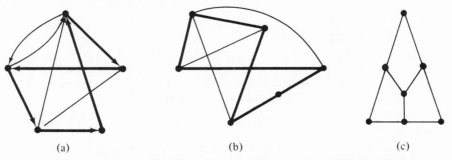

(a)   (b)   (c)

Figure 9.4   A directed graph (a) and an undirected graph (b) in which
Hamiltonian cycles are shown in boldface; (c) is an undirected graph
containing no Hamiltonian cycle.

Theorem 9.5

The problem of determining whether a digraph has a Hamiltonian cycle is NP-complete.

Proof: As in the other proofs, the power of nondeterministic algorithms makes it trivial to show that the problem is in $\mathcal{NP}$. To prove that it is $\mathcal{NP}$-complete, we will show that the problem of determining whether there is a vertex cover of size l is transformable to it. That is, given any undirected graph $G = (V, E)$, we will show how to construct a digraph $\hat{G} = (\hat{V}, \hat{E})$ in polynomial time so that G has a vertex cover of size l if and only if $\hat{G}$ has a Hamiltonian cycle.

$\hat{G}$ is defined as follows. The set of vertices is

$$\hat{V} = \{(v, e, \delta)|v \in V, e \in E \text{ such that } e \text{ is incident with } v, \delta = 0 \text{ or } 1\} \cup \{a_1, a_2, \ldots, a_l\},$$

where $a_1, a_2, \ldots, a_l$ are new symbols. That is, the vertices of $\hat{G}$ consist of l vertices $a_1, a_2, \ldots, a_l$, together with a pair of vertices for each edge and incident vertex. These pairs of vertices in $\hat{G}$ are shown in boxes in Figure 9.5 The edges of $\hat{G}$ are best described informally. To each edge of G, $e \in E$ between vertices u and v, there corresponds a subgraph of four vertices and six edges in $\hat{G}$, as shown in boldface in Figure 9.5. Furthermore, for each vertex v of G, the subgraphs corresponding to the edges incident with v are linked by edges, as shown in Figure 9.5. The order of the edges in this "edge-list" is arbitrary but fixed. The edge-lists for vertices u and v are shown in Figure 9.5. In addition to all these edges, there is an edge from each a_i, $1 \leq i \leq l$, to the first element of each edge-list and an edge to each a_i, $1 \leq i \leq l$, from the last element of each edge-list.

Suppose that G has a vertex cover $\{v_1, v_2, \ldots, v_l\}$. We must show that $\hat{G}$ has a Hamiltonian cycle. Let $e_{i,1}, e_{i,2}, \ldots, e_{i,d_i}$ be the edges of G incident with v_i, $1 \leq i \leq l$, in the order in which they appear in the "lists" of $\hat{G}$. Consider the cycle

$$a_1, (v_1, e_{1,1}, 0), (v_1, e_{1,1}, 1), (v_1, e_{1,2}, 0), (v_1, e_{1,2}, 1), \ldots, (v_1, e_{1,d_1}, 0), (v_1, e_{1,d_1}, 1),$$

$$a_2, (v_2, e_{2,1}, 0), (v_2, e_{2,1}, 1), (v_2, e_{2,2}, 0), (v_2, e_{2,2}, 1), \ldots, (v_2, e_{2,d_2}, 0), (v_2, e_{2,d_2}, 1),$$

$$\vdots$$

$$a_l, (v_l, e_{l,1}, 0), (v_l, e_{l,1}, 1), (v_l, e_{l,2}, 0), (v_l, e_{l,2}, 1), \ldots, (v_l, e_{l,d_l}, 0), (v_l, e_{l,d_l}, 1), a_1.$$

This cycle goes through every vertex in $\hat{V}$ except those in edge-lists corresponding to vertices in V other than v_i, $1 \leq i \leq l$. However, since $\{v_1, v_2, \ldots, v_l\}$ is a vertex cover of G, every vertex $w \in V - \{v_1, v_2, \ldots, v_l\}$ that has an incident edge is adjacent to one of the v_i; that is, there is an edge $e = (v_i, w)$ in E. Consider the vertices $(w, e, 0)$ and $(w, e, 1)$ in $\hat{V}$ in the edge-list for w. We can extend the cycle to include them by replacing the edge from $(v_i, e, 0)$ to $(v_i, e, 1)$ in the cycle with the path

$$(v_i, e, 0), (w, e, 0), (w, e, 1), (v_i, e, 1).$$

This can be done for every vertex in $\hat{V}$ that is not in the cycle. The result is a cycle in $\hat{G}$ that contains every vertex in $\hat{V}$, that is, a Hamiltonian cycle.

Conversely, suppose that $\hat{G}$ has a Hamiltonian cycle. Since the cycle contains every vertex in $\hat{V}$, it must contain every a_i, $1 \leq i \leq l$; and since there are no edges in $\hat{E}$ between any pair of a_i's, the Hamiltonian cycle can be broken into l nontrivial disjoint paths, each path beginning at some a_j and ending at some a_k with no other a_i's in between. Examination of the boldface configuration in Figure 9.5 shows that any Hamiltonian cycle entering the configuration at $(u, e, 0)$ must exit at $(u, e, 1)$; it cannot exit somewhere else and then return to $(u, e, 1)$ later (why not?). Consequently, with each of

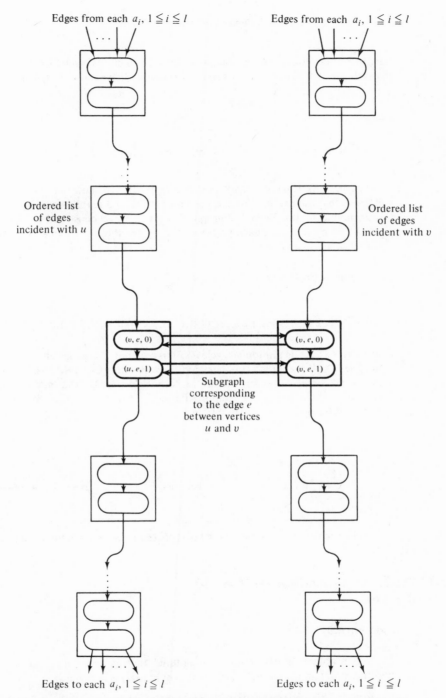

Figure 9.5 The subgraph of $\hat{G}$ corresponding to the edge $e = (u, v)$ in G is shown in boldface. The lists of edges incident with u and v are also shown; each box contains the pair of vertices in $\hat{G}$ corresponding to an edge and incident vertex in G.

the l paths between the a_i's we can associate a vertex $v \in V$ such that every vertex in $\hat{V}$ on the path not of the form (v, e, δ) is of the form (w, e, δ), where w is adjacent to v in G by the edge $e \in E$. These l vertices in V form a vertex cover of G.

Corollary

The problem of determining whether an undirected graph has a Hamiltonian cycle is $\mathcal{NP}$-complete.

Proof: As usual, the problem is clearly in $\mathcal{NP}$. To show that it is $\mathcal{NP}$-complete, we will prove that the problem of determining whether a digraph has a Hamiltonian cycle is transformable to it. Given a digraph $G = (V, E)$, we construct an undirected graph $\hat{G} = (\hat{V}, \hat{E})$ by replacing each vertex $v \in V$ by the three-vertex configuration shown in Figure 9.6. In other words,

$$\hat{V} = \{(v, \delta)|v \in V, \delta = 0, 1, \text{or } 2\}$$

and

$$\hat{E} = \{((u, 2), (v, 0))|(u, v) \in E\} \cup \{((v, 0), (v, 1))|v \in V\} \cup \{((v, 1), (v, 2))|v \in V\}.$$

Obviously, $\hat{G}$ has a Hamiltonian cycle if G has one, but G also has one if $\hat{G}$ does, since any such cycle in $\hat{G}$ must enter a configuration corresponding to v at $(v, 0)$ and leave it at $(v, 2)$; there is no other way for the cycle to include $(v, 1)$, the middle vertex of the configuration. Thus $\hat{G}$ has a Hamiltonian cycle if and only if G has one. $\hat{G}$ is easily constructed from G in polynomial time.

Figure 9.6   Construction of the part of $\hat{G}$ corresponding to the vertex v in G.

9.2.3 The Traveling Salesman Problem Revisited

Theorem 9.6

The problem of determining an optimal tour in the symmetric n-city traveling salesman problem (i.e., with a symmetric cost matrix) is $\mathcal{NP}$-hard.

Proof: The symmetric traveling salesman problem can be viewed as the problem of finding a minimum-cost Hamiltonian cycle in a weighted, undirected graph. With this point in mind, it is easy to prove that the problem of determining the existence of a Hamiltonian cycle in an undirected graph is transformable to the symmetric traveling salesman problem.

Given an undirected graph $G = (V, E)$, $|V| = n$, we construct a symmetric n-city traveling salesman problem as follows. Let $V = \{v_1, v_2, \ldots, v_n\}$. The cost of going between cities i and j is defined by

$$C_{ij} = \begin{cases} 1 & \text{if } (v_i, v_j) \in E, \\ 2 & \text{otherwise.} \end{cases}$$

Clearly, the cost of an optimal tour is n if G contains a Hamiltonian cycle; if G does not contain a Hamiltonian cycle, the cost of an optimal tour must be at least $n + 1$ ($n - 1$ edges of cost 1 and one edge of cost 2).

Corollary

The problem of determining an optimal tour in the general (i.e., not necessarily symmetric) n-city traveling salesman problem is 𝒩𝒫-hard.

Corollary

Given a bound b, the problem of determining whether there exists a tour of cost at most b in the symmetric traveling salesman problem is 𝒩𝒫-complete.

Proof: The problem is in 𝒩𝒫 from our discussion in Section 9.1. It is 𝒩𝒫-hard by an argument similar to that in the proof of Theorem 9.6.

Actually, matters are even worse than they seem from Theorem 9.6 and its corollaries. In Section 4.3 we presented the nearest insertion algorithm, which ran in polynomial time and computed a tour whose cost was guaranteed to be no more than twice the cost of an optimal tour. That proof depended heavily on the symmetry of the cost matrix and the assumption that the costs satisfied the triangle inequality. One might hope for a similar "approximate" algorithm in the general case—that is, when the triangle inequality does not hold; unfortunately, however, even this problem is 𝒩𝒫-hard. If we define a tour to be *ε-suboptimal* when its cost is no more than $1 + \varepsilon$ times the cost of an optimal tour, we have

Theorem 9.7

For any $\varepsilon > 0$, the problem of finding an ε-suboptimal tour for the symmetric n-city traveling salesman problem is 𝒩𝒫-hard.

Proof: The proof is similar to the proof of Theorem 9.6. Given an undirected graph $G = (V, E)$, $V = \{v_1, v_2, \ldots, v_n\}$, we construct a symmetric traveling salesman problem by defining

$$C_{ij} = \begin{cases} 1 & \text{if } (v_i, v_j) \in E \\ \lceil n\varepsilon \rceil + 2 & \text{otherwise.} \end{cases}$$

Clearly the cost of an optimal tour is n if and only if G contains a Hamiltonian cycle; if G does not contain a Hamiltonian cycle the cost of an optimal tour must be at least $n + \lceil n\varepsilon \rceil + 1$. Since

$$\frac{n + \lceil n\varepsilon \rceil + 1}{n} > 1 + \varepsilon,$$

a tour whose cost is more than n is ε-suboptimal if and only if it is optimal—that is, if and only if G has no Hamiltonian cycle.

Corollary

For any $\varepsilon > 0$, the problem of finding an ε-suboptimal tour for the general n-city traveling salesman problem is $\mathcal{NP}$-hard.

9.3 REMARKS AND REFERENCES

In this chapter we have examined a collection of combinatorial problems that are equivalent in the sense that either all of them can be solved by a deterministic polynomial-time algorithm or none of them can (Exercise 19). The class of such equivalent problems is extensive, including problems from diverse parts of computer science, mathematics, and operations research. At the end of this section we mention a few interesting $\mathcal{NP}$-complete and $\mathcal{NP}$-hard problems not covered in the text or exercises of this chapter.

Nondeterministic computation dates back to the early days of mathematical logic and computability theory: the models of computation based on what are called rewriting systems are all implicitly nondeterministic. The first explicit use of nondeterminism as a technique of computation was by Michael Rabin and Dana Scott in the summer of 1957. Their results were first published in

RABIN, M. O., and D. SCOTT, "Finite Automata and Their Decision Problems," *IBM J. Res. Develop.*, **3** (1959), 114–125.

This paper was reprinted in

MOORE, E. F., *Sequential Machines: Selected Papers*, Addison-Wesley, Reading, Mass., 1964.

The idea of using nondeterministic algorithms as a means of expressing backtrack searches is from

FLOYD, R. W., "Nondeterministic Algorithms," *J. ACM*, **14** (1967), 636–644.

Results proving that one problem can be transformed to another are common in mathematical logic and computability theory, but such results have a much different flavor than the ones presented here. The same type of results as given in this chapter occur in mathematical programming; for example, in

DANTZIG, G. B., "On the Significance of Solving Linear Programming Problems with Some Integer Variables," *Econometrica*, **28** (1960), 30–44 it is proved that a variety of combinatorial problems can be transformed into linear programming programs. Similarly, in

DANTZIG, G. B., W. O. BLATTNER, and M. R. RAO, "All Shortest Routes from a Fixed Origin in a Graph," *Theory of Graphs*, International Symposium, Rome, July 1966. Proceedings published by Gordon and Breach, New York, 1967, and by Dunold, Paris, 1967, 85–90

it is proved that the traveling salesman problem is transformable to the problem of determining a shortest path (without cycles) from one vertex in a weighted digraph to all vertices in the digraph when the weights can be positive, zero, negative, or ∞ (Exercise 17).

The key result of this chapter is Theorem 9.1—that satisfiability is $\mathcal{NP}$-complete. This result is from

COOK, S. A., "The Complexity of Theorem-Proving Procedures," *Proc. Third ACM Symposium on Theory of Computing* (1971), 151–158.

Cook's pioneering paper also contains the corollary to Theorem 9.1 that 3-satisfiability is $\mathcal{NP}$-complete, and a remark that 2-satisfiability is in $\mathcal{P}$ (Theorem 9.2) based on the results of

DAVIS, M., and H. PUTNAM, "A Computing Procedure for Quantification Theory," *J. ACM,* **7** (1960), 201–215.

Theorem 9.3 is implicit in Cook's proof that subgraph isomorphism is $\mathcal{NP}$-complete (see Exercise 11).

The richness of the class of $\mathcal{NP}$-complete problems was demonstrated in

KARP, R. M., "Reducibility Among Combinatorial Problems," *Complexity of Computer Computations*, R. E. Miller and J. W. Thatcher (Eds.), Plenum Press, New York, 1972, 85–104.

This influential paper contains Theorems 9.4, 9.5, and 9.6 in addition to many similar results. An important strengthening of Theorem 9.6 is given in

GAREY, M. R., R. L. GRAHAM, and D. S. JOHNSON, "Some $\mathcal{NP}$-Complete Geometric Problems," *Proc. Eighth ACM Symposium on Theory of Computing* (1976), 10–22,

where it is proved that the traveling salesman problem is $\mathcal{NP}$-complete even if it is restricted to sets of points with a Euclidean distance. Theorem 9.7 is from

SAHNI, S., and T. GONZALEZ, "P-Complete Approximation Problems," *J. ACM,* **23** (1976), 555–565.

Before going to the literature, the reader should be aware of the various terms that have been used for what we call *transformable* and $\mathcal{NP}$-*complete*. Karp used the term "reducible" as we have used "transformable," and Cook used the terms "polynomially reducible" or "$\mathcal{P}$-reducible" to mean something more general than "transformable." Karp used "polynomial complete" or simply "complete" for $\mathcal{NP}$-complete. The terms "transformable" and "$\mathcal{NP}$-

complete" have become standard because of the influence of

KNUTH, D. E., "A Terminological Proposal," *SIGACT News*, **6**, 1 (January 1974), 12–18

and

KNUTH, D. E., "Postscript about $\mathcal{NP}$-hard Problems," *SIGACT News*, **6**, 2 (April 1974), 15–16.

There are still many open questions for future research. In addition to the question of whether $\mathcal{P} = \mathcal{NP}$, there are unresolved questions about specific combinatorial problems. For example, the traveling salesman problem is known to be $\mathcal{NP}$-hard, but is it $\mathcal{NP}$-complete—that is, is it in $\mathcal{NP}$? The problem of determining whether two graphs are isomorphic is known to be in $\mathcal{NP}$, but is it $\mathcal{NP}$-complete?

The remainder of this section is devoted to describing a few $\mathcal{NP}$-hard and $\mathcal{NP}$-complete problems to show their diversity.

The *discrete multicommodity flow problem* is as follows. Given a graph and a set of k disjoint source-sink pairs of vertices (s_i, t_i), $1 \leq i \leq k$, we want to determine whether there is a set of k vertex disjoint paths p_i, $1 \leq i \leq k$, where p_i is from s_i to t_i. The proof that this problem is $\mathcal{NP}$-complete when the graph is undirected is due to Donald E. Knuth and appears in

KARP, R. M., "On the Computational Complexity of Combinatorial Problems," *Networks*, **5** (1975), 45–68.

(This survey paper is a superb introduction to the subject of $\mathcal{NP}$-complete problems.) Knuth's result is strengthened in

EVEN, S., and A. SHAMIR, "On the Complexity of Timetable and Integral Multicommodity Flow Problems," *SIAM J. Comput.*, to appear,

where the problem is proved to be $\mathcal{NP}$-complete for $k = 2$ in both directed and undirected graphs.

The paper by Even and Shamir also establishes the $\mathcal{NP}$-completeness of the *timetable problem*. Suppose that we are given a set of hours in the week, a collection of n teachers, each of whom is available to teach during some subset of the set of hours, a set of m classes, each of which can be scheduled only during certain hours of the week, and an $n \times m$ matrix of nonnegative integers in which the (i, j) element is the number of hours the ith teacher must teach the jth class. We wish to determine if it is possible to assign meeting times and teachers to the classes so that every class meets during the required hours, every teacher can be at his classes, there is only one teacher per class, and no teacher has two classes simultaneously.

A graph is k-*colorable* if all vertices can be colored using k different colors so that no two adjacent vertices have the same color. From Exercise 62 of Chapter 8 we know that every planar graph is 5-colorable. Also, it is easy to prove that the problem of testing 2-colorability is in $\mathcal{P}$ (Exercise 64 of Chapter 8), and, of course, 4-colorability of all planar graphs hinges on the famous four-color theorem. In

STOCKMEYER, L., "Planar 3-Colorability is Polynomial Complete," *SIGACT News* **5**, 3 (July 1973), 19–25

it is proved that determining whether a planar graph is 3-colorable is $\mathcal{NP}$-complete. This result is strengthened in

GAREY, M. R., D. S. JOHNSON, and L. STOCKMEYER, "Some Simplified $\mathcal{NP}$-Complete Problems," *Theoretical Comput. Sci.* **1** (1976), 237–267

to the problem of determining 3-colorability in planar graphs, each of whose vertices have at most four incident edges.

Given a sparse matrix (see Exercise 11 of Chapter 2) A and a vector b, we want to use Gaussian elimination to solve the system of linear equations

$$Ax = b,$$

taking advantage of the sparseness of A by not storing or performing arithmetic operations on some or all of the zeros. Alternatively, we can find x by applying Gaussian elimination to the equivalent permuted system

$$(PAQ)(Q^T x) = Pb,$$

where P and Q are permutation matrices (i.e., matrices of 0s and 1s containing exactly one 1 in every row and column). The problem of determining P and Q to minimize the time or storage required to apply Gaussian elimination to the permuted system is $\mathcal{NP}$-complete. See

ROSE, D. J., and R. E. TARJAN, "Algorithmic Aspects of Vertex Elimination in Directed Graphs," *SIAM J. Appl. Math.*, to appear

and

PAPADIMITRIOU, C. H., "The $\mathcal{NP}$-Completeness of the Bandwidth Minimization Problem," *Computing* **16** (1976), 263–270

for two different versions of this problem.

Given a table of names to be organized into a trie (see Section 6.3.4), it is not necessary to have the root test the first character, sons of the root test the second character, and so on. In fact, we may get a smaller trie (i.e., one with fewer internal nodes) by considering the characters in different orders. The problem of determining a trie with the minimum possible number of internal nodes is $\mathcal{NP}$-complete. Similarly, if each leaf of the trie is equally probable, determining the trie with minimum average search time is $\mathcal{NP}$-complete. Details of these and related results can be found in

COMER, D. and R. SETHI, "Complexity of Trie Index Construction," *Proc. Seventeenth Annual Symposium on Foundations of Computer Science* (1976), 197–207.

Given a finite set of sparse complex polynomials of one variable with integer coefficients, represented as a sequence of ordered pairs (coefficient, exponent) for each nonzero coefficient, the problem of determining the degree of the least common multiple of the set is $\mathcal{NP}$-hard. Also, determining the number of distinct complex zeros of a product of such polynomials is an $\mathcal{NP}$-hard problem. These and other similar results are presented in

PLAISTED, D. A., "Sparse Complex Polynomials and Polynomial Reducibility," to appear

and in

PLAISTED, D. A., "Some Polynomial and Integer Divisibility Problems are $\mathcal{NP}$-Hard," *Proc. Seventeenth Annual Symposium on Foundations of Computer Science* (1976), 264–267.

9.4   EXERCISES

1. Examine some of the algorithms in earlier chapters and determine whether the problems they solve are in $\mathcal{P}$. Are there any problems posed in earlier chapters that are *not* in $\mathcal{NP}$?

2. Devise nondeterministic polynomially bounded algorithms for solving Exercises 1(b), 1(f), 1(k), 1(m), 3(a), 3(b), 3(c), 3(d), and 3(e) of Chapter 4 and Exercises 41, 60, 61, and 63 of Chapter 8.

3. (a) Prove that there is a nondeterministic polynomial-time algorithm to answer the question "Is n composite?"

   ★(b) Prove that there is a nondeterministic polynomial-time algorithm to answer the question "is n prime?" [*Hint:* Prove that n is a prime if and only if there exists an a such that $a^{n-1} \equiv 1 \pmod{n}$, but such that $a^{(n-1)/p} \not\equiv 1 \pmod{n}$ for any prime divisor p of $n - 1$. Combine this result with the use of **choice** to "guess" the prime factorization of $n - 1$ and a value of a.]

4. Prove that if $\mathcal{P} = \mathcal{NP}$, then all problems in $\mathcal{NP}$ are $\mathcal{NP}$-complete.

5. Rewrite the following boolean expressions in conjunctive normal form and in conjunctive normal form with at most three literals per clause.

   (a) $((X_1 \vee X_2) \wedge \overline{((\overline{X_1 \wedge X_2}) \vee (\bar{X}_1 \wedge X_3))}) \vee \bar{X}_3$

   (b) $(\bar{X}_1 \wedge \bar{X}_2 \wedge X_3) \vee (\bar{X}_1 \wedge X_2 \wedge \overline{(X_2 \vee \bar{X}_3)}) \vee (X_1 \wedge X_3)$

   (c) $(X_1 \wedge \bar{X}_2) \vee (\overline{(X_1 \vee X_2 \vee X_3 \vee X_4)} \wedge (\bar{X}_1 \vee X_2))$

   (d) $(\overline{\bar{X}_1 \vee (\bar{X}_2 \wedge \bar{X}_3)} \vee X_3) \wedge (\bar{X}_1 \vee \overline{(X_2 \wedge \bar{X}_3)})$

   (e) $(X_1 \vee \bar{X}_2 \vee X_3 \vee X_4) \wedge ((X_2 \wedge X_3) \vee \bar{X}_4)$

6. Determine which of the boolean expressions in Exercise 5 are satisfiable.

7. Complete the proof of the corollary to Theorem 9.1 by proving that
   (a) the expression $\hat{E}$ obtained from E is satisfiable if and only if E is satisfiable.
   (b) the expression $\hat{E}$ has length polynomial in the length E.

8. Prove Theorem 9.2 by finding a deterministic polynomial-time algorithm for the problem. [*Hint*: Prove that $(X_1 \lor X_2) \land (\bar{X}_1 \lor X_3) \land E$ is satisfiable if and only if $(X_2 \lor X_3) \land E$ is satisfiable, where E is an arbitrary boolean expression that does not contain X_1 or $\bar{X}_1$.]

9. As a consequence of Theorem 9.2 and the corollary of Theorem 9.1, what would happen if we could write the boolean expression $X_1 \lor X_2 \lor X_3$ as the "and" of a sequence of terms, each of which is the "or" of two literals—that is, in the form $(\alpha_1 \lor \alpha_2) \land (\alpha_3 \lor \alpha_4) \land \cdots \land (\alpha_{k-1} \lor \alpha_k)$, where α_i, $1 \leq i \leq k$, are literals? Prove that $X_1 \lor X_2 \lor X_3$ cannot be so written.

10. The $0 - 1$ *integer programming problem* is to determine whether there is a column vector x of 0s and 1s such that $Cx \geq d$ for an integer matrix C and an integer vector d. Prove that this problem is $\mathcal{NP}$-complete. (*Hint*: Show that satisfiability is transformable to the problem by considering the matrix

$$
C_{ij} = \begin{cases}
1 & \text{if } X_j \text{ is in the } i\text{th clause} \\
-1 & \text{if } \bar{X}_j \text{ is in the } i\text{th clause} \\
0 & \text{otherwise}
\end{cases}
$$

and the vector

$$d_i = 1 - (\text{number of complemented variables in the } i\text{th clause}).)$$

11. Prove that the problem of determining whether one undirected graph contains a subgraph isomorphic to a second undirected graph is $\mathcal{NP}$-complete. (*Hint*: Use Theorem 9.3.)

12. Prove that the *set packing* problem is $\mathcal{NP}$-complete: Given a family of sets S_j, $1 \leq j \leq n$, determine whether the family contains l mutually disjoint sets. (*Hint*: Use Theorem 9.3.)

13. Given a digraph $G = (V, E)$, a *feedback vertex set* is set $R \subseteq V$ such that every cycle of G contains a vertex in R. Prove that determining the existence of a feedback vertex set of size k is an $\mathcal{NP}$-complete problem. (*Hint*: Use Theorem 9.4.)

14. Given digraph $G = (V, E)$, a *feedback edge set* is a set $S \subseteq E$ such that every cycle of G contains an edge in S. Prove that determining the existence of a feedback edge set of size k is an $\mathcal{NP}$-complete problem. (*Hint*: Use Theorem 9.4.)

15. Prove that the problem of determining for a general (i.e., not necessarily symmetric) cost matrix whether a specified edge occurs on any ε-suboptimal traveling salesman tour is $\mathcal{NP}$-hard.

16. Prove that the problem of determining whether for a general (i.e., not necessarily symmetric) cost matrix there is more than one ε-suboptimal traveling salesman tour is $\mathcal{NP}$-hard.

17. Prove that problem of determining the shortest path without a cycle between two vertices s and f of a weighted digraph in which the weights can be positive, negative,

zero, or ∞ is $\mathcal{NP}$-hard. If $s = f$, we allow the path to contain exactly one cycle, namely the path itself. [*Hint*: Show that the general traveling salesman problem is transformable to the problem by considering the complete weighted digraph $G = (V, E)$ in which the weight w_{ij} of edge (v_i, v_j) is defined by

$$w_{ij} = C_{ij} - K,$$

where $K > \sum_i \sum_j C_{ij}$ and C is a traveling salesman cost matrix. What is the shortest path with exactly one cycle from a vertex back to itself?]

18. What is wrong with the following "proof" that the traveling salesman problem is in $\mathcal{NP}$?

   "To prove that the problem is in $\mathcal{NP}$, we need only give a nondeterministic polynomial-time algorithm to solve it. Such an algorithm works as follows. Given an $n \times n$ matrix C of the intercity distances in an n-city traveling salesman problem in which each matrix entry is k bits long, it is clear that the sum of all n^2 entries is an upper bound on the cost of any tour, and thus the cost of an optimal tour lies in the range $[0, n^2 2^k]$. Using a binary search-type procedure (see Section 6.3.1) on the range $[0, n^2 2^k]$, only $\lg(n^2 2^k) = k + 2 \lg n$ applications of Algorithm 9.2 suffice to determine the cost of an optimal tour. We conclude that the traveling salesman problem is in $\mathcal{NP}$."

19. A number of corollaries in this chapter state that some problem is $\mathcal{NP}$-hard, not $\mathcal{NP}$-complete, example, the problem of finding an optimal traveling salesman problem is $\mathcal{NP}$-hard but not known to be in $\mathcal{NP}$. However, prove that the problem of finding an optimal traveling salesman tour can be done by a deterministic polynomial-time algorithm *if and only if* $\mathcal{P} = \mathcal{NP}$. (*Hint*: See Exercise 18.) Establish similar results for the other problems that were proved to be $\mathcal{NP}$-hard.

20. The footnote on page 402 states that it would be unreasonable to allow an integer n to be encoded in more than $O(\log n)$ characters. What would happen if encodings in proportional to n bits were allowed?

Index